Theory and Design of Seismic Resistant Steel Frames

Theory and Design of Seismic Resistant Steel Frames

F.M. Mazzolani
Institute of Structural Engineering, University of Napoli, Italy

and

V. Piluso
Department of Civil Engineering, University of Salerno, Italy

CRC Press
Taylor & Francis Group
Boca Raton London New York

CRC Press is an imprint of the
Taylor & Francis Group, an **informa** business

A SPON PRESS BOOK

CRC Press
Taylor & Francis Group
6000 Broken Sound Parkway NW, Suite 300
Boca Raton, FL 33487-2742

First issued in paperback 2019

© 1996 F.M. Mazzolani and V. Piluso
CRC Press is an imprint of Taylor & Francis Group, an Informa business

No claim to original U.S. Government works

ISBN-13: 978-0-419-18760-8 (hbk)
ISBN-13: 978-0-367-86608-2 (pbk)

Visit the Taylor & Francis Web site at
http://www.taylorandfrancis.com

and the CRC Press Web site at
http://www.crcpress.com

A catalogue record for this book is available from the British Library

Library of Congress Catalog Card Number: 95–67903

Contents

Preface

It seems to be of paramount importance that structural engineers, at least once in their life, should have the opportunity to experience a seismic event as direct protagonists; otherwise they cannot imagine the strength of the actions that structures are expected to resist.

The most recent earthquake in Campania and Basilicata in the South of Italy on 23rd November 1980 (and all the other similar earthquakes occurring from time to time elsewhere in the world) was very important for local technical specialists – as the authors are – as it allowed them to compare the behaviour of constructional materials and to find the reasons for the damage that was caused.

It thus became easier for these lucky engineers to understand why the design of seismic-resistant structures must be based on two basic limit states. In fact, the classical design approach for seismic-resistant structures requires that, in the case of moderate earthquakes occurring many times during the mean structural life, elastic behaviour has to be assured and sufficient stiffness has to be provided in order to limit any disturbance to the activities for which the building has been designed. Conversely, in the case of destructive earthquakes having a high return period, the structure has to provide sufficient strength, ductility and energy dissipation capacity in order to resist the earthquake action, even if it is severely damaged, preventing failure and safeguarding human lives. Following this approach, it is evident that the analysis of seismic-resistant structures represents the most difficult problem of engineering optimization, because strength, stiffness and ductility have to be properly balanced under the action of a random ground motion.

Earthquake effects on structures systematically reveal the mistakes made in design and construction, even the minutest mistakes. Newmark said that 'earthquake engineering is to the rest of engineering disciplines what psychiatry is to the other branches of medicine: it is a study of pathological cases to gain insight into the mental structure of normal human beings'. All this makes earthquake engineering captivating and fascinating, with a high educational value for any civil engineer.

It is universally recognized that steel structures represent an excellent solution for buildings in high seismicity areas. In fact, structural steels exhibit a strength and ductility which cannot be compared to that provided by other constructional materials. In addition, the industrial production of steel profiles and sheets guarantees high quality assurance. At the same time, reliable connections can be realized both in the workshop and in the field, leading to very safe structural typologies. Dissipative zones can be properly detailed, assuring stable hysteresis loops able to dissipate the earthquake input energy with high efficiency. But the reverse of the coin also exists: this idyllic situation only arises if certain conditions are met; if they are not, as was the case in the recent earthquakes in Mexico and California, then major accidents happen which contribute to undermining the reputation of steel structures.

To prevent such situations, this book provides a state-of-the-art review of the most advanced issues in the behavioural analysis and design of seismic-resistant steel frames. The contents of the book provide postgraduate students, researchers and professionals in civil and structural engineering with the basic principles on which modern seismic codes are based. References to the ECCS Recommendations and to the chapter on steel structures in Eurocode 8 are continuously made in all sections of the text. In addition, the book is much more than a simple background volume, because it gives also the most recent results, which can be used in the near future to improve the already codified provisions for steel structures in seismic zones. All recent developments in seismic design of steel frames are dealt with, including new items which cannot be found in any standard reference book on seismic engineering.

The material has been organized into ten chapters, which are devoted to the main subjects dealing with theory and design of seismic-resistant steel structures.

Chapter 1 starts with the basic seismic design criteria and the definition of seismic actions, as given by modern seismic codes. Chapter 2 and Chapter 3 are devoted to the very delicate problem of evaluating local and global ductility. A new design procedure aimed at the control of the failure mode and of the available ductility and energy dissipation capacity is provided. The collapse interpretation through the low cycle fatigue approach is also dealt with. Chapter 4 discusses the crucial issue of evaluating the q-factor, on which all modern provisions are based. A comparative state-of-the-art of the simplified methods for its evaluation is provided. The influence of second order effects is evidenced in Chapter 5, where a simple formulation for its evaluation is derived from a statistical base. The problems concerning the design of beam-to-column joints and the influence of their behaviour on the seismic inelastic response of steel frames are faced both in Chapter 6 and Chapter 7, the latter giving proposals for a new approach for analysing semi-rigid steel frames. The influence of the geometrical configuration and the examination of 'irregular' cases leading to damage concentration are dealt with in Chapter 8. Finally, the influence of random material variability is investigated in Chapter 9, while Chapter 10 provides the first results of research on the influence of claddings on the dynamic behaviour of frames and the modelling of their contribution when they are made of trapezoidal sheet panels.

The authors hope that this book will serve as a guide to those who want to become familiar with the effects of earthquakes, even if they have not had any direct experience. In any case, its contents will supplement the education of all structural designers and civil engineers, as well as of people involved in codification and in research for which new gates are open to future developments.

The authors express their gratitude to all colleagues who have contributed to this work by means of profitable suggestions and fruitful discussions. In particular, they are grateful to Dr Mario De Stefano, who prepared most of the drawings in the section of Chapter 8 dealing with plan regularity, and gave many valuable suggestions thanks to his doctoral thesis subject.

Federico M. Mazzolani
Vincenzo Piluso

List of symbols

a coefficient, distance

b exponent of the fatigue law, width, coefficient, distance

b_{cf} column flange width

b_f flange width

c coefficient, shear panel flexibility

d depth, deterioration, distance

d_b beam depth

d_c column depth

d_w depth of the web

e eccentricity

e_a accidental eccentricity

e_d dynamic eccentricity

e_R strength eccentricity

e_s static eccentricity

f_u ultimate stress

f_y yield stress

$f(\ )$ function of

g gravity acceleration, distance

h section height, interstorey height

h_e height of the ideal two flange section

i index

i_m mechanism index

j index

k coefficient, index, distance, constant

k_x resisting element stiffness in x direction

k_y resisting element stiffness in y direction

m storey mass, stress increasing ratio

$\overline{m}$ nondimensional ultimate moment of a connection

n exponent

n_b number of bays

n_c number of columns

n_m number of storeys involved in a collapse mechanism

n_p number of purlins

n_s number of storeys, number of seam fasteners

n_{sc} number of side fasteners

p damage concentration factor

q actual value of the q-factor

q_d design value of the q-factor

$\overline{q}$ behaviour factor

r distance

s soil coefficient, shape of horizontal displacements, damage distribution coefficient, nondimensional buckling stress

s_p slip of sheet-to-purlin fasteners

s_s slip of seam fasteners

s_{sc} slip of side fasteners

t thickness

t_{bf} beam flange thickness

t_{cf} column flange thickness

t_f flange thickness

t_p panel zone thickness

t_w web thickness

u horizontal displacement

v vertical displacement, generic displacement

x displacement

$x_{p,i}$ amplitude of a plastic excursion

A peak ground acceleration, cross section area

A_w web area

B damage coefficient, width, coefficient of a Ramberg–Osgood type relation

C period dependent seismic coefficient

C_o seismic coefficient

C_1 coefficient

C_2 coefficient

C_3 coefficient

D normalized damage parameter, peak ground displacement, dead load

D_k stiffness degradation parameter

D_s Japanese structural coefficient

D_S strength degradation parameter

E energy, elastic modulus, earthquake load

E_h hysteretic energy, strain hardening modulus

E_r reduced modulus of elasticity

$\overline{E}$ nondimensional energy

E^* dissipated energy in one direction

F force, base shear, factor

F_e eccentricity factor

F_p design strength of sheet-to-purlin fasteners

F_s stiffness factor, design strength of seam fasteners

F_{sc} design strength of side fasteners

$\overline{F}$ nondimensional force

G tangential modulus of elasticity

G_h strain-hardening tangential modulus

H interstorey height

H_o sum of the interstorey heights of the storeys involved by a collapse mechanism

I importance factor, moment of inertia

I_b moment of inertia of a beam

I_c moment of inertia of a column

I_e moment of inertia of the ideal two flange section

I_ω warping constant

K stiffness, typology coefficient

K_i initial stiffness of a connection

K_l lateral stiffness of a frame

K_φ rotational stiffness of a connection

K_Θ torsional stiffness of a system

$\overline{K}$ nondimensional stiffness

L length, live load

L_e equivalent length

M total mass of the structural system, bending moment

M_b plastic moment of a beam

M_c plastic moment of a column reduced due to axial force

M_p plastic moment

M_{pb} plastic moment of a beam

M_u ultimate moment

N axial force

N_{lo} axial force in the lower flange

N_o external axial load

N_{rf} number of reversals to failure

N_{up} axial force in the upper flange

N_y squash load

O_s overstrength factor

Q storey shear

R normalized spectral acceleration, rotation capacity

R_e normalized elastic spectral acceleration

R_{st} stable part of the rotation capacity

R_w reduction factor

S soil factor, snow load

S_a spectral acceleration

$S_{a,d}$ design spectral acceleration

$S_{a,e}$ elastic spectral acceleration

S_v spectral velocity

T period of vibration

T_k period of vibration of a semirigid frame

T_∞ period of vibration of the ideal rigid frame

V peak ground velocity, total vertical load, shear force

V_c shear force of a column

V_o beam shear force produced by vertical loads

V_p shear force in the panel zone

W weight, work, elastic section modulus

W_p plastic work

Z seismic zone factor, plastic section modulus

α multiplier of horizontal forces, yield shear coefficient, amplification coefficient, coefficient

α_{cr} critical multiplier of vertical loads

α_o kinematically admissible multiplier

α_u ultimate multiplier of horizontal forces

α_y first yielding multiplier of horizontal forces

β coefficient

β_{i_m} mechanism function

β_o maximum normalized spectral acceleration

γ stability coefficient, coefficient

γ_{i_m} mechanism function

γ_p coefficient

γ_s slope of the softening branch of the $\alpha - \delta$ curve

δ top sway displacement

ε normal strain

ε_{lo} normal strain in the lower flange

ε_{up} normal strain in the upper flange

ε_y yield strain

ζ beam-to-column flexural stiffness ratio

η damping correction factor, storey ductility ratio, amplification factor, coefficient, design level

η_c cumulated ductility

θ rotation

θ_{i_m} mechanism function

θ_m rotation corresponding to the maximum flexural resistance

θ_p plastic rotation

θ_u ultimate rotation

θ_y yielding rotation

λ_f flange slenderness

λ_{i_m} mechanism function

λ_w web slenderness

λ_{LT} lateral torsional slenderness

μ kinematic ductility

μ_c cyclic ductility

μ_h hysteretic ductility

μ_F low cycle fatigue parameter

μ_{PA} Park and Ang damage parameter

ν viscous damping ratio, Poisson's ratio

ξ damage concentration index

ξ_{i_m} mechanism function

ρ nondimensional axial stress due to the axial load, parameter

ρ_i damage partition coefficient

ρ_{i_m} sum of column plastic moment ratio

ρ_R strength radius of gyration

ρ_s stiffness radius of gyration

ρ_M mass radius of gyration

σ normal stress

σ_0 axial stress due to the axial load

τ shear stress

φ connection rotation

φ_Δ reduction factor for second order effects

$\bar{\varphi}$ nondimensional rotation of a connection

χ curvature

ψ coefficient

ω frequency

Δ interstorey drift, displacement

Δ_e elastic interstorey drift

Δ_{i_m} mechanism function for second order effects

Δ_p inelastic interstorey drift

Φ geometrical irregularity index

Ω energy dissipated in the previous cycles

Subscripts

a accidental value, acceleration

b bottom, beam, bay, bolt

bf beam flange

c cyclic, column, connection, cumulated value

cf column flange

cr elastic critical value

d design value, dynamic

e elastic value, eccentricity, equivalent

ep end plate

f flange

h hysteretic, hardening value

i index

i_m mechanism index

j index

k index, semirigid, stiffness

lim limit value

lo lower flange

m mechanism, value corresponding to the maximum resistance

max maximum value

mec mechanism value

min minimum value

mon monotonic value

opt optimum value

p plastic value, purlin, panel zone

r	reduced
rf	reversals to failure
s	storey, seismic, static, seam fasteners, stiffness
sc	side fasteners
t	top, tentative value
y	first yielding value, property in y-direction
u	ultimate value
up	upper flange
v	velocity
w	web
x	property in x-direction
F	fatigue value
LT	lateral torsional
M	mass
R	resistance
S	strength
∞	rigid
θ	torsional

φ	connection
ω	warping
Δ	value related to second order effects

Superscripts

(g)	global mechanism
(t)	mechanism typology
(FS)	full strength
(PS)	partial strength
(1)	mechanism type 1, property in one direction
(2)	mechanism type 2
(3)	mechanism type 3
$+$	positive value
$-$	negative value

1

Basic seismic design criteria

1.1 DEVELOPMENT OF SEISMIC DESIGN CRITERIA

In order to understand the basis of the design strategy for seismic-resistant structures adopted in modern seismic practice (i.e. codes), it is useful to sketch out a brief history of structural design in seismic zones.

Seismic engineering has made great progress since the destructive earthquakes of the last century, when an engineering approach to seismic problems was first developed. This progress is ongoing, thanks to the vigorous research activity which ensures a continuous upgrade of the seismic codes.

Different periods can be identified along the path that has been followed by the structural design in areas of high seismicity. The first phase commenced after the severe earthquakes of the second half of the 19th century and of the beginning of the 20th century, when the first conscious attempts were made to understand the forces arising from seismic action and to provide rational design rules. In spite of these efforts, and due to the limited knowledge of the physical phenomenon, procedures for analysing structures subjected to seismic action were not developed; the provisions that were made were basically devoted to defining height and mass limitations of the buildings and to giving the minimum distance between them.

A second phase, of about five decades starting from the years between the two World Wars, has been characterized by the first, quite rough, definition of seismic forces as inertial forces equivalent to a reduced percentage of the dead and live weight.

These design forces were very small with respect to those arising during a severe earthquake. This limitation derived directly from the inability at that time to perform inelastic analyses; however, the greatest approximation was due to the fact that the design seismic forces were defined, for each case, without taking into account the structural typology and the nature of the foundation soil. Moreover, the structural analyses were limited to considering each storey separately from the whole skeleton, due to the difficulty in analysing highly redundant structural schemes by hand. Furthermore, due the lack of knowledge of structural dynamics, approximated laws of distribution along the building height were used for the seismic forces.

Since the early 1970s, a critical review of the previous seismic design approach has taken place. In these years, thanks to the availability of personal computers and the implementation of a great number of programs for structural engineering, structural analysis of highly redundant schemes, both in static and dynamic range, has become feasible for design purposes.

The most important achievement of this new phase, which is ongoing, is the clarification of mechanisms allowing the reduction of the design seismic forces. It has been established that the magnitude of these forces has to be assumed as a function of the deformation characteristics of the structure and of its energy dissipation capacity.

In addition, the possibility has been recognized of dissipating the earthquake input energy by means of plastic excursions, which have to be compatible with the plastic deformation capacity of the structure. As a consequence, the concepts of local and global ductility have been introduced and their importance in the design of seismic-resistant structures has been emphasized. The corresponding design approach, which is nowadays universally accepted, is characterized by the requirements of both strength and ductility, which together represent the 'seismic toughness' of a structure. This means that the reliability of the magnitude of the seismic forces assumed in design practice has to be justified by verifying the available structural ductility and the energy dissipation capacity.

The difficulties of this approach are illustrated by the effort required to ensure appropriate ductility levels. For this reason, the main objective of modern seismic codes is the introduction of the design procedures oriented towards the failure mode and structural ductility control. The art of 'detailing' is the most important content of seismic design recommendations.

The increase in earthquake safety achieved through the refinement of this design approach cannot be denied, and is widely demonstrated not only by theoretical and experimental studies, but also by the 'full scale tests' which occurred during recent earthquakes in areas already covered by modern seismic codes.

From the theoretical point of view, the checks required in order to guarantee the safety of a building in seismic areas can be examined in the light of the general methodology of 'limit states'. The 'ultimate limit state' corresponds to the prevention of structural collapse, while the 'serviceability limit state' imposes that the structure, together with non-structural components, should suffer no damage and that discomfort to the inhabitants should be reduced to a minimum.

Severe earthquakes, having a very small probability of occurrence, have to be examined in the safety check against the ultimate limit state. The 'return period', i.e. the timespan between such earthquakes, there-fore, is much greater than the design life of the structure.

The intensity of ground motion, which has to be considered in the check against the serviceability limit state, corresponds to frequent earth-quakes having a return period less than the life of the structure.

It is, therefore, clear that the methods allowing checks against the serv-iceability limit state are those based on the elastic structural analysis, while the safety against the ultimate limit state, which is associated with the collapse conditions of the structure, should be verified by the meth-ods of inelastic structural analysis and the plastic design.

The usual procedure for checking the fulfilment of the requirements, that a seismic-resistant structure has to possess, is based on elastic analy-ses under seismic horizontal forces. These are defined by reducing the base shear required in the structure in order to remain within the elastic range during the severest design earthquake, by means of a coefficient, namely the q-factor. This takes into account the structural ductility and the energy dissipation capacity. Under such horizontal forces, the struc-ture has to possess sufficient strength and stiffness in order to guarantee the fulfilment of the requirements associated to the serviceability limit state. Safety against the ultimate limit state is considered automatically verified, provided that the detailing rules and design procedures, sug-gested by seismic codes in order to control the failure mode and, therefore the energy dissipation capacity, are satisfied. Therefore it is usually assumed that, when the suggested design procedures and detail-ing rules are used, adoption of the design value of the q-factor given by the code provides safe results.

This procedure offers considerable simplification because it enables in a single shot, checks against both the serviceability limit state and the collapse limit state. At the same time, the limitations of such a procedure are evident; the suggested design procedures do not always lead to the foreseen failure mode and the expected ductility, so that the energy dis-sipation capacity of the structure can be less than that requested in order to prevent collapse under the severest design earthquake.

As a consequence, it seems that this design procedure should be in-tegrated by simple methods, to allow verification of the design assumptions regarding the earthquake input energy dissipation capacity of the structure and, therefore, the validity of the q-factor assumed in design.

Problems of the type described above represent typical items in the general field of earthquake engineering, and are independent of the constructional material.

In particular, for steel structures it has to be remembered that the European Convention for Constructional Steelwork (ECCS) has a consolidated tradition of more than 20 years in the problems of earthquake engineering [1]. Its Committee TC 13 'Seismic Design' followed this tradition by working as a consulting body of the drafting panel of Eurocode 8 (EC8) 'Structures in Seismic Regions', and in particular on Chapter 3 of Part 1.3, which is directly devoted to steel structures [2, 3]. The pressing deadlines for the issue of the first edition of EC8 forced TC 13 to accelerate the preparation of its recommendations, which have been adopted in EC8 by introducing only some small editorial changes [4, 5].

Only a few years have passed since the issue of the ECCS Recommendations for Steel Structures in Seismic Zones, but the need for a critical review is becoming increasingly urgent. In the meantime, research activity has led to the clarification and refinement of many problems concerning design of steel structures under seismic action. This book is mainly devoted to the presentation of results obtained by the authors in analysing the inelastic behaviour of moment-resisting frames under seismic loads.

In this context, it seems appropriate to begin this discussion with definitions representing the basics of modern seismic design procedures. After this, the inelastic behaviour exhibited by different structural typologies is examined and, in particular, the features of the seismic behaviour of moment-resisting frames are described. Next, the definition of seismic action is presented acccording to different seismic codes. Finally, the principal design criteria are introduced.

1.2 BASIC DEFINITIONS

Considering an elastic simple degree of freedom (SDOF) system subjected to a ground motion, the normalized elastic design response spectrum $R_e(T)$ is defined as the ratio between the maximum acceleration that the system has to withstand and the peak ground acceleration. This gives

$$R_e(T) = \frac{S_{a,e}(T)}{A}$$ (1.1)

where T is the period of the system, $S_{a,e}(T)$ is the maximum spectral acceleration that the indefinitely elastic system has to withstand, i.e. the linear elastic design response spectrum, and A is the peak ground acceleration.

A structural system could be designed to remain within the elastic range by adopting a design base shear force given by

$$F_e = M\, A\, R_e\,(T) = M\, S_{a,e}\,(T) \tag{1.2}$$

where M is the mass of the system.

As the above design criterion is extremely severe, leading to uneconomic solutions, structural ductility and energy dissipation capacity are usually taken into account. In this case the structural system is able to withstand severe earthquakes, preventing collapse even if it is designed by assuming a reduced base shear. The magnitude of this reduction depends on the features of the structural typology and the period of vibration. The design base shear is, therefore, given by

$$F_d = \frac{M\, A\, R_e(T)}{q_d\,(T)} = M\, S_{a,d}\,(T) \tag{1.3}$$

where F_d is the design base shear force, q_d is the design value of the q-factor and $S_{a,d}\,(T)$ is the design value of the spectral acceleration, i.e. the inelastic design response spectrum.

Therefore, the design value of the q-factor is defined as

$$q_d\,(T) = \frac{S_{a,e}\,(T)}{S_{a,d}\,(T)} \tag{1.4}$$

According to this relationship, the design value of the q-factor can be defined as the ratio between the linear elastic design response spectrum and the inelastic design response spectrum.

As will be discussed in Section 1.4, in modern European seismic codes, such as Eurocode 8 [6], the design value of the q-factor is expressed by means of a relationship in which, in general, two factors can be distinguished. The first, namely the behaviour factor, is related to the structural typology. It accounts for the ductility and energy dissipation capacity which can be considered intrinsic for that structural typology. The second factor takes into account the influence of the period of vibration and, therefore, is a function of T.

The general definition of the actual value of the q-factor, i.e. the one evaluated at the end of the design, is given by the ratio between the peak ground acceleration leading to collapse A_u and that which leads to the first yielding A_y [7, 8, 9]

$$q = \frac{A_u}{A_y} \tag{1.5}$$

It is evident that, according to the definition of the q-factor and its physical meaning, the structural typologies can be classified into two fundamental groups:

- non-dissipative structures
- dissipative structures

Non-dissipative structures have to be designed in order to withstand seismic action remaining within the elastic range, because they are not able to dissipate earthquake input energy by means of cyclic inelastic behaviour. Therefore, the corresponding q-factor is $q = 1$.

Dissipative structures are able to withstand severe earthquakes, thanks to their ductility and energy dissipation capacity due to ductile hysteretic behaviour of some of their components. These components are therefore referred to as dissipative zones. The q-factor of dissipative structures is always greater than one.

Non-dissipative parts of dissipative seismic structures and their connections to the dissipative ones have to be designed with sufficient overstrength in order to allow the cyclic yielding of the dissipative parts.

This means that non-dissipative parts have to be designed in order to remain in the elastic range and, therefore, they have to be proportioned on the basis of the maximum internal actions that the dissipative zones are able to transmit. Conversely, the dissipative zones have to be proportioned on the basis of internal actions arising from the seismic forces prescribed by the codes.

This is the general criterion for designing seismic-resistant dissipative structures and represents the so-called **capacity design approach**. The name means that non-dissipative parts have to be designed for the **capacity** of the fully yielded and strain-hardened dissipative zones.

The correct evaluation of the q-factor can be performed by a great number of dynamic inelastic analyses repeated for a great number of different ground motions. As a consequence, and because the corresponding computational effort is particularly cumbersome, simplified methods have been proposed by different researchers (Chapter 4).

The energy dissipation capacity of a structure and, therefore, the corresponding q-factor are strictly related to the mass, stiffness and strength distributions, both in plan and along the height. The term 'vertical regularity' refers to the ability of the structure to distribute ductility demands along its height as uniformly as possible. The term 'plan regularity' implies the ability of a building to vibrate separately in two orthogonal vertical planes without torsional coupling.

The influence of the structural regularity on the seismic behaviour of steel frames is discussed in Chapter 8.

The direct check against the ultimate limit state of structures subjected to destructive earthquakes requires the definition of parameters able to exhaustively characterize the structural damage. The 'admissible' value of such parameters depends on the materials, the structural details and the structural typology.

As a consequence, the seismic check can be performed by comparing the 'admissible' values of the damage parameters with the values requested to withstand the severest earthquake expected.

The traditionally used damage parameters are the maximum required ductility and the dissipated energy. On one hand, such parameters can be easily computed but, on the other, they can be criticized because they completely neglect the inelastic deformation history and fail to account for the combined action of the main damage mechanisms. In fact, the maximum ductility criterion considers the maximum displacement only, without providing any information on the number of plastic excursions and the ductility required by each of them. Conversely, the energy criterion takes into account all the plastic excursions, but it does not provide any information on how they accumulate plastic strains, giving rise to the total dissipated energy.

More realistic results are obtained by the combination of such parameters. Damage parameters based on the combination of the maximum required ductility criterion and of the energy criterion can be derived from experimental evidence [10, 11].

A very promising approach, which has been recently proposed by Krawinkler and Zohrei [12], is the use of a damage parameter based on the extension of the fatigue theory in the low-cycle range and on the linear damage cumulation assumption.

It is implicitly assumed that all deformation histories leading to the same value of a damage parameter give rise to similar structural damage. This assumption is fundamental to the practical use of parameters used for assessing the seismic resistance of structures.

The structural damage can be evaluated by means of one of the following parameters.

The **kinematic ductility**, i.e. the required ductility, is usually defined as the ratio between the maximum displacement occurring during the ground motion and the yielding displacement (Fig. 1.1)

$$\mu = \frac{x_{max}}{x_y} \tag{1.6}$$

Unlike the kinematic ductility, the **cyclic ductility** takes into account the possible changes of the origin of the plastic excursions (Fig. 1.1)

$$\mu_c = \frac{\max \{x_{p,i}\}}{x_y} + 1 \tag{1.7}$$

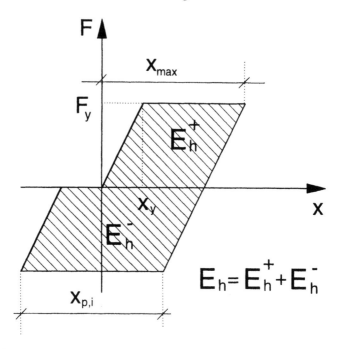

Fig. 1.1   Force versus displacement relation

For both kinematic ductility and cyclic ductility, the structural collapse is governed by the maximum plastic excursion, independently of the number of plastic excursions and on the plastic engagement required by each of them. Such an assumption is sufficiently close in the case of deformation histories characterized by only one cycle with a large plastic engagement and many cycles with small plastic excursions. On the contrary, for deformation histories characterized by many cycles having a similar plastic engagement, the use of an energy criterion seems more realistic. Two energy criteria have been adopted in the technical literature. The first criterion is represented by the maximum **hysteretic ductility in one direction** [13] which can be defined as

$$\mu_h^{(1)} = \frac{E^*}{F_y \, x_y} + 1 = \eta_c + 1 \tag{1.8}$$

where

$$E^* = \max\left\{ E_h^+ \, , \, E_h^- \right\} \tag{1.9}$$

and η_c is the cumulated plastic ductility in one direction.

The second criterion is represented by the **total hysteretic ductility** [14–16]. It represents the kinematic ductility required in the system so that it dissipates, under monotonic loading, the same amount of energy which is dissipated by the system subjected to the actual deformation history. As a consequence, it can be expressed by the relationship

$$\mu_h = \frac{E_h}{F_y \, x_y} + 1 \qquad (1.10)$$

In some cases, both the maximum plastic excursion and the dissipated energy can be significant. In such a case, the damage characterization can be rationally expressed through an opportune combination of the maximum displacement criterion and of the energy criterion.

On the basis of experimental data, it has been shown that, depending on material, structural details and structural typology, different combination criteria can be adopted. In particular, Park and Ang [9] have proposed a damage parameter based on a linear combination of the above criteria, given by

$$\mu_{P.A.} = \mu + \beta \, (\mu_h - 1) \qquad (1.11)$$

According to Cosenza *et al.* [17], the experimental values of β reported in the original work by Park [18], with reference to reinforced concrete structures, ranged between -0.3 and 1.2, with an average of about 0.15. Cosenza *et al.* found that the use of $\beta = 0.15$ gives results which closely correlate with the ones based on the damage model proposed by Krawinkler and Zohrei [12].

A very promising approach, proposed by Krawinkler and Zohrei [11], is represented by the application of the fatigue theory in the low-cycle range and of the linear damage cumulation according to the Miner rule. This approach leads to the following damage parameter

$$\mu_F = \sum_{i=1}^{n} (\mu_i - 1)^b \qquad (1.12)$$

where μ_i is the ductility required by the ith plastic excursion and n is the number of plastic excursions.

The exponent b, discussed in Chapter 3, depends on the slenderness of the member plates, the structural details and the material properties.

The above formulations provide a measure of the structural damage (local or global) from the absolute point of view, but they do not allow an immediate evaluation of the degree of safety with respect to collapse. For this reason, it is preferable to use normalized parameters, namely the damage functionals [16, 18]. The purpose of normalization is to obtain parameters which assume values equal to zero in the absence of plastic excursions, and unit values for collapse conditions. The normalization criterion should be based on experimental evidence by using limiting values that are easy to determine, generally by means of simple monotonic tests up to failure. Therefore, the normalization criterion should assure that the normalized parameter assumes a unit value under collapse conditions.

The following damage criteria (or collapse criteria) and the corresponding normalized damage parameters are widely used in the technical literature:

- normalized kinematic ductility

$$D_\mu = \frac{\mu - 1}{\mu_{u,mon} - 1} \tag{1.13}$$

- normalized cyclic ductility

$$D_{\mu_c} = \frac{\mu_c - 1}{\mu_{u,mon} - 1} \tag{1.14}$$

- normalized hysteretic ductility in one direction [13]

$$D_{\mu_h}^{(1)} = \frac{\mu_h^{(1)} - 1}{\mu_{u,mon} - 1} \tag{1.15}$$

- normalized total hysteretic ductility [17]

$$D_{\mu_h} = \frac{\mu_h - 1}{\mu_{u,mon} - 1} \tag{1.16}$$

- normalized Park-Ang's parameter [10]

$$D_{P.A.} = \frac{\mu + \beta\,(\mu_h - 1)}{\mu_{u,mon}} \tag{1.17}$$

- normalized low-cycle fatigue [16]

$$D_F = \frac{1}{(\mu_{u,mon} - 1)^b} \sum_{i=1}^{n} (\mu_i - 1)^b \tag{1.18}$$

where $\mu_{u,mon}$ represents the ultimate value of the kinematic ductility under monotonic loading conditions.

It is useful to note that the denominator of the second member term in equations (1.13)–(1.18) represents the 'admissible value' of the absolute damage. Moreover, it is interesting to note that two different assumptions can be made using the energy approach. The first assumption is that collapse is attained under cyclic loading conditions when the hysteretic ductility in one direction, positive or negative, is greater than the ultimate ductility under monotonic loading conditions [13]. This assumption corresponds to the use of the normalized damage parameter $D_{\mu_h}^{(1)}$ (1.15). The second assumption is that the available energy dissipation capacity under cyclic loads is equal to that under monotonic loads [17]. This assumption corresponds to the use of the normalized damage parameter D_{μ_h} (1.16). This collapse criterion is the most severe [17].

With reference to the low-cycle fatigue approach (1.18), it is important to point out that for $b = 1$ the energy criterion is obtained, while for $b \to \infty$ the maximum plastic excursion criterion is obtained.

Finally, it is easy to see that the normalized Park-Ang's parameter (1.17) does not provide a unit value at collapse under monotonic loading conditions. In spite of this inconvenience it is often used in seismic literature.

1.3 DISSIPATIVE STRUCTURAL TYPOLOGIES

The main typologies of seismic-resistant dissipative structures can be classified according to the type and the number of dissipative zones. Three different typologies can be recognized [20]:

- moment-resisting frames
- concentrically braced frames
- eccentrically braced frames

As this book is devoted to the analysis of seismic behaviour and to the design of moment-resisting frames, it is useful to compare this typology with concentrically braced frames and eccentrically braced frames, in order to emphasize the main differences between these systems. More detailed information on the latter typologies is available [21].

Moment-resisting frames (Fig. 1.2) are characterized by a large number of possible dissipative zones. These zones are represented by the ends of the members (i.e. beams and columns) which constitute the steel skeleton, where plastic hinges can develop. In this typology, the most important type of action is the bending moment, so that the energy dissipation, which takes place in plastic hinges, is due to the inelastic cyclic bending behaviour. In order to maximize the energy dissipation capacity of a moment-resisting frame, the structural design has to be conceived in a such a way that plastic hinges must lie in beams rather than in columns, except at the base of the frame. The failure mode corresponding to this condition is called the global collapse mechanism. In such a case,

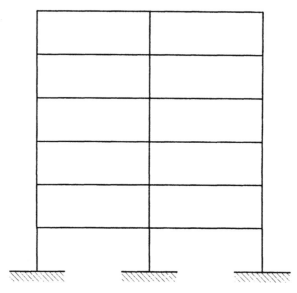

Fig. 1.2   Moment-resisting frames (MR frames)

dissipative zones are only located in beams near to beam-to-column connections.

This typology is widely used for low-rise buildings. In fact, they are able to provide a suitable energy dissipation mechanism, thanks to their large number of dissipative zones, thus leading to the fulfilment of the requirements which are necessary to prevent collapse under the severest design ground motion. On the other hand, fulfilment of the requirements necessary to guarantee the check against the serviceability limit state becomes more and more difficult as the height of the building increases. This is because framed structures are not able to provide sufficient stiffness in order to reduce sway deflection under moderate earthquakes or wind.

According to Eurocode 8 and to ECCS Recommendations [6, 22], the design value of the q-factor for rigid moment-resisting frames is provided by assigning to the behaviour factor the value $5\alpha_u/\alpha_y$, where α_u and α_y are, respectively, the maximum and first yielding values of the horizontal forces multiplier versus the top sway displacement behavioural curve (Fig. 1.3).

Concentrically braced frames can be designed in order to carry the total value of the seismic horizontal forces or only a part of them. In the first case, the beam-to-column connections have to be able to transfer shear force only; in the second case, they have to be able to transfer both

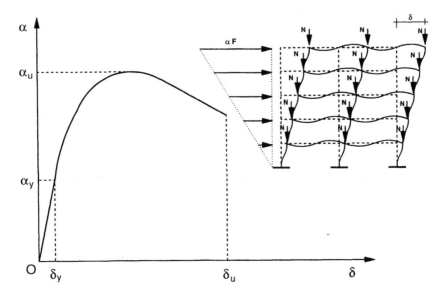

Fig. 1.3 Multiplier of horizontal forces versus top sway displacement curve

shear force and bending moment. In both cases connections must be of full strength type.

Dissipative zones of concentrically braced frames are represented mainly by the tensile diagonals, because of the assumption usually made that the compression diagonals are buckled. The inelastic cyclic performance of concentric bracings is rather unsatisfactory due to the repeated buckling of diagonal bars, which produces degradation of the energy dissipation capacity of the system as the number of cycles increases.

The bracing elements are located so that, irrespective of the type of beam-to-column connections, they are always mainly stressed by axial forces. The most common configurations, presented in Fig. 1.4, are

- X-braced frames
- V-braced frames
- inverted V-braced frames
- K-braced frames.

X-braced frames always have one diagonal in tension and the other in compression. The earthquake input energy is mainly dissipated by the tension diagonal, while the contribution of the compression diagonal decreases as its slenderness increases (Fig. 1.5).

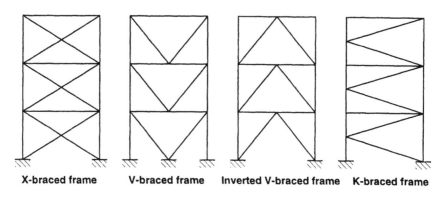

X-braced frame   **V-braced frame**   **Inverted V-braced frame**   **K-braced frame**

Fig. 1.4   Common configurations of concentrically braced frames (CB frames)

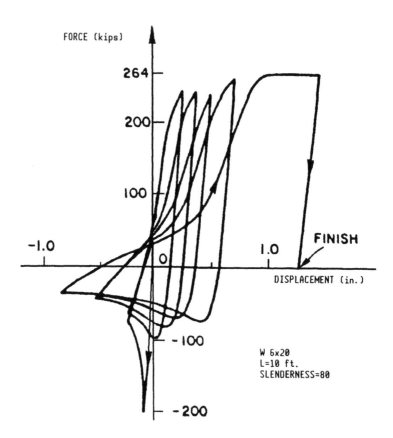

Fig. 1.5   Cyclic behaviour of axially loaded members

According to Eurocode 8 and ECCS Recommendations [6, 22], the design value of the q-factor of X-braced frames is given through the behaviour factor which is assumed equal to $4\alpha_u/\alpha_y$.

One tension and one compression diagonal are also present in V-braced and inverted V-braced frames. Due to the different behaviour of the tensile diagonal and the compression diagonal, there is a vertical action resulting from the axial forces in the bracing diagonals, producing a bending action in the beam, which has to be continuous in order to obtain a satisfactory behaviour. In any case, the post-buckling strength of the V-braced scheme is subjected to a quick degradation because, under load reversals, the previously buckled member probably could not return to its original alignment and the member which was in tension could exceed its capacity in compression. In this manner, both diagonal members could be in a buckled condition. For this reason, according to Eurocode 8 and ECCS Recommendations, the design value of the q-factor of V-braced frames is assigned by means of a behaviour factor equal to $2\alpha_u/\alpha_y$ [6, 22].

K-braced frames exhibit the same typical behaviour as V-braced frames, but a dangerous worsening of the seismic performances arises due to lateral displacement in the midspan of the column. This lateral displacement can undermine the structural ductility and failure mode control, leading to premature collapse due to the lateral buckling of the column. As a consequence, the reliability of K-braced frames, particularly in high-seismicity zones, is unsatisfactory. For this reason K-bracings cannot be considered as dissipative, due to the assumed cooperation of the column to the yielding mechanism, and the value $q_d = 1$ has to be assumed for the q-factor [6, 22].

Eccentrically braced frames constitute a suitable alternative structural typology with respect to moment-resisting frames and concentrically braced frames. They are characterized by the stiffening effect provided by the diagonals which are eccentrically located in moment-resisting frames. Due to the addition of these diagonals, the beam is divided in two or more parts. The shorter of these parts is called the 'link' and represents the dissipative element in eccentrically braced frames. In this typology, the earthquake input energy is dissipated by means of the inelastic cyclic shearing and bending of the link.

The common types of eccentrically braced frames, classified according to the location of the diagonal elements, are the D-brace, K-brace and V-brace (Fig. 1.6). They belong to the group of dissipative structures and are able to provide a ductility and energy dissipation capacity level similar to that of moment-resisting frames. For this reason, both structural typologies have, according to ECCS Recommendations [22], the same q-factor.

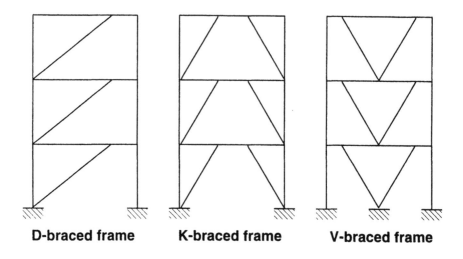

D-braced frame   **K-braced frame**   **V-braced frame**

Fig. 1.6   Typical configurations of eccentrically braced frames (EB frames)

The main differences between moment-resisting frames and all other most common steel structural typologies can be summarized on the basis of the requirements that a seismic-resistant structure has to satisfy: strength and stiffness against moderate ground motions with small return period and strength, plus ductility and energy dissipation capacity against severe earthquakes with a large return period.

With reference to the stiffness requirement, which is necessary to limit interstorey drifts under moderate seismic actions, it must be stated that moment-resisting frames are not very satisfactory, and can only be conveniently adopted for medium- and low-rise buildings. This defect is strongly reduced in eccentrically braced frames where the diagonal members allow an intermediate behaviour between the deformability of moment-resisting frames and the stiffness of concentrically braced frames.

Considering energy dissipation mechanisms, it can be shown that dissipative zones in moment-resisting frames are represented by the plastic hinges at the member ends, where the earthquake input energy is dissipated through inelastic cyclic bending. On the other hand, in eccentrically braced frames, the dissipative function is developed by the link beams where inelastic cyclic shearing and bending occur. Shear deformations play a very important role.

Finally, concentrically braced frames dissipate the earthquake input energy through the cyclic axial deformation of the diagonal members, mainly the tensile ones. In this case, buckling phenomena produce a great reduction in both stiffness and energy dissipation capacity as the number of cycles increases.

It can be concluded that the inelastic behaviour of moment-resisting frames leads to a better seismic performance under severe earthquakes with respect to concentrically braced frames, whereas, due to the great number of dissipative zones, to a ductility and to an energy dissipation capacity similar to that of eccentrically braced frames. Only the sway rigidity of moment-resistant frames is less than that of both concentrically and eccentrically braced frames.

1.4   SEISMIC ACTION

Seismic hazard maps are usually prepared in order to show the seismic zones of a region. They are based on a given annual probability of exceeding or, a given average return period of the quantity represented in the map (in general, the peak ground acceleration or the design seismic intensity is used). This allows the definition of different seismic zones, on the basis of a limiting value of the quantities represented in the map. Moreover, they are able to point out zones where effects of seismic action can be neglected. The seismic zonation of a region is, therefore, established in order to assure a uniform level of seismic protection to similar constructions throughout a particular region.

By means of seismic hazard maps, national territories are usually classified into different 'seismic zones' depending on the degree of local seismic activity, which is evaluated taking into account the general tectonic features of the region and seismic activity in the past.

Both ECCS Recommendations and Eurocode 8 [6, 22] consider three seismic zones: high seismicity, medium seismicity and low seismicity. These zones are defined through seismic hazard maps in which the peak ground acceleration is the determining parameter. Limiting values of $0.35g$, $0.25g$ and $0.15g$ (where g is the acceleration due to gravity) for the peak ground acceleration have been suggested for zones of high, medium and low seismicity, respectively.

Up to now, the most common type of dynamic analysis for the seismic design of buildings is the response spectrum approach. Therefore, elastic acceleration response spectra are the basic seismic input description used in most seismic codes.

The seismic action is defined by means of a normalized linear elastic design response spectrum and a standard value A of the peak ground acceleration (PGA). The peak ground acceleration defines the intensity of the earthquake according to the seismic zonation based on the seismic hazard maps.

For each value of the period T of a simple degree of freedom (SDOF) system

$$T = \frac{2\pi}{\omega} = 2\pi \sqrt{M/K} \tag{1.19}$$

where M is the mass of the system and ω the frequency; the normalized linear elastic design response spectrum $R_e(T)$ provides the ratio between the maximum acceleration that the system has to withstand and the peak ground acceleration.

According to this definition, the maximum acceleration that the system has to sustain is given by

$$S_{a,e}(T) = A\, R_e(T) \tag{1.20}$$

In Eurocode 8 [6] the linear elastic design response spectrum (LEDRS), corresponding to a damping ratio v, is expressed by means of the following relationships:

$$S_{a,e}(T) = A\, s \left[1 + \frac{T}{T_1}(\eta\, \beta_o - 1) \right] \qquad \text{for} \quad 0 \le T \le T_1 \tag{1.21}$$

$$S_{a,e}(T) = A\, s\, \eta\, \beta_o \qquad\qquad\qquad \text{for} \quad T_1 \le T \le T_2 \tag{1.22}$$

$$S_{a,e}(T) = A\, s\, \eta\, \beta_o \left(\frac{T_2}{T} \right)^{k_1} \qquad\quad \text{for} \quad T_2 \le T \le T_3 \tag{1.23}$$

$$S_{a,e}(T) = A\, s\, \eta\, \beta_o \left(\frac{T_2}{T_3} \right)^{k_1} \left(\frac{T_3}{T} \right)^{k_2} \qquad \text{for} \quad T \ge T_3 \tag{1.24}$$

where:

- β_o is the maximum normalized spectral acceleration, assumed constant between T_1 and T_2;
- T_1 and T_2 are the limits of the constant spectral acceleration branch;
- T_3 is the period value defining the beginning of the constant spectral displacement branch of the spectrum;
- k_1 and k_2 are exponents which influence the shape of the spectrum for a vibration period greater than T_2 and T_3, respectively;
- s is a soil parameter with a reference value of 1.0 for stiff soil conditions;

- η is a damping correction factor with reference value 1.0 for 5% viscous damping.

The value of the damping correction factor η is given by [6]

$$\eta = \sqrt{7/(2+v)} \geq 0.7 \qquad (1.25)$$

where v is the value of the viscous damping ratio of the structure, expressed as a percentage.

The values of the transition periods T_1 and T_2 essentially depend on the magnitude of the earthquake and on the ratios between the ground motion parameters, i.e. the peak ground values of acceleration A, velocity V and displacement D. As an example, the transition period T_1 is given approximately by $T_1 = 2\pi V/A$ [23].

The frequency content of the ground acceleration is influenced by the local soil conditions, so that it differs from that at the bedrock. In general, low-frequency components arise from softer soils. These frequency components are responsible for the absolute ground displacements, which, as a consequence, are higher in softer soils than in stiff soils.

The influence of the local soil condition is taken into account in Eurocode 8 by means of three different response spectra, each of which refers to a soil profile according to the following descriptions.

- Soil profile A – rock (shear wave velocity greater than 2000 m/s) or stable deposits of unconsolidated sediments as for B, with a depth of less than 50 m on a solid rock base.
- Soil profile B – stable deposits (compact sands or gravels or stiff clays) of depth exceeding 50 m on a solid rock base.
- Soil profile C – soft to medium-stiff deposits (sands, stiff clays) having a depth of 10 m or more.

As a guideline, possible values of the parameters defining the site-dependent response spectra suggested by Eurocode 8 are given in Table 1.1.

Table 1.1 Values of parameters defining elastic design response spectra

Soil profile	s	β_0	k_1	k_2	T_1 (s)	T_2 (s)	T_3 (s)
A	1.0	2.5	1.0	2.0	0.10	0.40	3.0
B	1.0	2.5	1.0	2.0	0.15	0.60	3.0
C	0.9	2.5	1.0	2.0	0.20	0.80	3.0

The response spectrum obtained for $s = 1$ represents the reference case, because it is used to model the seismic vibration on the bedrock or at the surface in rock or firm soil conditions. In this reference case the maximum spectral amplification is equal to β_0 which depends on the frequency content of the ground motion, on the ratio between the duration of the ground motion and the fundamental period of the system, on the viscous damping, and on the selected probability of exceedance. For typical earthquake durations of about 20–30 s and 5% damping, the value $\beta_0 = 2.5$ corresponds to a 50% probability of non-exceedance.

With reference to Eurocode 8, comparison of the corresponding spectra (Fig. 1.7) clearly shows the following properties of the site-dependent response spectra.

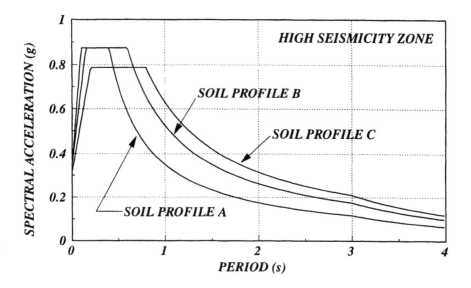

Fig. 1.7   Elastic design response spectra for high seismicity zones according to Eurocode 8

- The spectra for medium to firm soil sites have the highest spectral amplitudes (plateau) in the intermediate frequency range.
- At soft soil sites the frequency content is shifted to the lower frequency (higher period) range and the maximum spectral amplitude in the plateau area decreases when compared with firm soil sites.
- The rock site spectra are somewhat shifted to the higher frequency range and become narrower.

The linear elastic design response spectrum in equations (1.21) to (1.24) gives the acceleration which the structures should withstand to remain in the elastic range under the severest design earthquake.

Structures are usually not designed to support such orders of acceleration, but a reduced value is considered, taking into account that the earthquake input energy during severe ground motions can be dissipated through inelastic deformations. However, to prevent collapse, the value of these plastic deformations has to be limited according to the available local and global ductility. As a consequence, design spectra have to be derived in order to obtain the normalized pseudo-acceleration required for a specified level of inelastic response. As previously mentioned, these inelastic spectra are obtained in seismic codes by reducing the linear elastic design response spectrum by means of a coefficient, the q-factor, which is related to the structural typology, its ductility and energy dissipation capacity, and to the period of vibration.

The inelastic design response spectrum (IDRS) is therefore given by

$$S_{a,d}(T) = \frac{S_{a,e}(T)}{q_d(T)} \tag{1.26}$$

where $q_d(T)$ is the design value of the q-factor.

According to Eurocode 8, the inelastic design response spectra for the above three soil profiles are given by the following relationships:

$$S_{a,d}(T) = A\ s \left[1 + \frac{T}{T_1}\left(\frac{\eta\,\beta_o}{\bar{q}} - 1 \right) \right] \qquad \text{for} \quad 0 \le T \le T_1 \tag{1.27}$$

$$S_{a,d}(T) = \frac{A\ s\ \eta\ \beta_o}{\bar{q}} \qquad \text{for} \quad T_1 \le T \le T_2 \tag{1.28}$$

$$S_{a,d}(T) = \frac{A\ s\ \eta\ \beta_o}{\bar{q}} \left(\frac{T_2}{T} \right)^{k_{d_1}} \ge 0.20\ A \quad \text{for} \quad T_2 \le T \le T_3 \tag{1.29}$$

$$S_{a,d}(T) = \frac{A\ s\ \eta\ \beta_o}{\bar{q}} \left(\frac{T_2}{T_3} \right)^{k_{d_1}} \left(\frac{T_3}{T} \right)^{k_{d_2}} \ge 0.20\ A \quad \text{for} \quad T \ge T_3 \tag{1.30}$$

where $\bar{q}$ is the 'behaviour factor' given by the code which represents the intrinsic dissipative capacity of a given structural typology, and k_{d_1} and k_{d_2} are exponents which influence the shape of the spectrum for a vibration period greater than T_2 and T_3, respectively.

The guideline values of the parameters k_{d_1} and k_{d_2}, taken from Eurocode 8, are given in Table 1.2.

Table 1.2   Values of additional parameters defining inelastic design response
spectra

Soil profile	k_{d_1}	k_{d_2}
A	2/3	5/3
B	2/3	5/3
C	2/3	5/3

These spectra, which are shown in Fig. 1.8 for a value of the behaviour
factor equal to 6, take into account that in the short-period range the
advantages of the inelastic behaviour are less important. In addition, they
assume that for the theoretical case $T = 0$ (infinitely rigid structures) the
benefits of plastic behaviour cannot be exploited, so that $q_d = 1$.

Furthermore, in the case of tall and complex buildings, the increase
of vibration modes is accompanied by an increase in those modes which
can produce severe local damage. As a consequence, there is also a
greater likelihood that high ductility demands can be concentrated in
limited parts of buildings. Taking into account that in these cases there
is a greater risk of damage concentration and, on the other hand, that
these buildings have a greater sensitivity to second-order effects, the
European seismic codes adopt more conservative criteria for longer-pe-
riod structures.

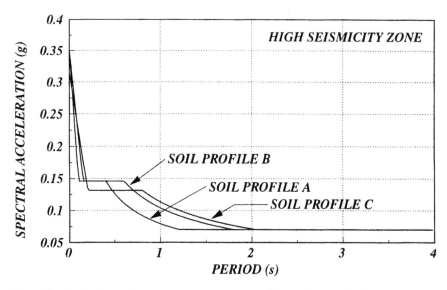

Fig. 1.8   Inelastic design response spectra according to Eurocode 8

In particular, in these cases it appears necessary to increase the lateral design forces and this has been quantified as an increment of 50% at a period of 2.0 s. This increase is gradually reduced as the building period decreases. This variation has been achieved by taking the parameter k_{d_1} equal to 2/3, instead of the value $k_1 = 1$ assumed for the corresponding parameter in the elastic design response spectrum [24].

The design value of the q-factor, resulting from the above spectra, can be derived from the direct application of equation (1.4). As a consequence, and with reference to Eurocode 8, the following relationships are obtained:

$$q_d(T) = \frac{1 + \dfrac{T}{T_1}\left(\eta\,\beta_o - 1\right)}{1 + \dfrac{T}{T_1}\left(\dfrac{\eta\,\beta_o}{\bar{q}} - 1\right)} \qquad \text{for} \quad 0 \le T \le T_1 \quad (1.31)$$

$$q_d(T) = \bar{q} \qquad \text{for} \quad T_1 \le T \le T_2 \quad (1.32)$$

$$q_d(T) = \bar{q}\left(\frac{T_2}{T}\right)^{k_1 - k_{d_1}} \le 5\,\eta\,s\,\beta_o\left(\frac{T_2}{T}\right)^{k_1} \qquad \text{for} \quad T_2 \le T \le T_3 \quad (1.33)$$

$$q_d(T) = \bar{q}\left(\frac{T_2}{T_3}\right)^{k_1 - k_{d_1}}\left(\frac{T_3}{T}\right)^{k_2 - k_{d_2}} \le 5\,\eta\,s\,\beta_o\left(\frac{T_2}{T_3}\right)^{k_1}\left(\frac{T_3}{T}\right)^{k_2} \quad \text{for } T \ge T_3 \quad (1.34)$$

These relationships show that the design value of the q-factor is dependent on the structural typology through the behaviour factor $\bar{q}$ and, due to the different shape of the inelastic design response spectrum with respect to that for the linear elastic case, on the period of vibration. The comparison between the elastic design response spectra, for high-seismicity zones, and the corresponding inelastic design response spectra is given in Fig. 1.9 for a value of the behaviour factor $\bar{q}$ equal to 6. For the same case, the resulting design values of the q-factor are given in Fig. 1.10.

For the determination of design loads to be used in calculations, the seismic action previously defined has to be multiplied by a partial safety factor, namely the importance factor I, which adapts the degree of seismic protection to the social and economic importance of the building. This objective is accomplished by identifying five categories of structure [22]:

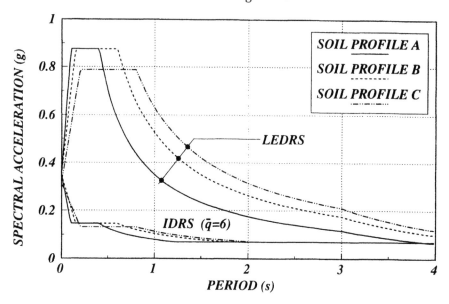

Fig. 1.9   Comparison between elastic and inelastic design response spectra for a
behaviour factor equal to 6

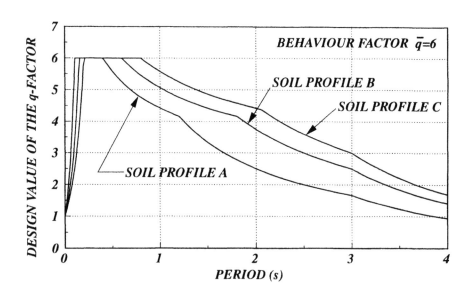

Fig. 1.10   Design values of the *q*-factor derived from Eurocode 8 spectra

- (I) structures having vital importance for civil protection, such as hospitals, fire stations, electricity plants, etc.;
- (II) structures where crowding has to be foreseen, such as schools, assembly halls, cultural institutions, churches, etc.;
- (III) structures which do not belong to either of the above two categories;
- (IV) structures of minor importance for public safety requirements, such as agricultural buildings;
- (V) provisional structures.

In the European codes, the suggested values of the importance factors for the first four classes I to IV are equal, respectively, to 1.4, 1.2, 1.0 and 0.8 [6, 22]. For provisional structures (category V) the importance factor can be stated as a function of the expected life of the structure. The values 0.4, 0.6, 0.8 and 1.0 have been suggested for an expected life of 2, 5, 10 and 20 years respectively [22].

1.5   COMPARISON BETWEEN SEISMIC CODES

It is useful to compare the inelastic design response spectra adopted by the European seismic codes with those adopted in the USA and Japan.

For this comparison of the seismic codes currently adopted in the USA, is chosen the 1990 edition of the seismic provisions of AISC-LRFD (American Institute of Steel Construction – Load and Resistance Factor Design) [25], which are based on document 7-88 of the ASCE (American Society of Civil Engineers) [26], and the UBC (Uniform Building Code) of 1991 [27], which is essentially coincident with the SEAOC (Structural Engineering Association Of California) Recommendations of 1990 [28].

With reference to moment-resisting frames, the main difference between these seismic codes and the European approach is represented by their classification by the former into two groups, namely ordinary moment-resisting frames (OMRF) and special moment-resisting frames (SMRF) depending on the design criteria.

Special moment-resisting frames are designed by considering some restrictive dimensioning and detailing rules, which are mainly intended to control the failure mode, the structural ductility and its energy dissipation capacity.

The fulfilment of all design rules requested for special moment-resisting frames is not necessary in the ordinary case.

In the AISC-LRFD seismic provisions of 1990 [25], the design base shear is expressed by means of the relationship

$$F_d = Z\,I\,K\,S\,C\,W \qquad (1.35)$$

where

- Z is the seismic zone factor, varying from 0.1875 to 1.0;
- I is the importance factor, varying from 1.0 to 1.5;
- K is the factor dependent on the structural typology (which is related to the q-factor adopted in the European codes), varying from 0.67 to 2.5 for structures designed and detailed according to the provisions;
- S is the soil factor, varying from 1.0 to 1.2;
- C is the period-dependent seismic coefficient which is provided by $C = 1/(1.5T^1/2)$ with the limitations $C \leq 0.12$ and $CS \leq 0.14$;
- W is the sum of the dead and partial live load.

By considering the severest seismic zone ($Z = 1$), buildings of normal importance ($I = 1$), special moment-resisting frames ($K = 0.67$) and the severest soil condition ($S = 1.2$), the design base shear coefficient is provided by

$$\frac{F_d}{W} = 0.536 \, \frac{1}{T^{1/2}} \leq 0.0938 \tag{1.36}$$

The AISC-LRFD code involves explicit consideration of limit states, load and resistance factor, and implicit probabilistic determination of reliability. The designation LRFD reflects the concept of factoring both loads and resistances. Taking into account that in the combination of the factored loads the earthquake load has a factor equal to 1.5, it is found that in the low-period range the design base shear coefficient is equal to $0.0938 \times 1.5 = 0.141$.

On the contrary, the seismic provisions given in UBC91 [27] are based on the allowable stress design (ASD). The design base shear is here expressed by means of the relationship

$$F_d = \frac{ZICW}{R_w} \tag{1.37}$$

where

- Z is the seismic zone factor, varying from 0.075 to 0.4;
- I is the importance factor, varying from 1 to 1.25;
- R_w is the factor dependent on the structural typology (which is related to the q-factor adopted in the European codes), varying from 4 to 12;
- C is the soil and period-dependent seismic coefficient, which is provided by $C = 1.25S/T^{2/3}$ with the limitation $C \leq 2.75$, where S is the soil factor, varying from 1.0 to 2.0;
- W is the sum of the dead and partial live load.

By considering the severest seismic zone ($Z = 0.40$), buildings of normal importance ($I = 1$), special moment-resisting frames ($R_w = 12$) and dense or stiff soil condition ($S = 1.2$), the design base shear coefficient is provided by

$$\frac{F_d}{W} = 0.05 \, \frac{1}{T^{2/3}} \le 0.0917 \tag{1.38}$$

Taking into account that in allowable stress design of UBC91 the safety factor is equal to 1.5, it is found that in the low-period range the first yielding design base shear coefficient is equal to $0.0917 \times 1.5 = 0.1375$.

It means that, when using both the UBC91 code and the LRFD recommendations for designing the same structural typology in the same seismic zone, they provide practically the same design base shear coefficient. Therefore, by comparing equations (1.35) and (1.37) it appears that the relationship $R_w \approx 8/K$ [29] can be used to correlate the two coefficients, playing the same role as the European behaviour factor.

In the Japanese seismic code [30, 31], the required base shear capacity is provided by

$$F_d = D_s \, F_{es} \, Z \, R(T) \, C_0 \, W \tag{1.39}$$

where

- D_s is the coefficient, varying from 0.25 to 0.45 for moment-resisting frames, which takes into account the ductility and energy dissipation capacity of the structure (it is the inverse of the European behaviour factor: $D_s = 1/\overline{q}$ [31]);
- $F_{es} = F_e F_s$ is a structural shape factor which considers the adverse effects due to eccentricity between the stiffness centre and the gravity centre in the plan (with F_e varying from 1.0 to 1.5) and the change of the building stiffness along the height (with F_s varying from 1.0 to 1.5) where, in the case of regular buildings, $F_e = F_s = F_{es} = 1$;
- Z is the seismic hazard zoning coefficient, varying from 0.7 to 1.0;
- $R(T)$ is the non-dimensional response spectrum given by

$$R(T) = 1 \qquad \text{for} \quad T \le T_o \tag{1.40}$$

$$R(T) = 1 - 0.2 \left(\frac{T}{T_o} - 1 \right)^2 \qquad \text{for} \quad T_o \le T \le 2\,T_o \tag{1.41}$$

$$R(T) = \frac{8T_o}{5T} \qquad \text{for} \quad T \ge 2\,T_o \tag{1.42}$$

with the limitation that $R(T) \ge 0.25$, where T_o is equal to 0.4, 0.6 and 0.8 for hard, medium and soft soil, respectively;

- C_0 is a coefficient which has to be assumed equal to 0.2 for the service-ability limit state and equal to 1.0 for the ultimate limit state.

Therefore, with reference to regular buildings made of ductile moment-resisting frames, the design base shear coefficient for the ultimate limit state analysis is equal to 0.25. The Japanese code is now under revision; however, the above magnitude of the design base shear coefficient has been confirmed in the draft of the new code [32, 33].

In Eurocode 8, with reference to the constant acceleration range of the inelastic design response spectrum, the design base shear is given by

$$F_d = \frac{M \, A \, s \, \beta_0}{\bar{q}} \qquad (1.43)$$

which provides a design base shear coefficient given by

$$\frac{F_d}{W} = 0.35 \frac{\beta_0}{\bar{q}} \qquad (1.44)$$

With respect to moment-resisting frames, if it is assumed $\alpha_u/\alpha_y \approx 1.2$, then the design value of the behaviour factor is equal to $5 \, \alpha_u/\alpha_y \approx 6$. As a consequence, for the low-period range with $\beta_0 = 2.5$, the design base shear coefficient is about 0.1458.

In the case of ECCS Recommendations, taking into account that the suggested maximum normalized spectral amplitude is equal to 3.0 [22] (against the value 2.5 suggested by Eurocode 8), the value 0.175 is obtained for the design base shear coefficient ($3 \times 0.1485/2.5 = 0.175$).

The values obtained for both ECCS Recommendations and Eurocode 8 are only qualitative, because the parameters defining the seismic action have to be provided by the competent national authorities on the basis of seismic hazard analyses [6, 22]. Due to this limitation, the values of the above parameters have to be considered as guideline values. Nevertheless the comparison among the design base shear coefficients adopted by the main seismic codes is interesting in order to understand the different safety approaches, even if they are referred to completely different seismic zones where, on the basis of the corresponding seismic hazard analyses, earthquakes of different magnitude and frequency content have to be expected for the same given probability of non-exceedance.

It can be seen that the Japanese code requires a design base shear about 1.75 times that adopted by the other codes.

The above comparison of seismic codes shows that the difference between corresponding design base shears is slightly influenced by the different elastic response spectra. As an example, for the severest design earthquake the maximum spectral accelerations are given by

$$\frac{S_{u,e}(T)}{g} = 5.33 \frac{S}{T^{1/2}} \leq 1.12 \qquad \text{for LRFD} \qquad (1.45)$$

$$\frac{S_{u,e}(T)}{g} = 0.40 \frac{1.25\,S}{T^{2/3}} \leq 0.40 \times 2.75 = 1.10 \qquad \text{for UBC91} \qquad (1.46)$$

$$\frac{S_{u,e}(T)}{g} \leq 1 \qquad \text{for the Japanese Code} \qquad (1.47)$$

$$\frac{S_{u,e}(T)}{g} \leq 0.35 \, \beta_o = 0.875 \qquad \text{for Eurocode 8} \qquad (1.48)$$

$$\frac{S_{u,e}(T)}{g} = 0.35 \frac{\beta_o}{(T/T_o)^{2/3}} \leq 1.05 \qquad \text{for ECCS} \qquad (1.49)$$

The maximum scatter is from 0.875 to 1.12.

The main reason for the difference is the values assumed for the parameters related to the q-factor. It is easy to recognize from equations (1.35), (1.37), (1.39) and (1.43) that the following relationships can be established between the behaviour factor, adopted in the European codes, and the similar coefficients K, R_w and D_s used in the 1990 AISC-LRFD, UBC91 and the Japanese seismic code, respectively:

$$\bar{q} = \frac{5.33}{K} \quad \text{for LRFD} \qquad (1.50)$$

$$\bar{q} = \frac{R_w}{1.5} \quad \text{for UBC88} \qquad (1.51)$$

$$\bar{q} = \frac{1}{D_s} \quad \text{for the Japanese code} \qquad (1.52)$$

For $K = 0.67$, $R_w = 12$ and $D_s = 0.25$ these give 7.96, 8 and 4 for LRFD, UBC91 and the Japanese code, respectively. It means that the Japanese code is the most conservative from this point of view.

Eurocode 8 provides a design base shear similar to that of the American seismic codes, because the value of the q-factor is less than the American one (6 against 8), but it has to be applied to reduce a spectral acceleration which is less than the American one (0.875 against 1.10).

As the spectral accelerations provided by the Japanese seismic code are comparable with the others, the greater magnitude of the design seismic forces is mainly due to the very restrictive value adopted for the behaviour factor.

This difference should be considered in the light of the unquestionable fact that the value of the behaviour factor has to be derived not only by engineering judgement, but also by cost–benefit considerations. In fact, seismic-resistant steel structures in Japan are very heavy in appearance compared to American and European ones, corresponding to higher safety levels against earthquakes.

1.6   DESIGN CRITERIA FOR MR FRAMES

The structural members, where dissipative zones are located, have to be designed to resist the actions produced by the design seismic forces, while the other members have to elastically withstand the maximum actions which the dissipative zones are able to transmit.

As already seen, the above statement represents the general design criterion of seismic-resistant dissipative structures, sometimes referred to as the 'capacity design approach'. According to this approach, the control of the failure mode of a seismic-resistant structure can be obtained by means of the correct selection of the dissipative zone location and by designing other zones so that they are able to elastically resist under actions equal to those that dissipative zones trasmit.

The optimum selection of the dissipative zone location no longer represents a great problem because the behaviours corresponding to the different failure modes for various structural typologies have been widely investigated, and those leading to the best seismic performance for a given structural typology have been recognized.

In the case of moment-resisting frames (MR frames), it is universally accepted that, in order to obtain the best seismic performance, they have to be designed so that the possible places of plastic hinge formation are located only in the beams near the beam-to-column connections, but never in the columns with the exception of the base of first-storey columns. The energy dissipation mechanism corresponding to the above location of dissipative zones is referred to as the global type mechanism (Fig. 1.11).

For this reason, the design rules provided by the modern seismic codes are mainly intended for fulfilment of the requirements necessary for obtaining the global-type energy dissipation mechanism. These provisions will be developed in detail in the following chapters; in this section it is sufficient only to assume that they mainly refer to the required flexural and shear strength of beam-to-column connections and to the column-to-beam moment ratio.

In particular, it is assumed that the flexural strength of the connections has to be greater than the plastic bending moment of the beam [6, 22, 25, 27] or the moment resulting from the panel zone nominal shear strength [25, 27]. The required shear strength of the connection has to be evaluated taking into account the actions transmitted by vertical loads and by considering the possibility that the flexural strength of the beam is reached at its ends.

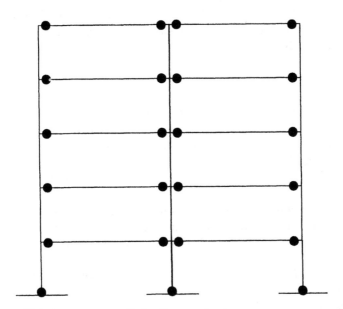

Fig. 1.11 Global-type energy dissipation mechanism

In addition to the above rules, which are intended to allow the complete development of the flexural strength of the beam, other design rules are provided in order to limit the possibility of column hinging. The most important of these states that at any beam-to-column intersection the sum of the plastic moment of the columns has to be greater than the sum of the plastic moment of the beams.

In both LRFD Recommendations and in the UBC91 code, MR frames meeting the above design criteria are referred to as 'special MR frames'; in the opposite case, they are referred to as 'ordinary MR frames'. According to the different energy dissipation capacity, in equation (1.37) the value $R_w = 12$ is assigned to special MR frames and the value $R_w = 6$ to ordinary MR frames [27], which correspond to European behaviour factors equal to $\bar{q} = 8$ and $\bar{q} = 4$, respectively.

In the European codes a similar distinction is not present, but by considering the design and detailing rules we can recognize that only special MR frames are allowed in European seismic zones according to the new generation codes.

A different approach is adopted in the Japanese seismic code [31], where the q-factor is assigned to the framed structure on the basis of the storey ductility ratio η. The background of the Japanese seismic code will be presented in detail in Chapter 5; in this section it is useful only to note that for MR frames the values D_s = 0.25, 0.30, 0.35 and 0.45 (q = $1/D_s$) are provided for the ductility classes $\eta \geq 4$, $4 > \eta \geq 2$, $2 > \eta \geq 0$ and $\eta < 0$, respectively.

1.7 REFERENCES

[1] *F.M. Mazzolani:* 'The ECCS activity in the field of recommendations for steel seismic resistant structures', 9th International Conference on Earthquake Engineering, Tokyo-Kyoto 1988.

[2] *F.M. Mazzolani:* 'The European Recommendations for steel structures in seismic areas', International Colloquium of Stability of Steel Structures, Budapest, April 1990.

[3] *F.M. Mazzolani:* 'A new generation of seismic codes: the European Recommendations for Steel Structures in Seismic Areas', 1st International Conference on Seismology and Earthquake Engineering, Teheran, May, 1991.

[4] *F.M. Mazzolani:* 'Design of Seismic Resistant Steel Structures: The ECCS approach', International Conference on Steel & Aluminium Structures, Singapore, 22–24 May, 1991.

[5] *F.M. Mazzolani:* 'The European Recommendations for Steel Structures in Seismic Areas: principles and design', Annual Technical Session of SSRC, Chicago, 15–17 April 1991.

[6] *Commission of European Communities:* 'Eurocode 8: Structures in Seismic Regions', Draft of October 1993.

[7] *G. Ballio:* 'ECCS approach for the design of steel structures against earthquakes', Proceedings of ECCS-IABSE Symposium, Luxembourg, 1985.

[8] *G. Ballio, C.A. Castiglioni, F. Perotti:* 'On the assessment of structural design factors for steel structures', IX World Conference on Earthquake Engineering, Tokyo-Kyoto, Japan, August, 1988.

[9] *G. Ballio, C.A. Castiglioni:* 'Le costruzioni metalliche in zona sismica: un criterio di progetto basato sulla accumulazione del danno', Giornate Italiane della Costruzione in Acciaio del C.T.A., Viareggio, 24–27 Ottobre 1993.

[10] *Y.J. Park, A.H.S. Ang:* 'Mechanic Seismic Damage Model for Reinforced Concrete', Journal of Structural Engineering, ASCE, April, 1985.

[11] *J.E. Stephens, J.T.P. Yao:* 'Damage Assessment using Response Measurements', Journal of Structural Engineering, ASCE, Vol. 113, April, 1987.

[12] *H. Krawinkler, M. Zohrei:* 'Cumulative Damage in Steel Structures Subjected to Earthquake Ground Motion', Computer & Structures, Vol. 16, N.1–4, 1983.

[13] *H. Akiyama:* 'Earthquake-Resistant Limit State Design for Buildings', University of Tokyo Press, 1985.

[14] *S. Mahin, V.V. Bertero:* 'An evaluation of inelastic seismic design spectra', Journal of the Structural Division, ASCE, September, 1981.

[15] *A.A. Nassar, H. Krawinkler:* 'Seismic Demands for SDOF and MDOF systems', J.A. Blume EEC Report No.95, Department of Civil Engineering, Stanford University, 1991.

[16] *P. Fajfar:* 'Equivalent ductility factor, taking into account low-cycle fatigue', Earthquake Engineering and Structural Dynamics, Vol. 21, pp. 837–848, 1992.

[17] *E. Cosenza, G. Manfredi, R. Ramasco:* 'An Evaluation of the Use of Damage Functionals in Earthquake Engineering Design', 9th European Conference on Earthquake Engineering, Moscow, September, 1990.

[18] *Y.J. Park:* 'Seismic Damage Analysis and Damage-Limiting of R/C Structures', PhD Thesis, Department of Civil Engineering, University of Illinois, Urbana, 1984.

[19] *E. Cosenza, G. Manfredi, R. Ramasco:* 'The Use of Damage Functionals in Earthquake Engineering: A Comparison between Different Methods', Earthquake Engineering and Structural Dynamics, Vol. 22, 1993.

[20] *F.M. Mazzolani:* 'Seismic Behaviour of Steel Structures', 1st National Conference on Steel Structures, Athens, June 1991.

[21] *F.M. Mazzolani, V. Piluso:* 'ECCS Manual on Design of Steel Structures in Seismic Zones', ECCS Document No.76, First Edition, January 1994.

[22] *ECCS-CECM-EKS:* 'European Recommendation for Steel Structures in Seismic Zones, Technical Working Group 1.3 (now TC13): Seismic Design', N.54, 1988.

[23] *Commission of European Communities:* 'Eurocode 8: Structures in Seismic Regions', Draft of May 1988.

[24] *S.A. Anagnostopoulos:* 'Methods of assessing seismic hazard for uniform selection of regional parameters in Eurocode 8', Background Document for Eurocode 8, Vol. 1, 'Seismic Input Data', May, 1988.

[25] *AISC-LRFD:* 'Seismic Provisions for Structural Steel Buildings – Load and Resistance Factor Design', American Institute of Steel Construction, November 15, 1990.

[26] *ASCE:* 'Minimum Design Loads for Buildings and Other Structures', American Society of Civil Engineers, ASCE 7-88, New York, 1990.

[27] *UBC:* 'Uniform Building Code', International Conference of Building Officials, Whittier, CA, 1988.

[28] *SEAOC:* 'Recommended Lateral Force Requirements and Commentary, Seismology Committee', Structural Engineers Association of California, Los Angeles, Sacramento, San Francisco, CA, 1990.

[29] *E.P. Popov:* 'Seismic Design' in 'Constructional Steel Design: An International Guide', Edited by P.J. Dowling, J.E. Harding and R. Bjorhovde, Elsevier Applied Science, London and New York, 1992.

[30] *IAEE:* 'Earthquake Resistant Regulations: A World List – 1980', International Association for Earthquake Engineering, August, 1980.

[31] **B. Kato:** Personal communication to Prof. F.M. Mazzolani, November, 1989.

[32] *AIJ:* 'Standard for Limit State Design of Steel Structures' (Draft), Architectural Institute of Japan, 1990 (English version, October 1992).

[33] **B. Kato:** 'Development and Design of Seismic-Resistant Steel Structures in Japan', International Workshop on Behaviour of Steel Structures in Seismic Areas, Timisoara, Romania, 26 June – 1 July, 1994.

2

Global ductility of MR frames

2.1 ANALYSIS METHODS

The most important problem which has to be faced by the engineer during the design process is the definition of the structural model. Structural modelling is related, on one hand to the analysis of the structure, and on the other hand to the verification of members: they represent two fundamental steps for evaluating the structural safety. The first step is performed in order to determine the stress distribution under the action of design loads; the second one is aimed at the control of the ability of members to withstand the internal forces. As a consequence, it is clear that the structural modelling is related to the use of a method of global analysis combined with a method of member verification.

Different types of global analysis can be performed (Fig. 2.1a):

1. linear elastic analysis;
2. elastic critical load analysis;
3. elastic analysis which includes the effects of displacements (geometric non-linearity);
4. elastic–plastic analysis which includes the effects of yielding (material non-linearity);
5. elastic–plastic analysis which includes both geometric and material non-linearities.

The above order corresponds to an increasing degree of difficulty, but the higher order methods are increasingly able to predict faithfully the global structural response. It is also important to establish that the type of analysis that can be performed is strictly related to the behavioural features of the member cross-sections.

As the type of structural analysis to be performed depends on the behaviour of the cross-sections of frame members, it is important to be

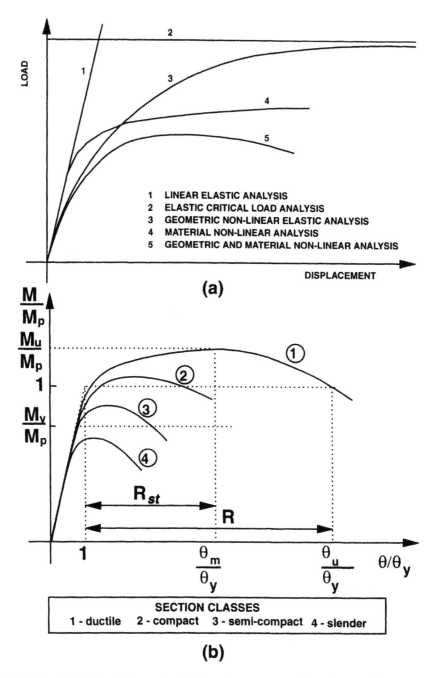

Fig. 2.1   Global analysis methods (a) and cross-section classification (b)

aware that, according to EC3, the cross-sections can be classified into four behavioural classes (Fig. 2.1b).

The term plastic or ductile cross-section (class 1) refers to sections which are able to develop a plastic hinge with a high rotation capacity, while the term compact cross-section (class 2) is used for those cases in which plastic hinges can develop, but a limited rotation capacity is available.

Semi-compact cross-sections (class 3) are characterized by the fact that the material yield strength can be attained in both extreme fibres, but local buckling in compressed zones prevents the development of the full plastic moment resistance.

Finally, in slender cross-sections (class 4) the premature development of local buckling of the compressed parts of the section prevents the attainment of the material yield strength.

As far as cross-section verification is concerned, it is clear that in the first two cases it can be performed by **plastic methods**, while in the second two cases **elastic methods** have to be used, referring to the full section in the case of semi-compact sections and to the effective section in the case of slender cross-sections.

According to the type of cross-section, different combinations of global analysis and member verification methods can be adopted.

The possibility of developing a plastic global analysis (plastic design) is restricted to the use of plastic cross-sections, so that plastic hinges can develop with a sufficient rotation capacity to allow a moment redistribution within the structure and to lead to the formation of a plastic mechanism. In this case the plastic global analysis is combined with a plastic verification method for members.

The use of compact sections can limit the possibility of reaching a plastic mechanism, because plastic hinges can form, but they do not guarantee a sufficient rotation capacity, so that moment redistribution is not likely to occur. As a consequence, in this case the elastic analysis has to be used to compute the internal forces which have to be compared with the cross-section plastic capacities. In the case of statically determined structures the load-bearing capacity is equal to that resulting from the adoption of plastic sections, because the attainment of the first plastic hinge leads to a failure mechanism.

For both semi-compact and slender sections, the plastic capacities cannot be fully developed; therefore, both global analysis and member verification have to be limited to the elastic range. In particular, for semi-compact sections, the ultimate limit state is represented by the occurrence of material yielding in the most stressed fibre, while in the case of slender sections it is represented by the local buckling of the compressed part, due to the inability of these sections to reach the material's yielding point.

As the term global ductility refers to the ability of a system to sustain large deformations in the plastic range, it is quite obvious that high-ductility moment-resisting frames require the adoption of plastic cross-sections which are able to provide sufficient rotation capacity, allowing the moment redistribution and the development of a plastic collapse mechanism.

The rotation capacity is a measure of the ability of a bent member to sustain plastic deformations. It represents a measure of the local ductility of bent members. It is worth stressing that there is, on one hand, a strict link between global and local ductility and, on the other hand, between ductility, either global or local, and the type of analysis to be performed.

In this chapter the parameters influencing the global ductility are examined, together with the dimensioning and detailing rules, which are provided by the seismic codes in order to improve the ductile behaviour of MR frames. The phenomena affecting the local ductility of bent members and the methods for its evaluation will be described in the next chapter.

2.2 THE ROLE OF GLOBAL DUCTILITY IN SEISMIC DESIGN

It has been already pointed out that the ability of a member or a system to sustain deformations beyond its yield point without significant loss of strength is usually referred to by the term ductility.

Ductility of structures plays a fundamental role in seismic design. In fact, structures are usually designed so that some of the energy input during severe earthquakes is dissipated through inelastic deformations. Therefore, seismic analysis of structures is usually divided, for the purpose of research activity, into two fundamental steps: the first step requires the evaluation of the available ductility, while the second is the computation of the ductility required by the severest design earthquake. It is clear that this methodology can be used to prevent collapse, provided that the available ductility is greater than that required.

For a given structure the global ductility is defined as the ratio between the values of a given displacement, in general the top sway displacement in the case of frames, evaluated at the ultimate and the yield conditions. This definition has been largely used in the past to extend to multi degree of freedom (MDOF) systems the large amount of data available for the inelastic behaviour of simple degree of freedom (SDOF) systems. The critical point of this extension is due to the number of parameters characterizing the pattern of yieldings of multi degree of freedom structures. In fact it is clear that only the inelastic response of SDOF systems can be characterized by using a simple response parameter such as the required global ductility, while in the case of MDOF systems different patterns of yieldings could correspond to the same maximum inelastic displacement. In this case the control of local ductility demands is, therefore, necessary. Moreover, it has to be noted that ductility itself does not represent a sufficient parameter with which to estimate the structural damage. Many researchers have pointed out that the concept of structural ductility fails to take into account the number of yield excursions and reversals; this knowledge may be useful to evaluate the amount of damage sustained by structures during severe earthquakes. Therefore, a more complete characterization of the inelastic response of structures requires the introduction of new damage parameters, such as the cyclic ductility, the hysteretic ductility, the Park and Ang damage index, and the low-cycle fatigue index [1–6].

2.3 DEFINITIONS OF GLOBAL DUCTILITY

The concept of global ductility refers to the ability of a structural system to sustain deformations beyond its yield point without significant loss of strength. In spite of the fact that the physical meaning of the ductility concept is universally recognized, its quantitative definition is not clearly established, because it depends on several parameters.

First of all, attention must be paid to the definition of the external actions. With reference to seismic design, the problem concerns the choice of the distribution of a horizontal force system, the intensity of which is defined by using a unique multiplier α. In this concept, different distribution patterns can be adopted depending, in general, upon the considered number of vibration modes. Once the distribution pattern of horizontal forces has been chosen, the global ductility can be evaluated as the ratio between the two top sway displacements δ, at ultimate and yield states. As different options can be adopted for the definition of ultimate and yield states, the definition for the ductility ratio is not unequivocally stated [7].

Alternative ways which have been used to define the yield state are shown in Fig. 2.2. They assume as reference parameter:

(a) the displacement at first yield;
(b) the elastic displacement under a load equal to the collapse load;
(c) the yield displacement provided by a bilinear approximation of the α–δ curve, based on the equivalence in the energy absorption capacity.

It has been pointed out that the ductility concept requires that deformations beyond the yield point have to be sustained without significant loss of strength. Therefore, it is clear that different alternatives are also possible for the definition of the ultimate displacement (Fig. 2.3). In particular, the ultimate displacement can be defined as the one corresponding to a fixed percentage reduction in load-carrying capacity [8, 9], or as the one corresponding to the attainment of the available rotational capacity R in the critical plastic hinge [10], or the attainment

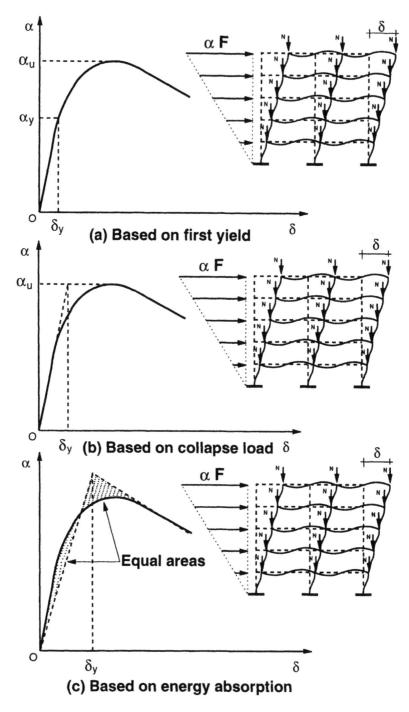

(a) Based on first yield

(b) Based on collapse load

Equal areas

(c) Based on energy absorption

Fig. 2.2   Alternative definitions of the yield displacement

of a given value of the required rotation capacity, so that it corresponds to a specified level of local damage.

2.4   RESPONSE PARAMETERS UNDER MONOTONIC LOADS

For any given distribution pattern of horizontal forces, the inelastic response of a structure under monotonic horizontal loads is completely described by the behavioural curve α–δ (Fig. 2.3), which relates the multiplier of horizontal forces α to the top horizontal displacement δ. This behavioural curve comprises two branches: an increasing branch and a softening branch.

The increasing branch can be divided into two parts (Fig. 2.3b). The first part, which represents the phase of elastic behaviour, extends from the origin until the point of first yielding is reached. The first-yielding multiplier and the corresponding displacement, α_y and δ_y, respectively, identify the beginning of the second part of the increasing branch. This part can develop due to the plastic redistribution capacity of the structure until the collapse (or ultimate) multiplier α_u and the corresponding displacement δ_{max} are reached.

The softening branch can be also divided into two parts (Fig. 2.3b). The first part is characterized by the fact that the structure is still indeterminate and the process of plastic hinge formation is in progress until a kinematic mechanism is formed. The displacement δ_{mec}, corresponding to the development of a collapse mechanism, provides the beginning of the last part of the softening branch. This part is strictly related to the type of collapse mechanism and to the magnitude of vertical loads. The ultimate displacement is attained when the maximum rotational capacity in the critical plastic hinge is reached.

The most important behavioural parameters characterizing the inelastic performance of a structure can be summarized as follows [10]:

- the global ductility $\mu = \delta_u / \delta_y$;
- the plastic redistribution parameter α_u / α_y;
- the slope parameter of the softening branch γ (i.e. the 'stability coefficient');
- the rotational capacity of the critical plastic hinge R;
- the type of collapse mechanism.

It is useful to clarify the meaning of the slope parameter of the α–δ behavioural curve. With reference to an elastic–perfectly plastic SDOF system, it can be observed that in the elastic range the equilibrium equation, including geometric non-linearity, is given by (Fig. 2.4)

$$F + \frac{N\delta}{h} = K\delta \qquad (2.1)$$

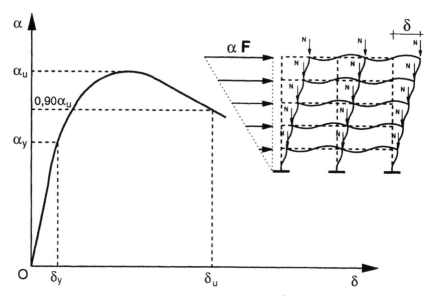

(a) Based on a fixed percentage reduction (10% as an example) in a load-carrying capacity

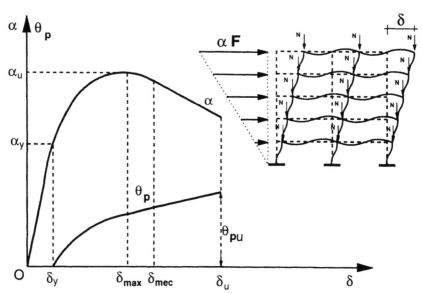

(b) Based on the ultimate plastic rotation θ_{pu} of the critical plastic hinge

Fig. 2.3   Alternative definitions of the ultimate displacement

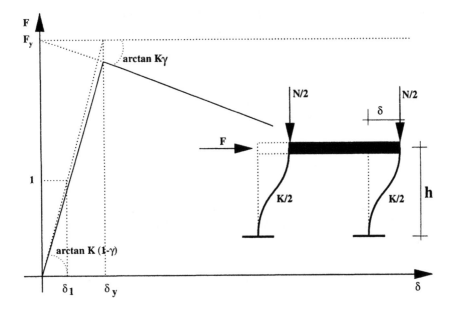

Fig. 2.4   Meaning of the slope parameter γ

where N is the total vertical load and K is the lateral stiffness. This gives

$$F = K \delta \left(1 - \frac{N}{K h} \right)$$ (2.2)

The equilibrium equation in the plastic range is given by

$$F + \frac{N \delta}{h} = F_y$$ (2.3)

where F_y is the yield shear resistance. With the introduction of the parameter

$$\gamma = \frac{N}{K h}$$ (2.4)

equation (2.2) can be written in the form

$$F = K \delta (1 - \gamma)$$ (2.5)

while equation (2.3) becomes

$$F = F_y - K \gamma \delta$$ (2.6)

which represents the softening branch of the F–δ curve.

The external force F can be expressed in the form

$$F = \alpha F_1 \qquad (2.7)$$

by means of the multiplier α which quantifies the magnitude of the external force F as a multiple of the reference force F_1. Taking into account that

$$F_1 = K \delta_1 \qquad (2.8)$$

where δ_1 represents the sway displacement under the force F_1 according to a linear elastic analysis (P–Δ effect not included), equation (2.6) can be written as

$$\alpha = \frac{F_y}{K \delta_1} - \gamma \frac{\delta}{\delta_1} \qquad (2.9)$$

Therefore, the absolute value of the slope of the softening branch of the α–δ curve is given by

$$\gamma_s = \frac{\gamma}{\delta_1} \qquad (2.10)$$

which gives

$$\gamma = \gamma_s \, \delta_1 \qquad (2.11)$$

Equation (2.9) states that the slope parameter γ represents the actual slope of the softening branch of the α versus δ/δ_1 non-dimensional behavioural curve, while the slope of the softening branch of the α–δ curve is given by γ_s. Moreover, equation (2.4) shows that the slope parameter γ provides a measure of the frame sensitivity to second-order effects, thus justifying the name of stability coefficient.

It must be stressed that plastic redistribution produces two effects. On one hand an increase of the load-carrying capacity with respect to first yielding is obtained; on the other hand the formation of plastic hinges is not contemporary so that some plastic hinges have to withstand higher inelastic rotations. This phenomenon can lead to a premature failure due to the limitations arising from local ductility of members.

Moreover, as the slope of the softening branch increases with increasing vertical loads, we can observe that global ductility can be limited by two situations:

- the required local ductility when the value of the critical elastic multiplier of vertical loads is high;
- the high slope of the softening branch in the opposite case.

In both cases the influence of the collapse mechanism type is decisive. More details on this subject are given later.

2.5 PARAMETERS AFFECTING GLOBAL DUCTILITY

It has been emphasized that the plastic redistribution leads, on one hand, to an increase of the load-carrying capacity with respect to first yielding and, on the other hand, to a more severe plastic engagement of some plastic hinges due to their premature formation. Furthermore, the frame sensitivity to second-order effects influences the slope of the softening branch of the behavioural curve α–δ and, as a consequence, it affects the value of the available global ductility.

Therefore, it can be concluded that the parameters affecting the value of the available global ductility are:

- the plastic redistribution parameter α_u / α_y;
- the member rotation capacity R;
- the stability coefficient γ.

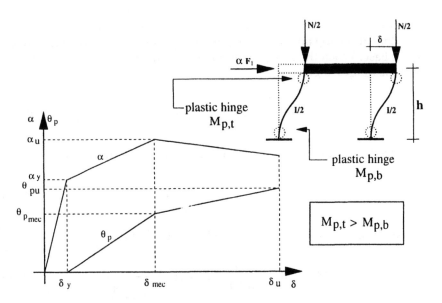

Fig. 2.5   Simplified model with plastic redistribution capacity

The most simple model that can be used to clarify the role of the above parameters for available global ductility is represented by a Grinter's portal frame in which, in order to represent the plastic redistribution characterizing actual frames, it is assumed that the plastic moments are different at the top and at the base of the columns (Fig. 2.5) [10].

With reference to the first-yielding condition, the following relationship can be stated:

$$\alpha_y \, F_1 + \frac{N \, \delta_y}{h} = \frac{12 \, E \, I}{h^3} \, \delta_y \tag{2.12}$$

which gives

$$\delta_y = \frac{h^3}{12 \, E \, I} \left(\alpha_y \, F_1 + \frac{N \, \delta_y}{h} \right) \tag{2.13}$$

When the collapse mechanism is attained, the equilibrium equation is

$$\alpha_u \, F_1 + \frac{N \, \delta_{mec}}{h} = \frac{12 \, E \, I}{h^3} \, \delta_y + \frac{3 \, E \, I}{h^3} \left(\delta_{mec} - \delta_y \right) \tag{2.14}$$

which, from equation (2.12), yields

$$\left(\delta_{mec} - \delta_y \right) = \frac{h^3}{3 \, E \, I} \left[\left(\alpha_u - \alpha_y \right) F_1 + \frac{N \left(\delta_{mec} - \delta_y \right)}{h} \right] \tag{2.15}$$

It is assumed that the plastic moment at the base of the columns is less than that at the top and, therefore, the first plastic hinges to be formed are at the base. When the collapse mechanism is attained, the plastic rotation of these hinges is given by

$$\theta_{p_{mec}} = \frac{h^2}{2 \, E \, I} \left[\left(\alpha_u - \alpha_y \right) F_1 + \frac{N \left(\delta_{mec} - \delta_y \right)}{h} \right] \tag{2.16}$$

By introducing the critical elastic multiplier of the vertical loads as

$$\alpha_{cr} = \frac{N_{cr}}{N} = \frac{\pi^2 \, E \, I}{N \, h^2} \tag{2.17}$$

equation (2.15) can be written in the following form

$$\left(\delta_{mec} - \delta_y \right) = \frac{h^3 \, F_1}{3 \, E \, I} \left(\alpha_u - \alpha_y \right) \frac{1}{1 - \dfrac{\pi^2}{3 \, \alpha_{cr}}} \tag{2.18}$$

Equations (2.13) and (2.17) provide

$$\frac{h^3 \, F_1}{3 \, E \, I} = \frac{4}{\alpha_y} \frac{\delta_y}{\alpha_y} \left(1 - \frac{\pi^2}{12 \, \alpha_{cr}} \right) \tag{2.19}$$

Therefore, equation (2.18) becomes

$$\left(\delta_{mec} - \delta_y \right) = 4 \frac{\alpha_u - \alpha_y}{\alpha_y} \, \delta_y \, \frac{1 - \dfrac{\pi^2}{12 \, \alpha_{cr}}}{1 - \dfrac{\pi^2}{3 \, \alpha_{cr}}} \tag{2.20}$$

Using equations (2.19) and (2.20), the relationship (2.16) can be written as

$$\theta_{p_{mec}} = 6 \frac{\alpha_u - \alpha_y}{\alpha_y} \, \frac{\delta_y}{h} \, \frac{1 - \dfrac{\pi^2}{12 \, \alpha_{cr}}}{1 - \dfrac{\pi^2}{3 \, \alpha_{cr}}} \tag{2.21}$$

Denoting with θ_{p_u} the ultimate value of the plastic rotation of the plastic hinges at the column base, the ultimate displacement is given by

$$\delta_u = \delta_y + \left(\delta_{mec} - \delta_y \right) + \left(\theta_{p_u} - \theta_{p_{mec}} \right) h \tag{2.22}$$

Using equations (2.20) and (2.21), relationship (2.22) provides

$$\delta_u = \delta_y - 2 \left(\frac{\alpha_u}{\alpha_y} - 1 \right) \delta_y \, f(\alpha_{cr}) + \theta_{p_u} \, h \tag{2.23}$$

where

$$f(\alpha_{cr}) = \frac{1 - \dfrac{\pi^2}{12 \, \alpha_{cr}}}{1 - \dfrac{\pi^2}{3 \, \alpha_{cr}}} \tag{2.24}$$

Therefore, the available global ductility is provided by [10]

$$\mu = \frac{\delta_u}{\delta_y} = 1 - 2 \left(\frac{\alpha_u}{\alpha_y} - 1 \right) f(\alpha_{cr}) + R \tag{2.25}$$

where

$$R = \frac{\theta_{p_u} \, h}{\delta_y} \tag{2.26}$$

is the rotation capacity of the plastic hinges at the column base. It is easy to recognize that, for the simplified model, the following is true:

$$\gamma = \frac{\pi^2}{12 \ \alpha_{cr}}$$ (2.27)

Therefore, equation (2.25) can be written as

$$\mu = 1 - 2 \ \frac{1-\gamma}{1-4 \ \gamma} \left(\frac{\alpha_u}{\alpha_y} - 1 \right) + R$$ (2.28)

This relationship states that the available global ductility decreases with increasing frame sensitivity to second-order effects, expressed by γ, and increasing plastic redistribution capacity, expressed by α_u / α_y. Moreover, it demonstrates that global ductility increases as local ductility, expressed by R, increases.

In the case of actual multistorey frames, the above-mentioned type of dependence is still valid but, from the quantitative point of view, it is modified due to different reasons. First of all, the type of collapse mechanism influences the frame sensitivity to second-order effects and equation (2.27) cannot be always applied. Secondly, for a given value of the available global ductility the required rotation capacity is influenced by the collapse mechanism type. These aspects will be discussed in the following section.

2.6   INFLUENCE OF THE COLLAPSE MECHANISM

2.6.1 General

It has been already shown that the collapse mechanism plays a very important role in the seismic design of structures, because it influences the value of the available global ductility and the sensitivity of the frame to second-order effects, thus leading to a reduction of the energy dissipation capacity of the structure.

It will be shown that the type of collapse mechanism affects the slope of the softening branch of the α–δ behaviour curve, which is a measure of the frame sensitivity to second order effects. The influence of frame sensitivity to second-order effects on the seismic inelastic response will be analysed in Chapter 5.

Moreover, the type of collapse mechanism influences the value of the required local ductility corresponding to a given value of the global duc-tility. As the available global ductility is strictly related to the frame sensitivity to second-order effects, through γ, and to the available rotation capacity R, it is clear that knowledge of the collapse mechanism's influence on the above parameters allows clarification of its influence on the available global ductility.

2.6.2 Influence on the stability coefficient

The collapse mechanisms of moment-resisting frames under seismic horizontal forces can be considered as belonging to three main typologies (Fig. 2.6). The collapse mechanism of the global type is a particular case of type-2 mechanisms.

The influence of the collapse mechanism on the slope of the softening branch of the α–δ behaviour curve can be analysed by evaluation of the equilibrium curves of the different mechanism typologies.

The vector of the design horizontal forces $\{F\}$ is given by

$$\{F\}^T = \{F_1, F_2, ..., F_k, ..., F_{n_s}\} \tag{2.29}$$

where F_k is the horizontal force applied to the kth storey and n_s the number of storeys.

The vector of the storey heights $\{h\}$ is given by

$$\{h\}^T = \{h_1, h_2, ..., h_k, ..., h_{n_s}\} \tag{2.30}$$

where h_k is the height of the kth storey.

The vector of the storey vertical loads $\{V\}$ is given by

$$\{V\}^T = \{V_1, V_2, ..., V_k, ..., V_{n_s}\} \tag{2.31}$$

where V_k is the total vertical load acting at the kth storey. In the following, it is assumed that vertical loads are concentrated at the nodes of the frame.

The equilibrium curve of the generic collapse mechanism can be obtained by the principle of virtual work

$$\alpha \{F\}^T \{du\} + \{V\}^T \{dv\} = \{M_p\}^T \{d\theta\} \tag{2.32}$$

where

- α is the multiplier of the horizontal forces;
- $\{du\}$ is the vector of the virtual horizontal displacements corresponding to the given mechanism;
- $\{dv\}$ is the vector of the virtual vertical displacements corresponding to the given mechanism;

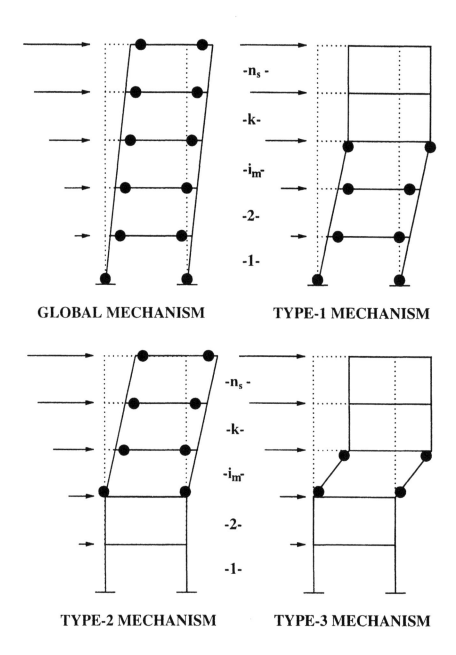

GLOBAL MECHANISM

TYPE-1 MECHANISM

$-n_s-$

$-k-$

$-i_m-$

$-2-$

$-1-$

TYPE-2 MECHANISM

TYPE-3 MECHANISM

$-n_s-$

$-k-$

$-i_m-$

$-2-$

$-1-$

Fig. 2.6 Collapse mechanism typologies of MR frames under seismic horizontal forces

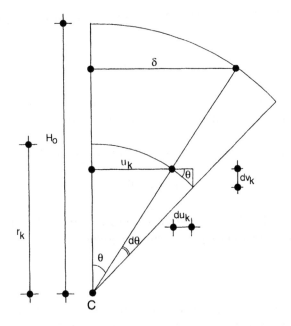

Fig. 2.7 Horizontal and vertical virtual displacements

- $\{d\theta\}$ is the vector of the virtual rotations of the plastic hinges of the given mechanism;
- $\{M_p\}$ is the vector of the plastic moments corresponding to the plastic hinges of the given mechanism.

With reference to Fig. 2.7, it can be seen that the horizontal displacement of the kth storey involved by the generic mechanism is given by

$$u_k = r_k \sin\theta \tag{2.33}$$

where r_k is the distance of the kth storey from the centre of rotation C, and θ is the angle of rotation. The top sway displacement is given by

$$\delta = H_o \sin\theta \tag{2.34}$$

where H_o is the sum of the interstorey heights of the storeys involved by the generic mechanism.

The relationship between vertical and horizontal virtual displacements is given by

$$dv_k = du_k \, \sin\theta \tag{2.35}$$

It shows that, as the ratio dv_k/du_k is independent of the considered storey, the vertical and horizontal virtual displacement vectors have the same shape.

In fact, the virtual horizontal displacements are given by

$$du_k = r_k \, d\theta \tag{2.36}$$

where r_k defines the shape of the virtual horizontal displacement vector, while the virtual vertical displacements are given by

$$dv_k = \frac{\delta}{H_o} \, r_k \, d\theta \tag{2.37}$$

and, therefore, they have the same shape r_k as the horizontal ones.

It can be concluded that, denoting with $\{s\}$ the shape vector of the horizontal displacements,

$$\{du\} = \{s\} \, d\theta \tag{2.38}$$

and

$$\{dv\} = \frac{\delta}{H_o} \{s\} \, d\theta \tag{2.39}$$

while

$$\{d\theta\} = \{I\} \, d\theta \tag{2.40}$$

because all plastic hinges are subjected to the same virtual rotation ($\{I\}$ is the unity vector).

As a consequence, equation (2.32) provides the following expression for the equilibrium curve of the generic collapse mechanism

$$\alpha = \frac{\{M_p\}^T \, \{I\} - \{V\}^T \, \{s\} \, \dfrac{\delta}{H_o}}{\{F\}^T \, \{s\}} \tag{2.41}$$

which can be written as

$$\alpha = \alpha_0 - \gamma_s \, \delta \qquad (2.42)$$

where

$$\alpha_0 = \frac{\{M_p\}^T \{I\}}{\{F\}^T \{s\}} \qquad (2.43)$$

is the kinematically admissible multiplier of the horizontal forces provided, for the generic collapse mechanism, by the limit design theory under the assumption of rigid–plastic behaviour of the material. Also,

$$\gamma_s = \frac{\{V\}^T \{s\} \dfrac{1}{H_0}}{\{F\}^T \{s\}} \qquad (2.44)$$

is the absolute value of the slope of the softening branch of the α–δ curve.

Expressions (2.43) and (2.44) are now explicated for the different collapse mechanism typologies, using the following notation:

- n_s is the number of storeys;
- n_c is the number of columns;
- n_b is the number of bays;
- k is the storey index;
- i is the column index;
- j is the bay index;
- i_m is the mechanism index;
- $M_{c,ik}$ is the plastic moment, reduced for the presence of the axial internal force, of the ith column of the kth storey;
- $M_{b,jk}$ is the plastic moment of the jth beam of the kth storey;
- $\alpha^{(g)}$ and $\gamma_s^{(g)}$ are, respectively, the kinematically admissible multiplier of the horizontal forces (rigid–plastic theory) and the slope of the softening branch of the α–δ curve, corresponding to the global type mechanism;
- $\alpha_{i_m}^{(t)}$ and $\gamma s_{i_m}^{(t)}$ have the same meaning as the previous symbols, but they refer to the i_mth mechanism of the tth typology ($t=1$–3).

In the case of the global-type mechanism (Fig. 2.6), the shape vector of the horizontal displacements is given by

$$\{s\} = \{h\} \qquad (2.45)$$

and

$$H_o = h_{n_s} \qquad (2.46)$$

because all storeys participate in the mechanism.

Therefore, the kinematically admissible multiplier is given by

$$\alpha^{(g)} = \frac{\sum\limits_{i=1}^{n_c} M_{c,i1} + \sum\limits_{k=1}^{n_s} \left(\sum\limits_{j=1}^{n_b} 2 M_{b,jk} \right)}{\sum\limits_{k=1}^{n_s} F_k \, h_k} \qquad (2.47)$$

while the slope of the softening branch is given by

$$\gamma^{(g)}_s = \frac{\dfrac{1}{h_{n_s}} \sum\limits_{k=1}^{n_s} V_k \, h_k}{\sum\limits_{k=1}^{n_s} F_k \, h_k} \qquad (2.48)$$

With reference to the i_mth mechanism of type 1 (Fig. 2.6), the shape vector of the horizontal displacements can be written as:

$$\{s\}^T = \{h_1, h_2, h_3, ..., h_{i_m}, h_{i_m}, h_{i_m}\} \qquad (2.49)$$

where the first element equal to h_{i_m} corresponds to the i_mth component. Moreover, for this case,

$$H_o = h_{i_m} \qquad (2.50)$$

Therefore, the kinematically admissible multiplier corresponding to the i_mth mechanism of type 1 is given by

$$\alpha^{(1)}_{i_m} = \frac{\sum\limits_{i=1}^{n_c} M_{c,i1} + \sum\limits_{k=1}^{i_m-1} \left(\sum\limits_{j=1}^{n_b} 2 M_{b,jk} \right) + \sum\limits_{i=1}^{n_c} M_{c,ii_m}}{\sum\limits_{k=1}^{i_m} F_k \, h_k + h_{i_m} \sum\limits_{k=i_m+1}^{n_s} F_k} \qquad (2.51)$$

while the corresponding slope of the softening branch of the α–δ curve is given by

$$
\overset{(1)}{\gamma_{s_{i_m}}} = \frac{\dfrac{1}{h_{i_m}} \left(\displaystyle\sum_{k=1}^{i_m} V_k\, h_k + h_{i_m} \sum_{k=i_m+1}^{n_s} V_k \right)}{\displaystyle\sum_{k=1}^{i_m} F_k\, h_k + h_{i_m} \sum_{k=i_m+1}^{n_s} F_k}
\tag{2.52}
$$

With reference to the i_mth mechanism of type 2 (Fig. 2.6), the shape vector of the horizontal displacements can be written as:

$$
\{s\}^T = \{0,\ 0,\ 0,\ \ldots,\ 0,\ h_{i_m} - h_{i_m-1},\ h_{i_m+1} - h_{i_m-1},\ \ldots,\ h_{n_s} - h_{i_m-1}\}
\tag{2.53}
$$

Which yields

$$
H_o = h_{n_s} - h_{i_m-1}
\tag{2.54}
$$

As a consequence, the kinematically admissible multiplier corresponding to the i_mth mechanism of type 2 is given by

$$
\overset{(2)}{\alpha_{i_m}} = \frac{\displaystyle\sum_{i=1}^{n_c} M_{c,ii_m} + \sum_{k=i_m}^{n_s} \left(\sum_{j=1}^{n_b} 2\, M_{b,jk} \right)}{\displaystyle\sum_{k=i_m}^{n_s} F_k\, (h_k - h_{i_m-1})}
\tag{2.55}
$$

while the corresponding slope of the softening branch of the α–δ curve is

$$
\overset{(2)}{\gamma_{s_{i_m}}} = \frac{\dfrac{1}{h_{n_s} - h_{i_m-1}} \displaystyle\sum_{k=i_m}^{n_s} V_k\, (h_k - h_{i_m-1})}{\displaystyle\sum_{k=i_m}^{n_s} F_k\, (h_k - h_{i_m-1})}
\tag{2.56}
$$

Finally, with reference to the i_mth mechanism of type 3 (Fig. 2.6), the shape vector of the horizontal displacements can be written as

$$\{s\}^T = \{0, 0, \ldots 0, 1, 1, 1, \ldots, 1\} \, (h_{i_m} - h_{i_m - 1}) \tag{2.57}$$

where the first non-zero term is the i_mth one, and

$$H_o = h_{i_m} - h_{i_m - 1} \tag{2.58}$$

The kinematically admissible multiplier of the i_mth mechanism of type 3 is therefore given by

$$\alpha_{i_m}^{(3)} = \frac{2 \displaystyle\sum_{i=1}^{n_c} M_{c,ii_m}}{\left(h_{i_m} - h_{i_m - 1}\right) \displaystyle\sum_{k=i_m}^{n_s} F_k} \tag{2.59}$$

and the corresponding slope of the softening branch of the α–δ curve is

$$\gamma_{s_{i_m}}^{(3)} = \frac{\displaystyle\sum_{k=i_m}^{n_s} V_k}{\left(h_{i_m} - h_{i_m - 1}\right) \displaystyle\sum_{k=i_m}^{n_s} F_k} \tag{2.60}$$

In the case of moment-resisting frames having a constant value h for the interstorey height

$$h_k = k \, h \tag{2.61}$$

In addition, if such a frame is subjected to the same vertical loads at each storey and to a triangular distribution of the horizontal forces

$$V_k = V \tag{2.62}$$

and

$$F_k = k \, F_1 \tag{2.63}$$

Under the above assumptions, the slope of the softening branch of the α–δ curve is given, for the different collapse-mechanism typologies, by the following simplified relationships:

$$\gamma_s^{(g)} = \frac{V}{F_1 \, h} \; \frac{3}{n_s \, (2 \, n_s + 1)} \tag{2.64}$$

$$\gamma_{s_{i_m}}^{(1)} = \frac{V}{F_1 \, h} \; \frac{3 \, (2 \, n_s + 1 - i_m)}{i_m \, (i_m + 1) \, (2 \, i_m + 1) + 3 \, (n_s - i_m) \, (n_s + i_m + 1)} \tag{2.65}$$

$$\gamma_{s_{i_m}}^{(2)} = \frac{V}{F_1 \, h} \; \frac{3}{(n_s - i_m + 1)} \; \frac{n_s \, (n_s + 1) - (i_m - 1) \, [\, 2 \, (n_s + 1) - i_m]}{n_s \, (n_s + 1) \, (2 \, n_s - 3 \, i_m + 4) + i_m \, (i_m - 1) \, (i_m - 2)} \tag{2.66}$$

$$\gamma_{s_{i_m}}^{(3)} = \frac{V}{F_1 \, h} \; \frac{2}{n_s + i_m} \tag{2.67}$$

The above are obtained by substituting the conditions (2.61), (2.62) and (2.63) into equations (2.48), (2.52), (2.56) and (2.60), and taking into account the following

$$\sum_{k=1}^{n_s} k = \frac{n_s \, (n_s + 1)}{2} \quad ; \quad \sum_{k=1}^{n_s} k^2 = \frac{n_s \, (n_s + 1) \, (2 \, n_s + 1)}{6} \tag{2.68}$$

The case in which a framed structure fails according to a global type mechanism (Fig. 2.6) can be adopted as a reference case, because it is generally considered able to exhibit enough ductility to withstand severe earthquakes.

The previous equations show that, for a given value of the total vertical load, collapse mechanisms other than the global type determine an increase of the slope of the softening branch of the α–δ curve.

This increase can be illustrated by means of the following amplification factors

$$\eta_{i_m}^{(t)} = \frac{\gamma_{s_{i_m}}^{(t)}}{\gamma_s^{(g)}} \quad t = 1\text{–}3 \quad i_m = 1, 2, ..., n_s \tag{2.69}$$

which represent the ratio between the slope of the softening branch of the α–δ curve of the i_mth mechanism of the tth type and the same slope corresponding to the global type mechanism.

For frames without set-backs and/or off-sets, it has been demonstrated [10] that values of the amplification factors corresponding to the failure modes shown in Fig. 2.6 are given by

$$\overset{(1)}{\eta_{i_m}} = \frac{n_s \, (2 \, n_s + 1) \, (2 \, n_s + 1 - i_m)}{i_m \, [\, (i_m + 1) \, (2 \, i_m + 1) + 3 \, (n_s - i_m) \, (n_s + 1 + i_m)]} \tag{2.70}$$

for type-1 mechanisms,

$$\overset{(2)}{\eta_{i_m}} = \frac{n_s \, (\, 2 \, n_s + 1)}{(n_s - i_m + 1)} \quad \frac{n_s \, (n_s + 1) - (i_m - 1 \,) \, [\, 2 \, (n_s + 1) - i_m]}{n_s \, (n_s + 1) \, (2 \, n_s - 3 \, i_m + 4) + i_m \, (i_m - 1) \, (i_m - 2)} \tag{2.71}$$

for type-2 mechanisms, and

$$\overset{(3)}{\eta_{i_m}} = \frac{2 \, n_s \, (2 \, n_s + 1)}{3 \, (n_s + i_m)} \tag{2.72}$$

for type-3 mechanisms.

Obviously, these relationships are valid for a triangular distribution pattern of the horizontal forces, according to the first vibration mode, together with a uniform distribution of vertical loads along the height.

In Figs 2.8–2.10 the influence of the collapse mechanism on the slope of the softening branch of the α–δ behavioural curve and, therefore, on the frame sensitivity to second-order effects is shown by means of the previously defined amplification factors. These amplification factors are given for the three considered mechanism typologies as a function of the mechanism index i_m.

Type-1 and type-3 mechanisms provide the same result for $i_m = 1$, because, for this case, they are coincident. The same thing happens for the type-2 and type-3 mechanisms for the case $i_m = n_s$. Finally, type-1 and type-2 mechanisms obviously provide a unit value for $i_m = n_s$ and $i_m = 1$ respectively.

Furthermore, it can be shown that the type-3 mechanism, namely the storey mechanism, always gives rise to a very high value of the amplification factor leading to intolerable values of γ, with a strong reduction of the available ductility. Moreover, for type-1 and 2 mechanisms, the value of the amplification factor is less than 2.0 only if the number of storeys involved by the collapse mechanism n_m ($n_m = i_m$ for type-1, $n_m = n_s - i_m + 1$ for type-2 and $n_m = 1$ for type-3) is greater than $(2/3)n$ [10].

In reference 10, it is shown that the slope parameter of the softening branch γ, for framed structures failing according to the global mechanism, can be obtained with good approximation by the relationship

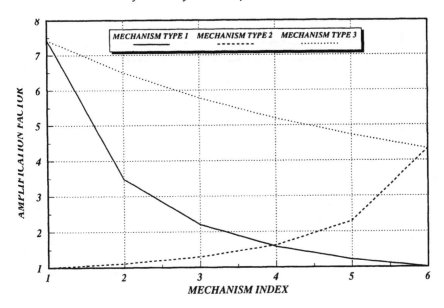

Fig. 2.8   Six-storey frames: influence of the collapse mechanism

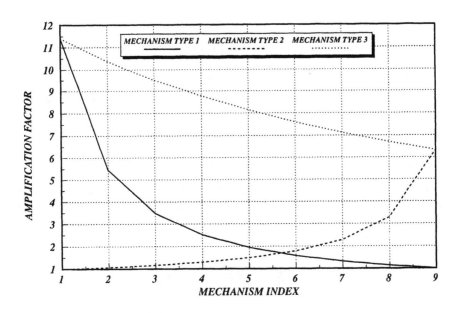

Fig. 2.9   Nine-storey frames: influence of the collapse mechanism

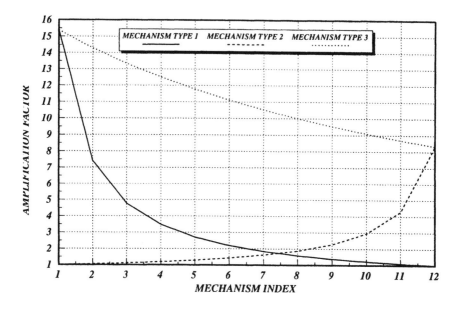

Fig. 2.10   Twelve-storey frames: influence of the collapse mechanism

$$\gamma \approx \frac{1}{\alpha_{cr}} \qquad\qquad (2.73)$$

where α_{cr} is the critical elastic multiplier of vertical loads (see also equation (2.27)).

For collapse mechanisms other than the global case, the stability coefficient γ can be quickly evaluated by means of equation (2.73), taking into account the previously defined amplification factors $(\gamma_{i_m}^{(t)} = \eta_{i_m}^{(t)} \gamma)$.

2.6.3 Influence on required rotation capacity

The type of failure mode influences the value of the rotation capacity required to obtain a given value of the available global ductility. After attainment of the collapse mechanism, the top horizontal displacement can be expressed as

$$\delta = \delta_{mec} + \left(\theta_p - \theta_{p_{mec}} \right) H_o \qquad\qquad (2.74)$$

The increase of required rotation capacity, as the top horizontal displacement increases, is therefore characterized by the gradient

$$\frac{d\,\theta}{d\,\delta} = \frac{1}{H_o} \tag{2.75}$$

which is dependent on the type of collapse mechanism, where H_o is provided by equations (2.46), (2.50), (2.54) and (2.58) for global, type-1, type-2 and type-3 mechanisms, respectively. These relationships, combined with equation (2.75), point to the negative influence of collapse mechanisms other than the global mode.

2.6.4 Influence on damage distribution

It has been shown that the type of collapse mechanism plays an important role in seismic performances of structures. However, the need to ensure a failure mode of the global type is not sufficiently understood. In fact, the failure mode also plays a very important role in damage distribution. This is illustrated by the energy design method proposed in reference 11 (Chapter 4) and is based on the damage distribution law

$$\frac{W_{p,i}}{W_p} = \frac{s_i\ p_i^{-\xi}}{\displaystyle\sum_{i=1}^{n_s} s_i\ p_i^{-\xi}} \tag{2.76}$$

where

- W_p is the total plastic work due to earthquake input energy;
- $W_{p,\,i}$ is the plastic work at the ith storey;
- ξ is a coefficient dependent upon the structural typology.

The coefficients s_i and p_i are given by

$$s_i = \left(\sum_{j=i}^{n_s} \frac{m_j}{M} \right)^2 \left(\frac{\alpha_{opt,\,i}}{\alpha_1} \right)^2 \left(\frac{K_1}{K_i} \right) \tag{2.77}$$

and

$$p_i = \frac{\alpha_i}{\alpha_{opt\,i}} \tag{2.78}$$

where

- M is the total mass of the structure;
- m_j is the mass at the jth storey;
- K_i is the shear stiffness at the ith storey;
- α_i is the yield shear coefficient of the ith storey

$$\alpha_i = Q_{y,i} \Big/ \sum_{j=i}^{n_s} m_j \, g$$

where $Q_{y,i}$ is the yield shear force;

- $\alpha_{opt,i}$ is the value of α_i corresponding to the optimum yield shear coefficient distribution under which a uniform damage distribution is obtained [11,12].

The coefficient p_i takes into account the inevitable scatter between the actual yield shear coefficient distribution and the optimum one and, in this sense, it is called the **damage concentration factor**. It is defined so that $p_1 = 1$ (Chapter 4).

The coefficients s_i take into account mass and stiffness distributions. It is easy to show that

$$a_1 = \frac{W_p}{W_{p,1}} = 1 + \sum_{i=2}^{n_s} s_i \left(\frac{p_i}{p_1} \right)^{-\xi} \qquad (2.79)$$

Equation (2.79) clarifies the physical meaning of the coefficient ξ, which is called **damage concentration index**. When ξ becomes sufficiently large, a_1 becomes unity. It means that damage concentration takes place in the first storey. A better damage distribution is obtained as far as the value of ξ decreases. The value of ξ ranges between 2.0 and 12.0 [11, 12]. Weak-column strong-beam structures are susceptible to damage concentration so that a maximum value of $\xi = 12$ has to be adopted. A very favourable mitigation of damage concentration is obtained in strong-column weak-beam structures due to the elastic action of columns; in this case, therefore, a value $\xi = 6$ can be used (Fig. 2.11) [11, 12].

It is important to realize that when weak column structures behave as shown in Fig. 2.11 they are also able to withstand strong earthquakes. In this case the failure mode is still a storey mechanism, but it involves all storeys and the energy dissipation capacity of the structure is not undermined.

Finally it is interesting to note that Akiyama has proposed a generalized form of weak beam type structure characterized by the presence of an elastic column (Fig. 2.11) [12]. The elastic column plays the role of damage distributor, while ordinary frames pin-jointed to the elastic column have the task of dissipating the earthquake input energy. By using this structural typology, a value of $\xi = 2$ can be reached leading to the best damage distribution capacity [12].

The above considerations, regarding the possibility to design structures able to resist severe earthquakes even if global failure mode is not attained, seem to be very important because, as discussed in the following, simple design rules able to lead to structures failing in global mode

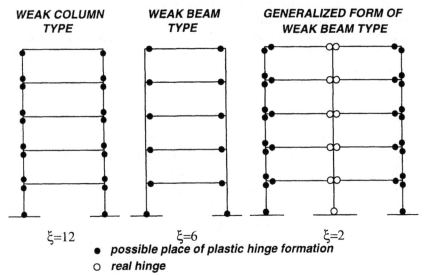

WEAK COLUMN TYPE $\xi=12$

WEAK BEAM TYPE $\xi=6$

GENERALIZED FORM OF WEAK BEAM TYPE $\xi=2$

● *possible place of plastic hinge formation*

○ *real hinge*

Fig. 2.11   Values of damage concentration index for different structural schemes

are not yet available and the use of sophisticated procedures is requested in order to obtain this design goal.

2.7   EVALUATION OF GLOBAL DUCTILITY

In the previous sections the parameters affecting the value of the available global ductility have been examined. An understanding of the role of these parameters is very important from the design point of view, because it allows design criteria to be established for the control of the structural ductility. Design criteria directed at failure mode control, which have been suggested by the modern seismic codes, are covered in the next section.

An evaluation of the available ductility is essential in order to verify the effectiveness of the adopted design procedure.

In practice, structural ductility evaluation can be performed by

• static inelastic analysis
• the mechanism curve method
• approximate relationships.

Static inelastic analysis, including both geometric and material non-linearities, allows the determination of the α–δ behavioural curve and, therefore, is the most accurate evaluation of structural ductility according to one of the alternative quantitative definitions given in Section 2.3. However, the use of static inelastic analysis in the design process can be excessively cumbersome so that, for practical purposes, the use of simplified methods can be justified.

The mechanism curve method [13, 14] was originally devoted to the approximate estimation of the ultimate multiplier, but it can be usefully adopted also for evaluating the structural ductility. This method uses the linear elastic versus top displacement relationship OA and the mechanism curve BD (Fig. 2.12), which can be determined by means of the equations derived in Section 2.6.1, to obtain an estimate of the ultimate multiplier α_u. Let the displacement at K be denoted by δ_K, where K is the point on the mechanism curve BD corresponding to the ultimate multiplier α_u, the true one. Now, if an estimate could be made of δ_K, the softening mechanism curve BD could be used to provide an estimate of the ultimate multiplier α_u. A convenient tool for estimating δ_K is to relate it empirically to δ_E, the linear elastic top displacement under horizontal forces corresponding to the rigid–plastic limit multiplier α_0. The study of a great number of frames [13] has shown that, by assuming $\delta_K = 2.5\,\delta_E$, the point K is located with sufficient accuracy to obtain an acceptable estimate of α_u. Moreover, depending on the definition adopted for the yield state (Section 2.3), it is also possible to evaluate the linear elastic top displacement $\delta_{E'}$ under horizontal forces corresponding to α_u or the first yielding displacement $\delta_{E''}$.

If the ultimate displacement is defined as the one corresponding to a fixed percentage reduction in load-carrying capacity (Section 2.3), then it can be immediately evaluated by means of the mechanism curve.

In the alternative case in which the ultimate displacement is defined as the one corresponding to the attainment of the available rotation capacity in the critical plastic hinge, an approximate evaluation can be performed.

The first step is represented by the evaluation of the collapse mechanism type by means of equations (2.51), (2.55) and (2.59), and by taking into account that it corresponds to the minimum kinematically admissible multiplier according to the kinematic theorem of plastic collapse. A knowledge of the collapse mechanism will establish whether the failing member, in which the plastic hinge attaining the maximum plastic rotation is located, is a beam or a column. It is usually a beam in the case of type-1 and type-2 mechanisms, while it is always a column in the case of type-3 mechanisms.

The available rotation capacity R of the critical member is easily computed by the method proposed in references [15–17], which will be described in the following chapter; the point of zero moment of the member at collapse can be evaluated taking into account that both ends of the member have yielded. As a consequence, the ultimate plastic rotation θ_{p_u} of the critical member can be computed from the relationship

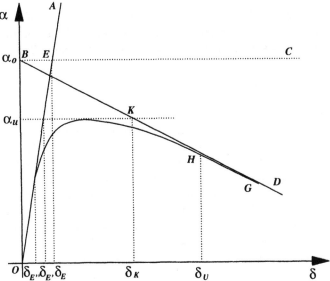

Fig. 2.12 The 'mechanism curve method'

$$\theta_{p_u} = R \ \theta_y \tag{2.80}$$

where θ_y can also be calculated as described in references 15–17, starting from a knowledge of the member bending moment diagram at collapse.

The ultimate top sway displacement can be now estimated by

$$\delta_u \approx \theta_{p_u} H_o \tag{2.81}$$

which takes into account, through H_o, the type of collapse mechanism.

Finally, the available global ductility is evaluated as the ratio between the ultimate displacement and that defining the yield state.

The last method for estimating the available global ductility is represented by the use of simplified relationships, which have been proposed, with different degrees of accuracy, by different researchers [10, 18]. Among these, the one proposed in reference 10 seems to allow a satisfactory degree of accuracy. Here, global ductility is defined as

$$\mu = 1 - 2 \left(\frac{\alpha_u}{\alpha_y} - 1 \right) + \psi \ R \tag{2.82}$$

where ψ is a coefficient dependent on the number of storeys, the beam-to-column stiffness ratio and the distribution of the column stiffness along the height. Its average value can be assumed equal to 2/3.

2.8   DESIGN CRITERIA

In modern seismic codes the reduction of the available ductility due to undesirable collapse mechanisms is taken into account through design criteria which aim to exploit the plastic reserves of the structural scheme. For this reason much attention has been directed towards the provision of simple design rules able to lead to structures failing in global mode. These rules are mainly devoted to the definition of the design actions and the corresponding checks of columns, beams and, as a consequence, beam-to-column connections. The design rules regarding columns and beams are discussed in this section, while those referring to beam-to-column connections will be described in Chapter 6.

In order to obtain a collapse mechanism of global type, modern codes introduce the concept universally recognized, that the flexural strength of columns has to be greater than the flexural strength of beams. Therefore, at each beam-to-column joint the following condition has to be satisfied

$$\sum M_{R,c} > \sum M_{R,b} \tag{2.83}$$

where $\sum M_{R,c}$ is the sum of the resisting moments of the columns connected to the joint, and $\sum M_{R,b}$ is the sum of the resisting moments of the beams connected to the joint.

These modern seismic codes are, therefore, aimed at providing amplification of the bending resistance of columns, following different design criteria.

In the European Convention for Constructional Steelwork (ECCS) design criterion [19] the amplification applies only to bending moments arising from seismic action. Therefore, the amplification coefficient α is computed by

$$\sum M_{c,o} + \alpha \sum M_{c,s} = \sum M_{R,b} \tag{2.84}$$

where $M_{c,o}$ is the column bending moment due to non-seismic loads, and $M_{c,s}$ is the column bending moment due to horizontal seismic forces.

The value of the amplification coefficient is therefore

$$\alpha = \frac{\sum M_{R,b} - \sum M_{c,o}}{\sum M_{c,s}} \tag{2.85}$$

and the design value of the column bending moment is given by

$$M_{c,d} = M_{c,o} + \alpha\, M_{c,s} \qquad (2.86)$$

According to ECCS Recommendations, the amplification coefficient α has to be set equal to 1.0 at the top floor of multistorey frames and equal to 1.20 at the base of the frame.

In Eurocode 8 [20] the fulfilment of requirement (2.83) is required, but a column-bending-moment amplification coefficient is not introduced. Equation (2.86) is provided, but with reference to the column-to-foundation connection for which the amplification coefficient α is assumed equal to 1.2.

A different approach is adopted by the CNR-GNDT code [21], in which the amplification of the total bending moment of the columns is proposed. Therefore, the following condition has to be satisfied

$$\alpha \sum M_c = \sum M_{R,b} \qquad (2.87)$$

where

$$M_c = M_{c,o} + M_{c,s} \qquad (2.88)$$

is the total bending moment due to seismic and nonseismic loads. The amplification coefficient α is therefore given by

$$\alpha = \frac{\sum M_{R,b}}{\sum M_c} \qquad (2.89)$$

and the design column bending moment is computed as

$$M_{c,d} = \alpha M_c \qquad (2.90)$$

Moreover, the CNR-GNDT code requires a 20% amplification of the α value in case of first-storey columns. Finally, when biaxial bending is neglected in performing safety checks of members, an ulterior 30% amplification of the α value is requested.

The above design rules aim to control the location of dissipative zones, which are expected at the beam ends rather than in the columns. As a consequence, the design actions of the beams have to be compatible with the above assumption. For this reason, according to ECCS Recommendations [19], the design shear strength of the beam has to satisfy the following condition

$$\frac{V_R}{\gamma_p} \geq V_o + \frac{M_{R,A} + M_{R,B}}{L} \tag{2.91}$$

where:

- V_0 is the shear force produced by vertical loads;
- L is the length of the beam;
- $M_{R,A}$ and $M_{R,B}$ are the resisting moments of the beam ends, so that the second term on the right hand side represents the additional shear force under the assumption that plastic hinges at member ends are developed;
- $\gamma_p \geq 1$ is a coefficient which has the task to assure that the shear force has a negligible influence on the plastic moment of the beam; the control of the influence of the shear force on the value of the plastic moment of the beam is not requested in the case in which the value $\gamma_p = 3$ is assumed.

These provisions are given also in the CNR-GNDT code [21]. The provision given by equation (2.91) is also covered in Eurocode 8 [20], but the value of γ_p is assumed equal to 2.0.

The shear–moment interaction can be accounted for in the method now described.

The point A of the shear–moment interaction diagram (Fig. 2.13) of a double T section, under the assumption of elastic–perfectly plastic behaviour and in the absence of axial forces, is given by

$$V_p^* = \tau_y \, (d - t_f) \, t_w \tag{2.92}$$

$$M_p^* = f_y \, (b_f - t_w) \, (d - t_f) \, t_f \tag{2.93}$$

where

- d is the beam depth
- t_f is the flange thickness
- b_f is the flange width
- t_w is the web thickness
- f_y is the yield stress in pure tension
- τ_y is the yield stress in pure shear ($\tau_y = f_y / \sqrt{3}$ for the von Mises yield criterion).

Neglecting the small increase in shear capacity which, in the interaction diagram happens for points to the left of A, the shear–moment interaction curve can be approximated as

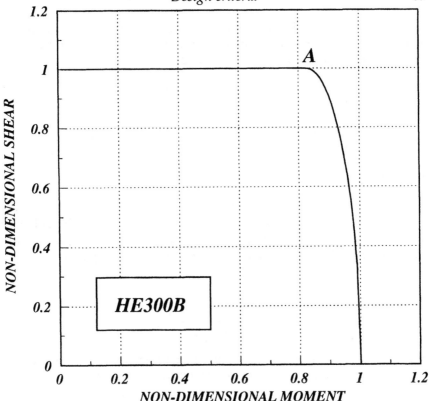

Fig. 2.13 A typical shear–moment interaction diagram

$$\left(\frac{|M| - M_p^*}{M_p - M_p^*}\right)^2 + \left(\frac{V}{V_p^*}\right)^2 = 1 \quad \text{for} \quad M_p^* \le |M| \le M_p \qquad (2.94)$$

$$V = V_p^* \quad \text{for} \quad |M| \le M_p^* \qquad (2.95)$$

It is interesting to note that the use of $\gamma_p = 2$ in equation (2.91) assures that the reduction of the flexural strength of the beam due to the influence of the shear force is less than 5%.

The inelastic response of steel frames, proportioned according to the above design criteria, has been investigated [22, 23] by using static inelastic analyses, leading to the characterization of the structural response by means of the α–δ behavioural curve and the pattern of yieldings. A wide number of structural schemes have been considered and the following conclusions have been derived [23].

- The codified design methods based upon the amplification of column bending moments are inadequate at obtaining frames failing in global mode.
- The establishment of global failure mode is not a sufficient condition to obtain a q-factor greater than that assumed in preliminary design.
- The tapering of columns can prevent the formation of the global-type mechanism.

Both in LRFD-AISC seismic provisions [24] and in UBC90 [25], the requirement given by equation (2.83) is recommended, but the influence of the axial load on the plastic capacity of columns is explicitly considered. The following design condition is therefore required:

$$\frac{\sum M_{R,c}\left(1 - \dfrac{N}{N_{R,c}}\right)}{\sum M_{R,b}} \geq 1 \tag{2.96}$$

where N is the axial load and $N_{R,c}$ the axial resistance of the column.

The requirement (2.96) is accompanied by the condition

$$\frac{\sum M_{R,c}\left(1 - \dfrac{N}{N_{R,c}}\right)}{V_n\, d_b\, \dfrac{H}{H - d_b}} \geq 1 \tag{2.97}$$

where V_n is the nominal strength of the panel zone of the beam-to-column joint, d_b is the average overall depth of the beams framing into the connection and H is the average of the storey heights above and below the connection.

The design shear strength V_n of the panel zone can be computed as

$$V_n = 0.55\, f_y\, d_c\, t_p\left(1 + \frac{3\, b_{cf}\, t_{cf}^{2}}{d_b\, d_c\, t_p}\right) \tag{2.98}$$

where

- t_p is the thickness of the panel zone including doubler plates;
- d_c is the overall depth of the column section;
- b_{cf} is the width of the column flanges;

- t_{cf} is the thickness of the column flanges;
- d_b is the overall depth of the beam.

Equation (2.98) takes into account the contribution of the column flanges to the panel zone shear capacity in the inelastic range.

The combination of the two provisions (2.96) and (2.97) is intended to assure that at a beam-to-column connection yielding occurs in the beam or in the panel zone rather than in the column.

In the above recommendations, therefore, the shear yielding of the panel zone is not precluded. In fact, well detailed panel-zone joints have exhibited a stable hysteretic behaviour and an excellent energy dissipation capacity after first yielding [26–28].

While permitting the panel zone to yield in shear, it is important to point out that large inelastic deformations of the beam-to-column panel zone can cause local kinks in the column flanges near the beam flanges, due to high curvature outside the panel zone. The beam flange welds often fracture during experimental tests, due to the action of tension in the beam flange and the high curvature near the edge of the distorted panel zone. In the case of beam-to-column connections made by web bolting and flange welding, the beam flange welds become even more vulnerable to the above phenomena once bolt slippage has occurred [26, 29].

In order to reduce the stress concentration in the beam flange welds, the use of doubler plates designed to develop at least 80% of the beam strength is suggested [30].

It has been pointed out [30] that, as the joint panel zone strength approaches the beam flexural strength, the magnitude of the maximum panel zone deformations under earthquake action is reduced, while the extent of the maximum beam and column plastic rotation is increased. In this case, particular attention has to be paid to the detailing of beam-to-column connections.

In other words, the choice of the pattern of yielding and, therefore, of the energy dissipation mechanism can follow two different approaches.

The first of these is based on the participation of the joint panel zone in dissipating energy in order to reduce, but not totally eliminate, the plastic engagement of beams and columns. In this case, the panel zone has to be able to develop sufficient strength in order to reduce its deformation and the consequent danger of the local kinking of column flanges and the fracture of the beam flange welds.

In the second approach, the participation of the joint panel zone is precluded. As a consequence, in this case, beam ends have to sustain severe plastic deformations; therefore beam-to-column connections have to be appropriately detailed. This second approach will be described in Section 2.10.2.

During the severest earthquake expected at any site, the column axial forces computed by using the specified design earthquake forces can be exceeded. This is a consequence of the lateral force reduction which is assumed for analysing an elastic model of the structure. Therefore, in such analysis, an underestimation of the overturning forces arises. In order to account for this underestimation, both in the LRFD-AISC seismic provisions and in UBC90, a limit to the required axial resistance of columns is provided.

As an example, and with reference to the LRFD-AISC seismic provisions, two additional load combinations are specified in order to define the minimum required column compressive strength and the minimum required tensile strength.

The minimum compression resistance of the columns has to be equal to or greater than the axial load computed under the following additional load combination

$$1.2\,D + 1.0\,E' + (\,0.5\,L \text{ or } 0.2\,S\,) \tag{2.99}$$

where D is the dead load, L is the live load, S is the snow load and E' is an amplified earthquake load defined as $E' = 3ZICSW$ corresponding to $K = 3$ (Chapter 1).

Taking into account that the corresponding normal load combination engaging the seismic forces is given by $1.2\,D + 1.5\,E + (0.5\,L \text{ or } 0.2\,S)$, where $E = KZICSW$ is the earthquake load (Chapter 1), it can be concluded that in order to define the minimum required compressive strength of columns the axial load N_0 computed under the design horizontal forces $1.5E$ has to be amplified according to a factor given by $1.0E'/1.5E = 3/1.5K = 2/K$. Therefore, the minimum required compressive strength is given by

$$N_{R,min} = N_v + \frac{2}{K}\,N_o \tag{2.100}$$

where N_v is the axial load computed under the vertical loads ($1.2D + 0.5L$ or $0.2S$).

Regarding the minimum required tensile strength of columns, the following additional load combination has to be considered

$$0.9D - 1.0\,E' \tag{2.101}$$

This integrates the normal load combination $0.9D - 1.5\dot{E}$ so that, for this case, it is sufficient to consider an amplification of the axial load computed under the design horizontal forces in the measure of the $2/K$ factor.

The two above additional load combinations are intended only to define the minimum required axial resistance of columns; therefore, they have to be applied without consideration of any concurrent flexural action on the members. Fulfilment of the above requirement is specified both for ordinary moment-resisting frames and for special moment-resisting frames.

Since the European seismic codes have no similar provision for the minimum required axial resistance of columns, it is useful to note that the amplification factor $2/K$ corresponds to $2q/5.33$ (where the q-factor is $q \approx 5.33/K$) leading to about 2.25 for $q = 6$.

2.9   FAILURE MODE CONTROL

2.9.1 Theoretical background of the design method

The previously described simple design criteria, suggested by modern seismic codes, do not always lead to frames failing in global mode. A more sophisticated design procedure, assuring the development of a collapse mechanism of global type, has been proposed [31, 32] and successively extended [33], where the influence of second-order effects has been included. The reliability of the method has been verified on a great number of structural schemes, leading in all cases to the fulfilment of the design object.

In the plastic design of structures attention is concentrated on the collapse state of the structure, while the elastic and partly-plastic preliminary stages are ignored. This assumption is justified when the primary interest is the ultimate resistance of a structure and its corresponding failure mode.

However, even when considering the collapsed state only, the complete investigation for a complicated structure might still be very cumbersome. As an alternative, the theory of limit design provides the way for avoiding many of the difficulties arising from a step by step analysis of the complete structure, by using 'bound theorems'.

It is well known that there are two classical approaches to the problem of calculating collapse loads, conventionally called the 'static' and the 'kinematic' approaches. In the equilibrium (static) approach, by satisfying the equilibrium equations and yield conditions, the collapse load is evaluated without considering the mode of deformation. In the geometric (kinematic) approach, by considering the mode of deformation and the energy balance, evaluation of the collapse load is achieved without considering the equilibrium equations.

The answers given by the two methods differ. In fact, a collapse load predicted by the equilibrium method is always on the 'low' side of the exact collapse load (i.e. the lower bound), while the collapse load predicted by the geometric approach is always on the 'high' side (i.e. the upper bound).

The static method is based on the lower-bound theorem, which states that if a stress distribution throughout the structure can be found so that it is everywhere internally in equilibrium and balances certain external loads and at the same time does not violate the yield conditions, then those loads will be carried safely by the structure. The corresponding load multiplier is a so-called statically admissible multiplier. As a consequence, the true collapse multiplier can be found as the maximum statically admissible multiplier.

The kinematic approach is based on the upper-bound theorem, which states that if an estimate of the plastic collapse load or the corresponding multiplier is made by equating the internal rate of dissipation of energy to the rate at which external forces do work in any postulated kinematically admissible mechanism of deformation, then the estimate will be either high, or correct. As a consequence, the true collapse multiplier can be found as the minimum kinematically admissible multiplier.

The theory of limit design has been mainly used in order to compute the collapse multiplier of a given structure. In the case of frames, the best known methods for achieving this are the elementary mechanism combination method (Neal and Symonds method) and the moment distribution method (Horne method). The moment distribution method has been also applied in order to search for structural solutions leading to the minimum weight.

The application of structural design to control the failure mode is a relatively recent problem, arising from seismic design needs, which up-to-now has been faced mainly through simplified rules provided by seismic codes. These rules were not derived by the limit design approach, and in most cases they do not allow the specified failure mechanism to be achieved.

2.9.2 Basic hypothesis

Starting from these considerations, a new design method has been proposed to control the failure mode of seismic-resistant steel frames. This method was first formulated in 1993, neglecting the P–Δ effect [31, 32], and subsequently it was revised and generalized by including second-order effects [33]. It is based on the observation that the collapse mechanisms of frames under horizontal forces can be considered as belonging to three main typologies (Fig. 2.6). The collapse mechanism of the global type is a particular case of the type-2 mechanism. As a consequence, the control of the failure mode can be performed through the analysis of $3n_s$ mechanisms (n_s being the number of storeys). Assuming that the beam sections are already designed to resist vertical loads, the values of the plastic section modulus of columns have to be defined in such a way that the kinematically admissible multiplier of the horizontal forces corresponding to the global mechanism is less than those corresponding to the other $3n_s-1$ kinematically admissibile mechanisms. According to the upper bound theorem, this means that the above stated multiplier is the true collapse multiplier and, therefore, that the true collapse mechanism is represented by the global failure mode.

2.9.3 Design conditions

In order to design a frame failing in global mode, the cross-sections of columns have to be dimensioned in such a way that, according to the upper bound theorem, the kinematically admissible horizontal force multiplier corresponding to the global type mechanism is the minimum among all kinematically admissible multipliers. Including the influence of second-order effects, through the equilibrium curves (2.42) of the mechanisms, the following design conditions are obtained

$$\alpha^{(g)} - \gamma_s^{(g)} \, \delta \le \alpha_{i_m}^{(1)} - \gamma_{s_{i_m}}^{(1)} \, \delta \qquad i_m = 1 - n_s \qquad (2.102)$$

$$\alpha^{(g)} - \gamma_s^{(g)} \, \delta \le \alpha_{i_m}^{(2)} - \gamma_{s_{i_m}}^{(2)} \, \delta \qquad i_m = 1 - n_s \qquad (2.103)$$

$$\alpha^{(g)} - \gamma_s^{(g)} \, \delta \le \alpha_{i_m}^{(3)} - \gamma_{s_{i_m}}^{(3)} \, \delta \qquad i_m = 1 - n_s \qquad (2.104)$$

where the displacement δ has to be the one corresponding to the complete development of the global mechanism [33].

Therefore, there are $3n_s$ design conditions to be satisfied in the case of a frame having n_s storeys. These conditions, which derive directly from the application of the upper bound theorem and of the equilibrium curves of the collapse mechanisms, will be integrated by conditions related to technological limitations.

2.9.4 Conditions to avoid type-1 mechanisms

The n_s conditions (2.102) can be explicated in the following form:

$$\frac{\displaystyle\sum_{i=1}^{n_c} M_{c,i1} + \sum_{k=1}^{n_s}\left(\sum_{j=1}^{n_b} 2M_{b,jk}\right) - \sum_{k=1}^{n_s} V_k \frac{h_k}{h_{n_s}}\,\delta}{\displaystyle\sum_{k=1}^{n_s} F_k\, h_k}$$

$$\leq \frac{\displaystyle\sum_{i=1}^{n_c} M_{c,i1} + \sum_{k=1}^{i_m-1}\left(\sum_{j=1}^{n_b} 2M_{b,jk}\right) + \sum_{i=1}^{n_c} M_{c,ii_m} - \frac{\delta}{h_{i_m}}\left(\sum_{k=1}^{i_m} V_k h_k + h_{i_m}\sum_{k=i_m+1}^{n_s} V_k\right)}{\displaystyle\sum_{k=1}^{i_m} F_k\, h_k + h_{i_m}\sum_{k=i_m+1}^{n_s} F_k} \tag{2.105}$$

It now has to be shown that a constant term and two known discrete functions can be recognized. The term

$$\theta_1 = \sum_{k=1}^{n_s}\left(\sum_{j=1}^{n_b} 2\,M_{b,jk}\right) \tag{2.106}$$

is a known constant, because the plastic moments of beams $M_{b,jk}$ are known, being the beams designed in order to resist vertical loads. For the same reason, the discrete function

$$\xi_{i_m} = \sum_{k=1}^{i_m-1}\left(\sum_{j=1}^{n_b} 2\,M_{b,jk}\right) \tag{2.107}$$

is a known function of the mechanism index i_m. Finally, the discrete function

$$\lambda_{i_m} = \frac{\displaystyle\sum_{k=1}^{n_s} F_k\, h_k}{\displaystyle\sum_{k=1}^{i_m} F_k\, h_k + h_{i_m}\sum_{k=i_m+1}^{n_s} F_k} \tag{2.108}$$

is a known function of the mechanism index i_m.

Therefore, the i_mth condition to be satisfied in order to avoid type-1 collapse mechanisms can be written as

$$
\sum_{i=1}^{n_c} M_{c,i1} + \theta_1 - \sum_{k=1}^{n_s} V_k \frac{h_k}{h_{n_s}} \, \delta \le \lambda_{i_m} \sum_{i=1}^{n_c} M_{c,i1} + \lambda_{i_m} \xi_{i_m} + \lambda_{i_m} \sum_{i=1}^{n_c} M_{c,ii_m}
$$

$$
- \frac{\lambda_{i_m}}{h_{i_m}} \delta \left(\sum_{k=1}^{i_m} V_k h_k + h_{i_m} \sum_{k=i_m+1}^{n_s} V_k \right) \tag{2.109}
$$

The influence of second-order effects is represented by the parameters

$$
\Delta_{i_m}^{(1)} = \frac{1}{h_{n_s}} \sum_{k=1}^{n_s} V_k h_k - \frac{\lambda_{i_m}}{h_{i_m}} \left(\sum_{k=1}^{i_m} V_k h_k + h_{i_m} \sum_{k=i_m+1}^{n_s} V_k \right) \tag{2.110}
$$

Now, the i_mth condition to be satisfied in order to avoid type-1 collapse mechanisms can be written as

$$
\sum_{i=1}^{n_c} M_{c,i1} + \theta_1 \le \lambda_{i_m} \sum_{i=1}^{n_c} M_{c,i1} + \lambda_{i_m} \xi_{i_m} + \lambda_{i_m} \sum_{i=1}^{n_c} M_{c,ii_m} + \Delta_{i_m}^{(1)} \, \delta \tag{2.111}
$$

It is convenient to introduce the parameter

$$
\rho_{i_m}^{(1)} = \frac{\displaystyle\sum_{i=1}^{n_c} M_{c,ii_m}}{\displaystyle\sum_{i=1}^{n_c} M_{c,i1}} \tag{2.112}
$$

which is the ratio between the sum of the reduced plastic moments of the columns of the i_mth storey and the same sum corresponding to the first-storey columns. By means of this parameter, the i_mth condition to be satisfied in order to avoid type-1 collapse mechanisms can be written in the form

$$\rho_{i_m}^{(1)} \geq \frac{(1 - \lambda_{i_m}) \sum_{i=1}^{n_c} M_{c,i1} + \theta_1 v_{i_m}^{(1)} - \lambda_{i_m} \xi_{i_m} - \Delta_{i_m}^{(1)} \delta}{\lambda_{i_m} \sum_{i=1}^{n_c} M_{c,i1}}$$

(2.113)

which has to be applied for $i_m = 1, 2, ..., n_s$.

2.9.5 Conditions to avoid type-2 mechanisms

The n_s conditions (2.103) can be explicated in the form

$$\frac{\sum_{i=1}^{n_c} M_{c,i1} + \sum_{k=1}^{n_s} \left(\sum_{j=1}^{n_b} 2 M_{b,jk} \right) - \sum_{k=1}^{n_s} V_k \dfrac{h_k}{h_{n_s}} \delta}{\sum_{k=1}^{n_s} F_k \, h_k}$$

$$\leq \frac{\sum_{i=1}^{n_c} M_{c,ii_m} + \sum_{k=i_m}^{n_s} \left(\sum_{j=1}^{n_b} 2 M_{b,jk} \right) - \dfrac{\delta}{h_{n_s} - h_{i_m - 1}} \sum_{k=i_m}^{n_s} V_k \left(h_k - h_{i_m - 1} \right)}{\sum_{k=i_m}^{n_s} F_k \left(h_k - h_{i_m - 1} \right)}$$

(2.114)

Now, a new discrete function can be introduced

$$\theta_{i_m} = \sum_{k=i_m}^{n_s} \left(\sum_{j=1}^{n_b} 2 M_{b,jk} \right)$$

(2.115)

This is still a known function, because plastic moments of beams are already established, being the beams designed in order to resist vertical loads. Moreover, it is useful to note that the constant θ_1 in equation (2.106) is the value of the function θ_{i_m} for $i_m = 1$.

It is convenient to introduce another known discrete function of the mechanism index i_m

$$\gamma_{i_m} = \frac{\displaystyle\sum_{k=1}^{n_s} F_k \, h_k}{\displaystyle\sum_{k=i_m}^{n_s} F_k \left(h_k - h_{i_m-1} \right)} \qquad (2.116)$$

The i_mth condition to be satisfied in order to avoid type-2 collapse mechanisms can now be written as

$$\sum_{i=1}^{n_c} M_{c,i1} + \theta_1 - \frac{\delta}{h_{n_s}} \sum_{k=1}^{n_s} V_k \, h_k \leq \gamma_{i_m} \sum_{i=1}^{n_c} M_{c,ii_m} + \gamma_{i_m} \, \theta_{i_m}$$

$$- \frac{\gamma_{i_m} \, \delta}{h_{n_s} - h_{i_m-1}} \sum_{k=i_m}^{n_s} V_k \left(h_k - h_{i_m-1} \right) \qquad (2.117)$$

The influence of second-order effects can be represented by the parameters

$$\Delta_{i_m}^{(2)} = \frac{1}{h_{n_s}} \sum_{k=1}^{n_s} V_k \, h_k - \frac{\gamma_{i_m}}{h_{n_s} - h_{i_m-1}} \sum_{k=i_m}^{n_s} V_k \left(h_k - h_{i_m-1} \right) \qquad (2.118)$$

Following the method adopted in the previous section, a new series of parameters is introduced, such as

$$\rho_{i_m}^{(2)} = \frac{\displaystyle\sum_{i=1}^{n_c} M_{c,ii_m}}{\displaystyle\sum_{i=1}^{n_c} M_{c,i1}} \qquad (2.119)$$

By means of these parameters, the i_mth condition to be satisfied in order to avoid type-2 collapse mechanisms can be written in the form

$$\rho_{i_m}^{(2)} \geq \frac{\sum\limits_{i=1}^{n_c} M_{c,i1} + \theta_1 - \gamma_{i_m} \theta_{i_m} - \Delta_{i_m}^{(2)} \, \delta}{\gamma_{i_m} \sum\limits_{i=1}^{n_c} M_{c,i1}} \qquad (2.120)$$

2.9.6 Conditions to avoid type-3 mechanisms

The n_s conditions (2.104), which have to be satisfied in order to avoid type-3 collapse mechanisms, can be written in the form

$$\frac{\sum\limits_{i=1}^{n_c} M_{c,i1} + \sum\limits_{k=1}^{n_s} \left(\sum\limits_{j=1}^{n_b} 2\, M_{b,jk} \right) - \sum\limits_{k=1}^{n_s} V_k \dfrac{h_k}{h_{n_s}} \delta}{\sum\limits_{k=1}^{n_s} F_k\, h_k} \leq \frac{2 \sum\limits_{i=1}^{n_c} M_{c,ii_m} - \delta \sum\limits_{k=i_m}^{n_s} V_k}{\sum\limits_{k=i_m}^{n_s} F_k\, \Delta h_{i_m}} \qquad (2.121)$$

By introducing a new known discrete function of the mechanism index i_m

$$\beta_{i_m} = \frac{\sum\limits_{k=1}^{n_s} F_k\, h_k}{\sum\limits_{k=i_m}^{n_s} F_k\, \Delta h_{i_m}} \qquad (2.122)$$

the above design conditions provide the relationship

$$\sum\limits_{i=1}^{n_c} M_{c,i1} + \theta_1 \leq 2\,\beta_{i_m} \sum\limits_{i=1}^{n_c} M_{c,ii_m} + \frac{\delta}{h_{n_s}} \sum\limits_{k=1}^{n_s} V_k\, h_k - \delta\, \beta_{i_m} \sum\limits_{k=i_m}^{n_s} V_k \qquad (2.123)$$

In this case, the influence of second-order effects can be represented through the parameters

$$\Delta_{i_m}^{(3)} = \frac{1}{h_{n_s}} \sum\limits_{k=1}^{n_s} V_k\, h_k - \beta_{i_m} \sum\limits_{k=i_m}^{n_s} V_k \qquad (2.124)$$

A final series of new parameters is now introduced:

$$\rho_{i_m}^{(3)} = \frac{\sum\limits_{i=1}^{n_c} M_{c,ii_m}}{\sum\limits_{i=1}^{n_c} M_{c,i1}} \tag{2.125}$$

By means of these parameters, the i_mth condition to be satisfied in order to avoid type-3 collapse mechanisms is expressed by the relationship

$$\rho_{i_m}^{(3)} \geq \frac{\sum\limits_{i=1}^{n_c} M_{c,i1} + \theta_1 - \Delta_{i_m}^{(3)} \delta}{2 \beta_{i_m} \sum\limits_{i=1}^{n_c} M_{c,i1}} \tag{2.126}$$

2.9.7 Technological conditions

According to the above formulations, the $3n_s$ design conditions have been derived directly from the application of the upper bound theorem. In particular, for each storey, there are three design conditions to be satisfied because three collapse mechanism typologies have been considered. As these design conditions have to be contemporaneously satisfied for each storey, the ratio

$$\rho_{i_m} = \frac{\sum\limits_{i=1}^{n_c} M_{c,ii_m}}{\sum\limits_{i=1}^{n_c} M_{c,i1}} \tag{2.127}$$

between the sum of the reduced plastic moments of columns of the i_mth storey and the same sum corresponding to the first-storey columns allows the above design conditions to be satisfied if the following relationship is verified:

$$\rho_{i_m} = \max \left\{ \rho_{i_m}^{(1)}, \rho_{i_m}^{(2)}, \rho_{i_m}^{(3)} \right\} \tag{2.128}$$

As the section of columns can only decrease along the height of the frame, the values of ρ_{i_m} (with $i_m = 1, 2, ..., n_s$) obtained by means of the conditions derived through the application of the upper bound theorem have to be modified in order to satisfy the following technological limitation

$$\rho_1 \geq \rho_2 \geq \rho_3 \geq, ..., \geq \rho_{n_s} \qquad (2.129)$$

2.9.8 Evaluation of the axial load in columns at the collapse state

If the sum of the reduced plastic moment of columns for the first storey is known, then the previously explained design conditions allow the definition, through the ratios ρ_k ($k = 1, 2, ..., n_s$), of the same sum corresponding to the kth storey, which guarantees that failure does not occur according to mechanisms belonging to the three examined typologies. In order to define the plastic section modulus of the columns, evaluation of the axial load in the columns at the collapse state is required.

Evaluation of the column axial forces is particularly easy, because they can be derived taking into account that the shear forces in the beams, at the collapse state, are due to the vertical loads and to the fact that plastic hinges are located at both ends of beams, so the plastic moments are applied there. The sum of these shear forces transmitted by the beams at each storey, above the considered one, provides the axial forces in the columns of the considered storey.

2.9.9 Design algorithm

It has been shown that the upper bound theorem allows the assessment of a condition for avoiding each undesired collapse mechanism, by considering the ratio between the sum of the reduced plastic moments of the kth storey column and the same sum corresponding to the first storey. As three different collapse mechanism typologies have been considered, there are $3n_s$ design conditions to be satisfied, which are provided by relationships (2.113), (2.120) and (2.126). These design conditions have to be integrated by the technological condition (2.129). These relationships can be used to design a frame failing in global mode and, therefore, having a mechanism equilibrium curve given by $\alpha = \alpha^{(g)} - \gamma_s^{(g)}\,\delta$, where $\alpha^{(g)}$ is the kinematically admissible multiplier given by equation (2.47) and $\gamma_s^{(g)}$ is the slope of the mechanism equilibrium curve given by equation (2.48). The algorithm to solve this problem is now presented. The following steps have to be performed.

a. Selection of the maximum displacement up to which it is desired to assure that the collapse mechanism cannot be different from the global one; this displacement has to be the ultimate displacement and therefore it can be evaluated as $\delta = \theta_{p_u} h_{n_s}$, where θ_{p_u} is the ultimate value of the beam plastic rotation which can be evaluated starting from the beam plastic rotation capacity. The computation of the beam plastic rotation capacity can be performed by means of a method proposed in reference [16] (Chapter 3).

b. Computation of the storey functions θ_{i_m}, β_{i_m}, γ_{i_m}, λ_{i_m} and ξ_{i_m}, which are provided by equations (2.115), (2.122), (2.116), (2.108) and (2.107), respectively.

c. Computation of the parameters $\Delta_{i_m}^{(t)}$ (with $t = 1, 2, 3$, related to the influence of second-order effects, given by equations (2.110), (2.118) and (2.124).

d. Computation of the slopes $\gamma_{s_{i_m}}^{(t)}$ (with $t = 1, 2, 3$) of the equilibrium curves of the considered mechanisms, as provided by equations (2.52), (2.56) and (2.60).

e. Computation, for the selected displacement δ, of a tentative value $\alpha_t = \alpha^{(g)} - \gamma_s^{(g)} \delta$ of the multiplier corresponding to the global mechanism by imposing that the reduced plastic moment of the first-storey columns is not less than the plastic moment of the beams.

f. Computation of the limit values $\rho_{i_m}^{(1)}$, $\rho_{i_m}^{(2)}$ and $\rho_{i_m}^{(3)}$ provided by equations (2.113), (2.120) and (2.126), respectively.

g. Computation of the value of ρ_{i_m} which avoids failure modes corresponding to the three examined collapse mechanism typologies

$$\rho_{i_m} = \max \left\{ \rho_{i_m}^{(1)}, \rho_{i_m}^{(2)}, \rho_{i_m}^{(3)} \right\} \qquad (2.130)$$

h. Modification of the computed values of ρ_{i_m} in order to satisfy the following technological condition

$$\rho_1 \geq \rho_2 \geq \rho_3 \geq ,...., \geq \rho_{n_s} \qquad (2.131)$$

i. Computation of the corresponding kinematically admissible multipliers $\alpha_{i_m}^{(1)}$, $\alpha_{i_m}^{(2)}$ and $\alpha_{i_m}^{(3)}$ provided by equations (2.51), (2.55) and (2.59), respectively.

j. Computation of the ultimate multiplier as the minimum among all the kinematically admissible multipliers, but including the influence of second-order effects for the selected displacement

$$\alpha_u = \min\left\{ \alpha_{i_m}^{(t)} - \gamma_{s_{i_m}}^{(t)}\, \delta;\ \text{with}\ i_m = 1, 2, 3, ..., n_s\ \text{and}\ t = 1, 2, 3 \right\} \quad (2.132)$$

k. If the condition

$$\left| \alpha_u - \alpha_t \right| > tolerance \quad (2.133)$$

is verified, then the value of $\sum_{i=1}^{n_c} M_{c,i1}$ corresponding to $\alpha_u = \alpha_o^{(g)} - \gamma_s^{(g)} \delta$ has to be computed and the procedure has to be repeated starting from step **f**. In the opposite case, it can be assumed that $\alpha^{(g)} = \alpha_o^{(g)} - \gamma_s^{(g)} \delta = \alpha_u$ and the column sections can be derived according to the following steps.

l. Computation of the axial force in the columns at the collapse state.

m. Computation, for each storey, of the sum of the reduced plastic moments by means of equation (2.127). The reduced plastic moment of each column can be now obtained by assuming that each column provides the same contribution to the above mentioned sum.

n. Definition of the section of the *i*th column of the *k*th storey, by assuming that the point (N_{ik} , $M_{c, ik}$) belongs to the yielding surface.

The described design method has been applied to different structural schemes; the number of bays varied from 2 to 6 and the number of storeys varied from 2 to 8. The design results have been verified by means of static inelastic analyses including second-order effects. Structures failing in global mode have been obtained in all the examined cases, pointing out the reliability of the design method.

2.10 DUCTILITY CONTROL

2.10.1 Ductility evaluation

The method described in Section 2.9 allows the design of frames failing in global mode, so that the object of failure mode control is achieved. Starting from this result, a design oriented to the ductility control can also be performed. In addition, the mechanism curve method described in Section 2.7 can be used to evaluate the available global ductility. In fact, in a frame designed according to the method of Section 2.9, the plastic hinge location is known *a priori* so that we can state that the ultimate top sway displacement is achieved when the maximum rotation capacity is attained in a beam end. As already mentioned, the available rotation capacity R can be easily computed by means of the method proposed in references 15–17, because the point of zero moment of the beam, at collapse, can be evaluated taking into account that both ends of the beam have yielded. As a consequence, the limit plastic rotation θ_p can be computed by equation (2.80) and the ultimate top sway displacement can be estimated by equation (2.81), where $H_o = h_{n_s}$ because the designed frame fails in global mode. Finally, the available global ductility is evaluated as the ratio between the ultimate displacement and the one defining the yield state (Section 2.7).

2.10.2 Design of the beam-to-column panel zone

Under the assumption that the shear stress is uniformly distributed in the panel zone, the shear stress developed within the panel zone is given by

$$\tau_p = \frac{V_p}{\left(d_c - 2\ t_{cf} \right) t_p} \tag{2.134}$$

where V_p is the shear force in the panel zone. With reference to Fig. 2.14,

$$V_p = \frac{\sum M_b}{\left(d_b - t_{bf} \right)} - V_c \tag{2.135}$$

where

$$V_c = \frac{\sum M_c}{H - d_b} \tag{2.136}$$

is the shear force in the columns evaluated by assuming that the zero-moment points are located in the middle section of the columns.

Therefore, taking into account the equilibrium condition, $\sum M_b = \sum M_c$, equation (2.135) provides

$$V_p = \frac{\sum M_b}{\left(d_b - t_{bf} \right)} \left(1 - \frac{d_b - t_{bf}}{H - d_b} \right) \tag{2.137}$$

The average shear stress in the panel zone is given by

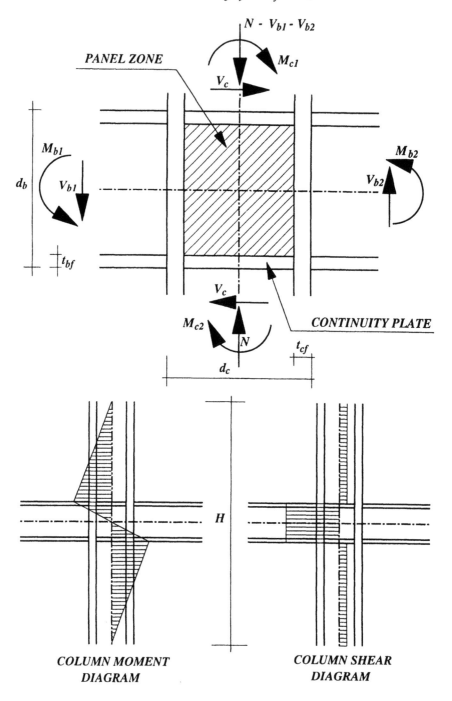

Fig. 2.14   Actions on the joint panel zone

$$\tau_p = \frac{\sum M_b}{t_p \left(d_c - 2\, t_{cf}\right) \left(d_b - t_{bf}\right)} \left(1 - \frac{d_b - t_{bf}}{H - d_b}\right) \tag{2.138}$$

In order to assure that yielding occurs in the beam rather than in the panel zone, according to the von Mises yield criterion, the following condition has to be satisfied

$$\frac{\sum M_{R,b}}{t_p \left(d_c - 2\, t_{cf}\right) \left(d_b - t_{bf}\right)} \left(1 - \frac{d_b - t_{bf}}{H - d_b}\right) \le \tau_y \left(1 - \left(\frac{N}{N_y}\right)^2\right)^{\frac{1}{2}} \tag{2.139}$$

This provides the design value t_p of the panel zone thickness, including doubler plates,

$$t_p \ge \frac{1}{\tau_y \left(1 - \left(\frac{N}{N_y}\right)^2\right)^{\frac{1}{2}}} \; \frac{\sum M_{R,b}}{\left(d_c - 2\, t_{cf}\right) \left(d_b - t_{bf}\right)} \left(1 - \frac{d_b - t_{bf}}{H - d_b}\right) \tag{2.140}$$

where $\sum M_{R,b}$ is the sum of the resisting moments of the beams connected to the joint and N_y is the squash load of the column.

In equation (2.140) the column axial force N has to be evaluated with reference to the collapse state, as explained in Section 2.9.8. This requirement can be significant in the case of panel zones belonging to exterior columns.

The assumption of uniformly distributed shear has been shown to be acceptable in estimating the stress field developed within the joint panel zone from numerous experimental investigations [26, 28]. However, by means of numerical simulations, it has been demonstrated [30] that the average shear stress computed by means of equation (2.128) is about 20% lower than the stress shown in contour near the centre of the joint, while it is 20% higher than the stress shown at the edges of the joint. As a consequence, attaching the centre of the doubler plates to the column web by using plug welds seems a good practice in order to obtain a better participation of the doubler plates in resisting the panel zone shear [29]. Moreover, it is useful to note that the stress contours in the joint panel zone seem almost identical whether or not the doubler plates are extended beyond the continuity plates [30].

It is useful to note that in the previous equations the term $(d_b - t_{bf})$ represents the height of the panel zone. In fact, if the distance between the stiffeners on the column web is practically equal to the distance between the beam flanges, i.e. continuity plates are adopted, such a distance is assumed as the height of the panel zone. It has been pointed out that, in the case of end plate bolted joints [34], stiffeners can be located away from the tension bolts. In this case, the height of the panel zone should be evaluated as the distance from the centre of tension bolts to the compression centre of the end plate. The above value should also be taken as the height of the panel zone when stiffeners are absent [34].

The exclusion of the panel zone from the participation in dissipating energy under earthquake action increases the plastic engagement of the beam ends, which have to withstand severe plastic rotations, but also reduces the frame sensitivity to second-order effects. As a consequence, the beam-to-column connections have to be appropriately detailed and constructed. As an example, fully welded beam-to-column connections can sometimes provide the required deformation capacity.

2.11 REFERENCES

[1] *H. Krawinkler, M. Zohrei:* 'Cumulative Damage in Steel Structures Subjected to Earthquake Ground Motion', Computer & Structures, Vol. 16, n.1–4, 1983.

[2] *H. Banon, D. Veneziano:* 'Seismic Safety of Reinforced Concrete Members and Structures', Earthquake Engineering and Structural Dynamics, Vol. 10, 179–193, 1982.

[3] *J.E. Stephens, J.T.P. Yao:* 'Damage Assessment Using Response Measurements', Journal of Structural Engineering, ASCE, Vol. 113, April, 1987.

[4] *S. Mahin, V.V. Bertero:* 'An Evaluation of Inelastic Seismic Design Spectra', Journal of the Structural Division, ASCE, September 1981.

[5] *E. Cosenza, G. Manfredi, R. Ramasco:* 'An Evaluation of the Use of Damage Functionals in Earthquake Engineering Design', 9th European Conference on Earthquake Engineering, Moscow, September 1990.

[6] *E. Cosenza, G. Manfredi, R. Ramasco:* 'The Use of Damage Functionals in Earthquake Engineering: A Comparison between Different Methods', Earthquake Engineering and Structural Dynamics, Vol. 22, 1993.

[7] *G.H. Powell, R. Allahabadi:* 'Seismic Damage Prediction by Deterministic Methods: Concepts and Procedures', Earthquake Engineering and Structural Dynamics, Vol.16, 719–734, 1988.

[8] *A.S. Elnashay, M. Chryssanthoupoulos:* 'Effect of Random Material Variability on Seismic Design Parameters of Steel Frames', Earthquake Engineering and Structural Dynamics, Vol. 20, 101–114, 1991.

[9] *F.M. Mazzolani, E. Mele, V. Piluso:* 'On the Effect of Randomness of Yield Strength in Steel Framed Structures under Seismic Loads', ECCS TC 13, Document N.TC13.01.91.

[10] *E. Cosenza:* 'Duttilità Globale delle Strutture Sismo-Resistenti in Acciaio', PhD Thesis, Università di Napoli, 1987.

[11] *H. Akiyama:* 'Earthquake Resistant Limit State Design for Buildings', University of Tokyo Press, 1985.

[12] *H. Akiyama:* 'Earthquake Resistant Design Based on the Energy Concept', Proceedings of the 9th World Conference on Earthquake Engineering, Tokyo, Kyoto, paper 8-1-2, Vol. V, August 2–9, 1988.

[13] *M.R. Horne, L.J. Morris:* 'Optimum Design of Multi-Storey Rigid Frames', Chapter 14 of 'Optimum Structural Design – Theory and Application', edited by R.H. Gallagher and O.C. Zienkiewicz, Wiley, 1973.

[14] *M.R. Horne, L.J. Morris:* 'Plastic Design of Low-Rise Frames', Constrado, Collins Professional and Technical Books, London 1981.

[15] *V. Piluso:* 'Il Comportamento Inelastico dei Telai Sismo-Resistenti in Acciaio', Tesi di Dottorato (PhD Thesis) in Ingegneria delle Strutture, IV Ciclo, Università di Napoli, 1992.

[16] *F.M. Mazzolani, V. Piluso:* 'Evaluation of the Rotation Capacity of Steel Beams and Beam-Columns', 1st COST C1 Workshop, Strasbourg, 28–30 October 1992.

[17] *F.M. Mazzolani, V. Piluso:* 'Member Behavioural Classes for Steel Beams and Beam-Columns', XXVI CTA, Collegio dei Tecnici dell'Acciaio, Giornate Italiane della Costruzione in Acciaio, Italian Conference on Steel Construction, Viareggio, 24–27 Ottobre 1993.

[18] *J. Sakamoto, A. Miyamura:* Critical Strength of Elastoplastic Steel Frames under Vertical and Horizontal Loading, Trans. Architectural Institute of Japan, Vol. 124, 1–7, 1966

[19] *ECCS (European Convention for Constructional Steelwork):* 'European Recommendations for Steel Structures in Seismic Zones', 1988.

[20] *Commission of the European Communities:* 'Eurocode 8: European Code for Seismic Regions – Design', Part 1.3: Buildings Draft of April 1993.

[21] *CNR-GNDT:* 'Norme Tecniche per le Costruzioni in Zone Sismiche', Dicembre, 1984.

[22] *R. Landolfo, F.M. Mazzolani, M. Pernetti:* 'L'Influenza dei Criteri di Dimensionamento sul Comportamento Sismico dei Telai in Acciaio', IV Convegno Nazionale, L'Ingegneria Sismica in Italia, Milano, Ottobre 1989.

[23] *R. Landolfo, F.M. Mazzolani:* 'The Consequences of the Design Criteria on the Seismic Behaviour of Steel Frames', 9th European Conference on Earthquake Engineering, Moscow, September, 1990.

[24] *AISC-LRFD:* 'Seismic Provisions for Structural Steel Buildings – Load and Resistance Factor Design', American Institute of Steel Construction, November 15, 1990.

[25] **UBC:** 'Uniform Building Code', International Conference of Building Officials, Whittier, CA, 1990.

[26] **H. Krawinkler, V.V. Bertero, E.P. Popov:** 'Inelastic Behaviour of Steel Beam-to-Column Subassemblages', Report No. UCB/EERC–71/7, Earthquake Engineering Research Center, University of California, Berkeley, 1971.

[27] **V.V. Bertero, E.P. Popov, H. Krawinkler:** 'Further Studies on Seismic Behaviour of Steel Beam-Column Subassemblages', Report No. UCB/EERC–73/27, Earthquake Engineering Research Center, University of California, Berkeley, 1973.

[28] **H. Krawinkler:** 'Shear in Beam-Column Joints in Seismic Design of Steel Frames', Engineering Journal, AISC, Vol. 15, No. 3, 1978.

[29] **E.P. Popov, N.R. Amin, J.J.C. Louis, R.M. Stephen:** 'Cyclic Behaviour of Large Beam-Column Assemblies', Earthquake Spectra, 1, No.2, 1985.

[30] **K. Tsai, E.P. Popov:** 'Steel Beam-Column Joints in Seismic Moment-Resisting Frames', Report No. UCB/EERC–88/19, Earthquake Engineering Research Center, University of California, Berkeley, 1988.

[31] **F.M. Mazzolani, V. Piluso:** 'Failure Mode and Ductility Control of Seismic Resistant MR frames', Italian Conference on Steel Construction, Viareggio, 24–27 Ottobre, 1993.

[32] **F.M. Mazzolani, V. Piluso:** 'Dimensionamento a collasso dei telai sismo-resistenti in acciaio', VI Convegno Nazionale, L'Ingegneria Sismica in Italia, Perugia, 13–15 Ottobre, 1993.

[33] **F.M. Mazzolani, V. Piluso:** 'A new method to design steel frames failing in global mode including P-Δ effects', International Workshop on Behaviour of Steel Structures in Seismic Areas, Timisoara, Romania, 26 June – 1 July, 1994.

[34] **G. Ballio, Y. Chen:** 'The Assessment of the Resistance of Shear Panel in Beam-to-Column Connections', Italian Conference on Steel Construction, Viareggio, 24–27 Ottobre, 1993.

3

Local ductility of beams and beam-columns

3.1 INTRODUCTION

In the previous chapter, it was shown that plastic deformation capacity is a key parameter in the assessment of the ultimate load-carrying capacity of steel moment-resisting frames subjected to seismic horizontal forces.

In limit design of structures it is requested that plastic hinges must have a sufficient rotation capacity without losing the bending capacity of the sections, so that the complete development of a collapse mechanism is allowed. In seismic design, a much greater plastic deformation capacity is sometimes required to dissipate the earthquake input energy.

The rotation capacity of steel members is undermined by the occurrence of local buckling of the plate elements which constitute the member cross-section and, if torsional restraints are not provided, by the occurrence of lateral torsional buckling.

In order to design cross-sections able to provide sufficient rotational capacity, the local buckling phenomenon has to be controlled. In particular the occurrence of local buckling within the elastic range has to be strictly avoided; therefore, geometrical properties of the cross-sections have to guarantee the attainment of plastic range before buckling.

The problem of correlating the plastic deformation capacity with the geometrical and mechanical properties of the cross-section has a primary importance, as it is demonstrated by the research efforts in this field and by the role of the problem in the modern seismic and non-seismic codes.

Haaijer and Thürlimann in 1958 [1] proposed that a section suitable for plastic design should not buckle before the occurrence of strain hardening. Their design philosophy became the guideline for the following research work. They investigated separately the effects of flange and web buckling without considering the restraining effect of the adjacent plate parts. Test results have pointed out that buckling of plates constituting double T-sections under bending moment do not develop independently, but they are geometrically compatible.

Lay [2] examined mainly the effect of flange buckling, by establishing a mathematical model for compressed flanges under yield stress supported by the web by means of a fictitious torsional spring. Starting from the above studies, Lay and Galambos [3, 4] provided a relationship for evaluating deformation capacity up to the occurrence of local buckling.

Both Haaijer [5] and Lay [2] assumed the plate-buckling problem to be solved as bifurcation of equilibrium.

Ben Kato [6] approached the problem from a different aspect. In the case of plates with rather high width-to-thickness ratios, the beginning of the yield is accompanied by fine 'crumplings' (waviness). The plate does not immediately lose its load-bearing capacity, but it can maintain the load during a certain deformation depending on the plate geometry and on the steel characteristics. This 'crumpling' can be considered as a kind of yield mechanism where not constant but increasing moments develop in plastic hinges due to strain hardening.

Experimental results of Lukey and Adams [7, 8] have shown that section failure does not immediately occur after local buckling and an important amount of plastic deformation is available before the bending capacity of the section falls below the plastic moment. Therefore, the postbuckling behaviour forms an important part of the calculation of the rotation capacity.

One of the main limitations of Kato's model is that the buckling of the plates forming the cross-section develops in a geometrically independent manner.

Climenhaga and Johnson [9] have described the postbuckling behaviour of bent steel members by means of geometrically compatible yield-line models which have been adapted to the form of real experimental buckling shapes. In this approach, the kinematic theorem of plasticity is applied, giving an upper bound.

Ivanyi [10, 11] also undertook a postbuckling investigation of compressed and bent members with the aid of the yield mechanism curve. Recently, Gioncu *et al.* [12] considered a suitable yield mechanism for cyclic bending.

The theoretical studies, briefly recalled above, have been accompanied by experimental research which has led to the formulation of empirical relationships for evaluating the rotation capacity. This approach, adopted by Nakamura [13], Mitani and Makino [14] and Kato and Akiyama [15], also takes into account the unstable part of the deformation capacity, due to postbuckling behaviour.

A different approach, introduced by Kemp [16], is based on the use of simplified structural models allowing a mathematical procedure leading to closed-form solutions. In this approach, the deformation capacity is evaluated with reference to its stable part only; postbuckling behaviour is neglected. Kemp evaluated the conditions leading to the local buckling of the compressed flange by means of a theoretical formulation.

Due to the difficulties arising in the theoretical evaluation of the ultimate stress leading to the local buckling of the compressed flange, a modification to the previous approach has been introduced by Ben Kato [17–19]. This modification has led to a new approach for evaluating rotation capacity, namely semiempirical methods, in which the theoretical evaluation of the moment versus curvature relationship is integrated by the use of empirical relations, derived from experimental data, for defining local buckling conditions. The same approach has been followed by Mazzolani and Piluso [20, 21].

The last possibility for evaluating rotation capacity is represented by the use of sophisticated numerical simulations based on the finite strip method [22–24] or on the finite element method [25]. In both cases, the analysis of the postbuckling behaviour is the most delicate point of the numerical approach. An empirical method based on the best fitting of both experimental and numerical simulation data has been proposed by Spangemacher and Sedlacek [26, 27].

In this chapter, the role of the parameters affecting the deformation capacity of steel beams and beam-columns will be examined and the different formulations for evaluating rotation capacity, provided in technical literature, will be presented and critically compared.

3.2 CROSS-SECTION CLASSIFICATION

A very important concept in design of steel structures, which has been introduced for the first time in Eurocode 3 [28], is represented by the subdivision of structural sections into different behavioural classes (Fig. 3.1). They are:

- class 1 – plastic sections
- class 2 – compact sections
- class 3 – semi-compact sections
- class 4 – slender sections.

Sections belonging to the first class are characterized by the ability to develop plastic hinges with high rotation capacity.

Class 2 sections are able to provide their maximum plastic flexural strength, but they have a limited deformation capacity.

Structural sections fall into class 3 when the bending moment leading to first yielding can be attained but, due to local buckling phenomena, plastic redistribution is not possible.

Finally, sections belonging to class 4 are not able to develop their total elastic flexural strength due to the premature attainment of local buckling of the compressed parts of the section, so that failure occurs in the elastic range.

It is clear that the parameter governing the above behaviour, and

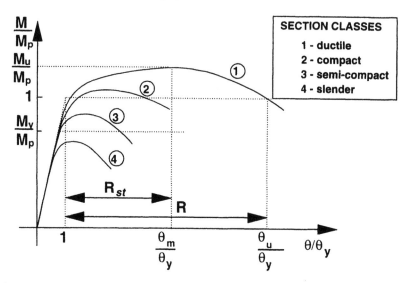

Fig. 3.1   Cross-section behavioural classes

therefore, which defines the class of the structural sections, is the width-to-thickness ratio b/t of the compressed plates which constitute the section. The limiting values of the b/t ratios defining the different behavioural classes according to Eurocode 3 [28] are given in Tables 3.1–3.4.

The problem has also been faced in the ECCS Recommendations for steel structures in seismic zones [29], which provide a limitation to the value of the q-factor depending upon the class of cross-sections in dissipative zones.

It has been pointed out [17–19] that the theoretical and experimental backgrounds of b/t limitations for defining classes of structural sections are not completely assessed. In particular, width-to-thickness ratio limitations are prescribed independently for flange and web. It seems that such an independent limitation is unreasonable, because the flange is restrained by the web and the web is restrained by the flange.

3.3   EVALUATION OF ROTATION CAPACITY

Beams and columns of rigid frames subjected to horizontal forces have to withstand double curvature bending, which can be simulated by an assembly of configurations of cantilever beams. Moreover, the rotation capacity of cantilever beams can be compared to the one of centrally loaded beams which are usually adopted as test specimen (Fig. 3.2). Rotation capacity can be defined as the ratio between the plastic rotation at the collapse state $\theta_p = \theta_u - \theta_y$ and the elastic limit θ_y (Fig. 3.1)

Table 3.1　Limiting values of the width-to-thickness ratios for different behavioural classes – webs

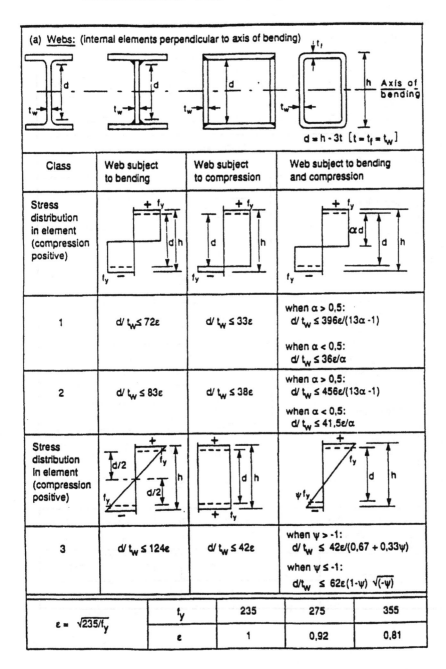

| | (a) Webs: (internal elements perpendicular to axis of bending) |
| | $d = h - 3t$ $[t = t_f = t_w]$ |

Class	Web subject to bending	Web subject to compression	Web subject to bending and compression
Stress distribution in element (compression positive)			
1	$d/t_w \leq 72\varepsilon$	$d/t_w \leq 33\varepsilon$	when $\alpha > 0,5$: $d/t_w \leq 396\varepsilon/(13\alpha - 1)$ when $\alpha < 0,5$: $d/t_w \leq 36\varepsilon/\alpha$
2	$d/t_w \leq 83\varepsilon$	$d/t_w \leq 38\varepsilon$	when $\alpha > 0,5$: $d/t_w \leq 456\varepsilon/(13\alpha - 1)$ when $\alpha < 0,5$: $d/t_w \leq 41,5\varepsilon/\alpha$
Stress distribution in element (compression positive)			
3	$d/t_w \leq 124\varepsilon$	$d/t_w \leq 42\varepsilon$	when $\psi > -1$: $d/t_w \leq 42\varepsilon/(0,67 + 0,33\psi)$ when $\psi \leq -1$: $d/t_w \leq 62\varepsilon(1-\psi)\sqrt{(-\psi)}$

$\varepsilon = \sqrt{235/f_y}$	f_y	235	275	355
	ε	1	0,92	0,81

Local ductility of beams and beam-columns

Table 3.2 Limiting values of the width-to-thickness ratios for different behavioural classes – internal flange elements

(b) Internal flange elements: (internal elements parallel to axis of bending)

Class	Type	Section in bending		Section in compression	
	Stress distribution in element and across section (compression positive)				
1	Rolled Hollow Section	$(b-3t_f)/t_f$	$\leq 33\varepsilon$	$(b-3t_f)/t_f$	$\leq 42\varepsilon$
	Other	b/t_f	$\leq 33\varepsilon$	b/t_f	$\leq 42\varepsilon$
2	Rolled Hollow Section	$(b-3t_f)/t_f$	$\leq 38\varepsilon$	$(b-3t_f)/t_f$	$\leq 42\varepsilon$
	Other	b/t_f	$\leq 38\varepsilon$	b/t_f	$\leq 42\varepsilon$
	Stress distribution in element and across section (compression positive)				
3	Rolled Hollow Section	$(b-3t_f)/t_f$	$\leq 42\varepsilon$	$(b-3t_f)/t_f$	$\leq 42\varepsilon$
	Other	b/t_f	$\leq 42\varepsilon$	b/t_f	$\leq 42\varepsilon$

$\varepsilon = \sqrt{235/f_y}$	f_y	235	275	355
	ε	1	0,92	0,81

Table 3.3 Limiting values of the width-to-thickness ratios for different behavioural classes – flanges

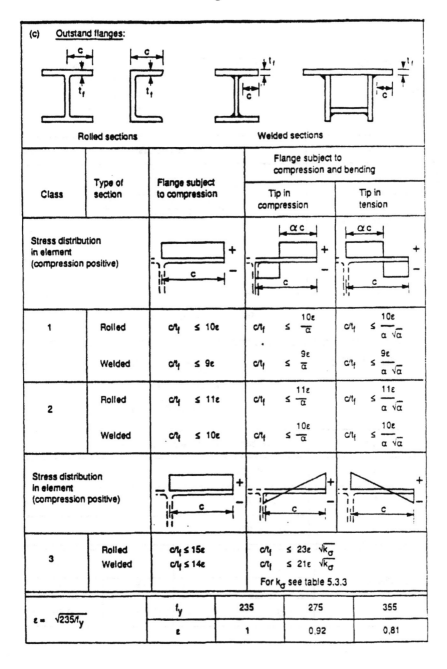

(c) Outstand flanges:

Rolled sections   Welded sections

Class	Type of section	Flange subject to compression	Flange subject to compression and bending	
			Tip in compression	Tip in tension
Stress distribution in element (compression positive)				
1	Rolled	$c/t_f \leq 10\varepsilon$	$c/t_f \leq \dfrac{10\varepsilon}{\alpha}$	$c/t_f \leq \dfrac{10\varepsilon}{\alpha\sqrt{\alpha}}$
	Welded	$c/t_f \leq 9\varepsilon$	$c/t_f \leq \dfrac{9\varepsilon}{\alpha}$	$c/t_f \leq \dfrac{9\varepsilon}{\alpha\sqrt{\alpha}}$
2	Rolled	$c/t_f \leq 11\varepsilon$	$c/t_f \leq \dfrac{11\varepsilon}{\alpha}$	$c/t_f \leq \dfrac{11\varepsilon}{\alpha\sqrt{\alpha}}$
	Welded	$c/t_f \leq 10\varepsilon$	$c/t_f \leq \dfrac{10\varepsilon}{\alpha}$	$c/t_f \leq \dfrac{10\varepsilon}{\alpha\sqrt{\alpha}}$
Stress distribution in element (compression positive)				
3	Rolled	$c/t_f \leq 15\varepsilon$	$c/t_f \leq 23\varepsilon\sqrt{k_\sigma}$	
	Welded	$c/t_f \leq 14\varepsilon$	$c/t_f \leq 21\varepsilon\sqrt{k_\sigma}$	
			For k_σ see table 5.3.3	

$\varepsilon = \sqrt{235/f_y}$	f_y	235	275	355
	ε	1	0,92	0,81

Table 3.4   Limiting values of the width-to-thickness ratios for different behavioural classes – angles and tubular sections

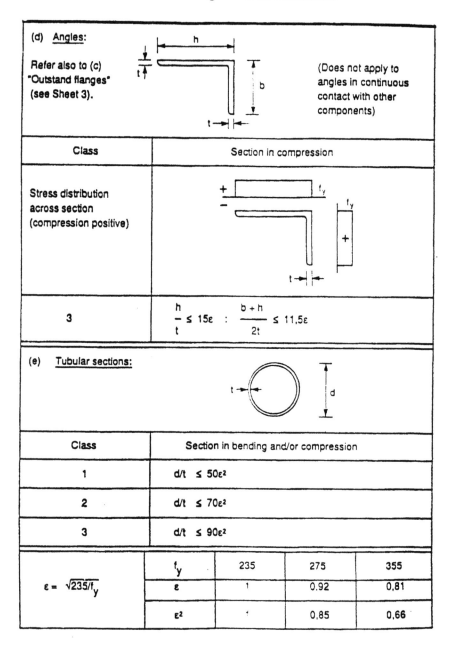

(d) **Angles:**			
Refer also to (c) "Outstand flanges" (see Sheet 3).			(Does not apply to angles in continuous contact with other components)

Class	Section in compression		
Stress distribution across section (compression positive)			
3	$\dfrac{h}{t} \le 15\varepsilon$: $\dfrac{b+h}{2t} \le 11,5\varepsilon$		

(e)   **Tubular sections:**

Class	Section in bending and/or compression		
1	$d/t \le 50\varepsilon^2$		
2	$d/t \le 70\varepsilon^2$		
3	$d/t \le 90\varepsilon^2$		

$\varepsilon = \sqrt{235/f_y}$	f_y	235	275	355
	ε	1	0,92	0,81
	ε^2	1	0,85	0,66

$$R = \frac{\theta_p}{\theta_y} = \frac{\theta_u - \theta_y}{\theta_y} = \frac{\theta_u}{\theta_y} - 1 \qquad (3.1)$$

It is useful to point out that rotation capacity can be divided into two parts. The first is related to the stable increasing branch of the moment versus rotation diagram, while the second one is related to the unstable softening branch describing the postbuckling behaviour. The stable part of the rotation capacity is given by

$$R_{st} = \frac{\theta_m}{\theta_y} - 1 \qquad (3.2)$$

In order to evaluate the rotation capacity of steel members by means of closed form relationships, different methods have been proposed. They can be divided into the three groups of theoretical methods, semi-empirical methods and empirical methods.

Theoretical methods are based upon the approximated theoretical evaluation of the moment versus curvature relationship of member cross-section and upon the theoretical analysis of buckling phenomena.

Semi-empirical methods differ from theoretical ones due to the fact that local buckling phenomena are taken into account by means of

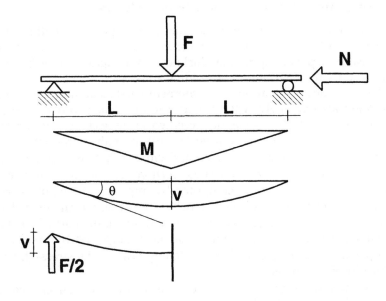

Fig. 3.2   Structural schemes adopted in the experimental evaluation of the rotation capacity

relationships provided by experimental evidence. In this context stub column tests can be performed.

Both theoretical and semi-empirical methods are able to clarify the
role of the parameters affecting the rotation capacity, but they are limited
to the analysis of the stable part of the plastic deformation capacity.

Empirical methods are based upon the direct analysis of experimental
data of full-scale member tests or upon the analysis of the numerical
simulation data. In both cases they can take into account, by means of
experimental or numerical evidence, both the precritical and the postbuckling behaviour of steel members.

In the following, methods belonging to the different groups are
presented and the corresponding results are examined and discussed on
the basis of some existing test data.

First of all, the theoretical basis for the evaluation of rotation capacity
is given in Section 3.4. Some theoretical and semi-empirical methods are
presented in Sections 3.5 and 3.6, respectively. Empirical methods are
examined in Section 3.7. A comparison between the different methods
and the experimental data is provided in Section 3.8.

3.4   THEORETICAL BASES

3.4.1 Simplified approach for closed-form solutions

Both theoretical and semi-empirical methods require evaluation of the
moment versus curvature relationship. For H cross-sections, this
relationship depends upon the thickness of flanges and web and is
influenced by the distribution of residual stresses.

The theoretical evaluation of the moment versus curvature law can be
performed appoximately with reference to an ideal two-flange section.
This approach, which is generally accepted in the technical literature [3,
4, 16], has also been used by Kato [17–19] to analyse the deformation
capacity of square hollow sections and circular hollow sections. In all
cases the equivalent ideal two-flange section has the same area as the
real item, while its depth is computed by imposing the equivalence of
plastic moment for both sections. Finally it must be remembered that in
the theory developed in [17–19], the rotation due to the part of the member which remains in elastic range has been neglected by assuming a
rigid-plastic-hardening or a rigid-hardening law for the stress–strain relationship of the material.

In the following, more general relationships [20, 21] which take into
account also the elastic deformability are presented, and the theory of
Kato is derived as a particular case corresponding to the limit condition
$E \rightarrow \infty$, where E is the elastic modulus.

3.4.2 Moment–curvature relationship

In this Section the moment–curvature relationship of an ideal two-flange section is obtained by considering an elastic–plastic flow–strain hardening model for the stress–strain relationship of the material (Fig. 3.3). Let:

- N_o be the external axial load;
- N_{up} the force in the upper flange;
- N_{lo} the force in the lower flange;
- M the external bending moment;
- ε_{up} the strain in the upper flange;
- ε_{lo} the strain in the lower flange;
- ε_y the yield strain;
- ε_h the strain corresponding to the beginning of the strain hardening;
- E the elastic modulus;
- E_h the hardening modulus;
- σ_y the yield stress;
- $\sigma_{cr} = s\sigma_y$ the critical stress due to the local buckling of the compressed flange or to the lateral–torsional buckling (where, with reference to local buckling, $s \geq 1$ in the case of ductile and compact sections, $W/Z < s < 1$ in the case of semi-compact sections and $s \leq W/Z$ in the case of slender sections; where W is the elastic modulus of the section and Z the plastic modulus);
- $\sigma_o = \rho\sigma_y$ the stress due to the external axial load ($\rho \leq 1$; $\rho = 0$ for pure bending);
- A the area of the cross-section;
- h_e the depth of the two-flange section.

According to the definition of rotation capacity represented in Fig. 3.1, it is found that for $s < 1$, i.e. for semi-compact and slender sections, $R = 0$.

Stresses, strains and forces are assumed positive when a compression results.

The equilibrium equation in the longitudinal direction requires

$$N_{up} + N_{lo} = N_o \tag{3.3}$$

while the rotation equilibrium equation can be written as

$$M = \frac{N_{up} - N_{lo}}{2} h_e \tag{3.4}$$

The general expression of the curvature is given by

$$\chi = \frac{\varepsilon_{up} - \varepsilon_{lo}}{h_e} \tag{3.5}$$

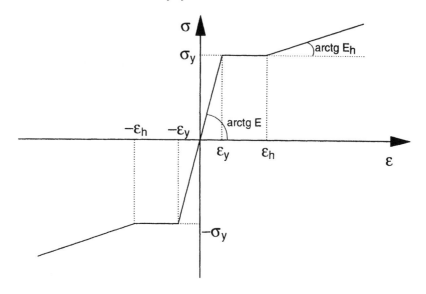

Fig. 3.3 Stress–strain relationship

Different behavioural phases can be identified depending upon the strain state of the flanges: elastic, plastic or strain hardening.

Case 1: both flanges in the elastic range
In this case the bending moment is given by

$$M = \chi\, E\, I_e \tag{3.6}$$

where

$$I_e = \frac{A\, h_e^2}{4} \tag{3.7}$$

the moment of inertia of the two-flange section.
   Equation (3.6) can be used until the yielding of the compressed flange is reached for a bending moment given by

$$M = (1 - \rho)M_p \tag{3.8}$$

where

$$M_p = \frac{A\, \sigma_y}{2}\, h_e \tag{3.9}$$

is the plastic moment of the two-flange section.

Equation (3.8) can be obtained from equations (3.3) and (3.4) by considering that in this case the upper-flange force is

$$N_{up} = \frac{A\,\sigma_y}{2} \qquad (3.10)$$

The limit value of the curvature corresponding to the attainment of the bending moment (3.8) is given by

$$\chi_1 = (1-\rho)\,\frac{M_p}{E\,I_e} = 2\,(1-\rho)\,\chi_y \qquad (3.11)$$

where

$$\chi_y = \frac{\varepsilon_y}{h_e} \qquad (3.12)$$

Case 2: the strain state of the upper flange corresponds to the plastic flow branch
In this case the constant value of the bending moment is given by equation (3.8), while the plastic flow produces an increase of the curvature. Therefore, equation (3.8) has to be used until the following value of the curvature is reached

$$\chi_2 = 2\,(1-\rho)\,\chi_y + \chi_h - \chi_y = (1-2\,\rho)\,\chi_y + \chi_h \qquad (3.13)$$

where

$$\chi_h = \frac{\varepsilon_h}{h_e} \qquad (3.14)$$

Case 3: the upper flange is in strain-hardening range
The forces in the upper and lower flanges are given by

$$N_{up} = \frac{A\,\sigma_y}{2} + E_h\,\frac{A}{2}\,(\varepsilon_{up} - \varepsilon_h) \qquad (3.15)$$

$$N_{lo} = \rho\sigma_y\,A - \frac{A\sigma_y}{2} - E_h\,\frac{A}{2}\,(\varepsilon_{up} - \varepsilon_h) \qquad (3.16)$$

Combining equations (3.3), (3.4), (3.15) and (3.16) the following expression for the bending moment is obtained

$$M = (1-\rho)\,M_p + E_h\,\frac{A}{2}\,(\varepsilon_{up} - \varepsilon_h)\,h_e \qquad (3.17)$$

Now, two possibilities can occur for the lower flange:

a. the critical bending moment $(s - \rho) M_p$ is attained when the lower flange is still in elastic range;

b. the critical bending moment $(s - \rho) M_p$ is attained when the lower flange is in plastic flow or in hardening range.

The first situation is verified if the following condition is satisfied

$$N_{lo} > - \sigma_y \frac{A}{2} \tag{3.18}$$

By combining relationships (3.16) and (3.18), it can be shown that the following condition has to be satisfied

$$\varepsilon_{up} - \varepsilon_h < \frac{2 \rho \sigma_y}{E_h} \tag{3.19}$$

The condition corresponding to the attainment of the critical bending moment can be expressed by means of the relationship

$$(s - \rho) M_p = (1 - \rho) M_p + E_h \frac{A}{2} (\varepsilon_{up} - \varepsilon_h) h_e \tag{3.20}$$

which gives

$$\varepsilon_{up} - \varepsilon_h = (s - 1) \frac{\sigma_y}{E_h} \tag{3.21}$$

By combining (3.19) and (3.21) it can be concluded that the critical bending moment is attained with the lower flange in elastic range if the following condition is satisfied

$$\rho > \frac{s - 1}{2} \tag{3.22}$$

For the opposite case, the lower flange is in plastic flow or strain-hardening range when the critical bending moment is attained.

Case 3a: $\rho > (s - 1) / 2$

The force in the lower flange can be written in the form

$$N_{lo} = \varepsilon_{lo} E \frac{A}{2} \tag{3.23}$$

Equating (3.23) and (3.16) the following relationship between ε_{lo} and ε_{up} is obtained

$$\varepsilon_{lo} = (2\,\rho - 1)\,\varepsilon_y - \frac{E_h}{E}\,(\varepsilon_{up} - \varepsilon_h) \qquad (3.24)$$

By combining equations (3.5), (3.24) and (3.14) it can be shown that

$$\varepsilon_{up} - \varepsilon_h = \frac{E}{E + E_h}\, h_e\,[\,\chi + (2\,\rho - 1)\,\chi_y - \chi_h\,] \qquad (3.25)$$

By combining equations (3.17), (3.25) and (3.7), the following moment–curvature relationship is obtained

$$M = (1 - \rho)\,M_p + \frac{2\,E\,E_h}{E + E_h}\,I_e\,[\,\chi - (1 - 2\rho)\,\chi_y - \chi_h\,] \qquad (3.26)$$

By introducing the **reduced modulus of elasticity**

$$E_r = \frac{2\,E\,E_h}{E + E_h} \qquad (3.27)$$

the moment–curvature relation can be simplified as

$$M = (1 - \rho)\,M_p + E_r\,I_e\,[\,\chi - (1 - 2\rho)\,\chi_y - \chi_h\,] \qquad (3.28)$$

The value of the curvature χ_{cr} corresponding to the attainment of the critical bending moment $(s - \rho)\,M_p$ is obtained by

$$(s - \rho)\,M_p = (1 - \rho)\,M_p + E_r\,I_e\,[\,\chi_{cr} - (1 - 2\rho)\,\chi_y - \chi_h\,] \qquad (3.29)$$

Therefore, the critical curvature is

$$\chi_{cr} = \frac{(s - 1)\,M_p}{E_r\,I_e} + (1 - 2\,\rho)\,\chi_y + \chi_h \qquad (3.30)$$

Case 3b: $\rho \le (s - 1)\,/\,2$
In this case, equation (3.28) is valid until the condition $\varepsilon_{lo} = -\,\varepsilon_y$ is attained. For this condition, equation (3.24) provides

$$-\,\varepsilon_y = (2\rho - 1)\,\varepsilon_y - \frac{E_h}{E}\,(\varepsilon_{up} - \varepsilon_h) \qquad (3.31)$$

or

$$2\,\rho\,\varepsilon_y = \frac{E_h}{E}\,(\varepsilon_{up} - \varepsilon_h) \qquad (3.32)$$

therefore, from equation (3.25)

$$2 \rho \, \chi_y = \frac{E_h}{E + E_h} \, [\chi_3 - (1 - 2 \rho) \, \chi_y - \chi_h] \qquad (3.33)$$

By taking into account that

$$2 \chi_y = \frac{M_p}{E \, I_e} \qquad (3.34)$$

and by introducing the reduced modulus of elasticity, the following limit value of the curvature is obtained

$$\chi_3 = \frac{2 \rho \, M_p}{E_r \, I_e} + (1 - 2 \rho) \, \chi_y + \chi_h \qquad (3.35)$$

As soon as this value of the curvature has been attained, a plastic flow is obtained and the plastic increase $\Delta \chi = \chi_h - \chi_y$ is verified. Therefore, the bending moment assumes a constant value which can be obtained by introducing the limit curvature (3.35) in relation (3.28). This value,

$$M = (1 + \rho) \, M_p \qquad (3.36)$$

remains constant until the following limit value of the curvature is reached

$$\chi_4 = \frac{2 \rho \, M_p}{E_r \, I_e} + (1 - 2 \rho) \, \chi_y + 2 \, \chi_h - \chi_y \qquad (3.37)$$

When the value of curvature is greater than that provided by relationship (3.37), it means that the lower flange is also in the strain-hardening range. Equations (3.15), (3.16) and (3.17) are still valid. Moreover, the force in the lower flange can be expressed as

$$N_{lo} = - \sigma_y \, \frac{A}{2} + \frac{E_h \, A}{2} \, (\varepsilon_{lo} + \varepsilon_h) \qquad (3.38)$$

By equating (3.16) and (3.38) the following relationship between lower flange and upper flange strains is obtained

$$\varepsilon_{lo} + \varepsilon_{up} = \frac{2 \, \rho \, \sigma_y}{E_h} \qquad (3.39)$$

By combining equations (3.5) and (3.39) the upper flange strain can be related to the curvature as follows

$$\varepsilon_{up} = \frac{\rho \, \sigma_y}{E_h} + \chi \, \frac{h_e}{2} \tag{3.40}$$

Finally by using equations (3.9), (3.14), (3.17) and (3.40), the expression of the bending moment is obtained

$$M = M_p + E_h \, I_e \, (\chi - 2 \, \chi_h) \tag{3.41}$$

The critical value of the curvature is obtained from the condition

$$(s - \rho) M_p = M_p + E_h \, I_e \, (\chi_{cr} - 2 \, \chi_h) \tag{3.42}$$

and, therefore, is given by

$$\chi_{cr} = \frac{(s - \rho - 1) \, M_p}{E_h \, I_e} + 2 \, \chi_h \tag{3.43}$$

In Table 3.5 the moment versus curvature relationships are summarized for the three fundamental cases: $\rho > (s - 1)/2$ (Fig. 3.4), $\rho \leq (s - 1)/2$ (Fig. 3.5) and $\rho = 0$. Furthermore in Table 3.6 the limiting case $E \rightarrow \infty$ is derived by taking into account that

$$E_r \rightarrow 2 E_h \tag{3.44}$$

$$\chi_y \rightarrow 0 \tag{3.45}$$

Obviously, in this case the derived relationships defining the moment–curvature relationship are coincident with the ones provided in references [17–19].

Moreover it is useful to note that by assuming $\chi_h = \chi_y$ in the relationships given in Table 3.5, the moment versus curvature relationship for the case of elastic-strain hardening behaviour is obtained. The case of rigid-strain hardening behaviour can be analysed by means of the relationships provided in Table 3.6 by assuming $\chi_h = 0$.

3.4.3 Evaluation of the rotation θ_m

By using the moment–curvature relationships previously derived and by assuming the attainment of the critical bending moment $(s - \rho) \, M_p$ as the buckling condition, it is easy to compute the ultimate rotation of the cantilever beam or of the centrally loaded beam with an axial force (Fig. 3.2), by integrating the curvature diagram.

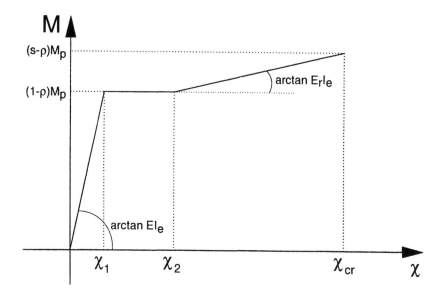

Fig. 3.4   Moment versus curvature relationship for $\rho > (s - 1)/2$

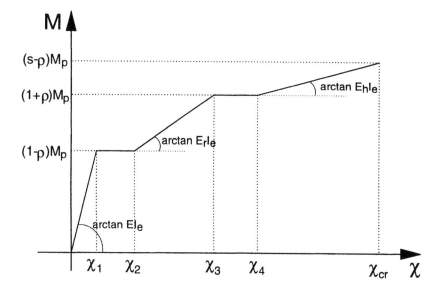

Fig. 3.5   Moment versus curvature relationship for $\rho \leq (s - 1)/2$

Table 3.5 Moment versus curvature relationships for an ideal two-flange section with elastic–plastic flow-strain hardening behaviour

CASE for $\rho > \dfrac{s-1}{2}$

$$M = \chi\, E\, I_e \qquad \text{for} \quad \chi \leq \chi_1$$

$$M = (1-\rho)M_p \qquad \text{for} \quad \chi_1 \leq \chi \leq \chi_2$$

$$M = (1-\rho)\,M_p + E_r\, I_e\,[\chi - (1 - 2\rho)\,\chi_y - \chi_h] \qquad \text{for} \quad \chi_2 \leq \chi \leq \chi_{cr}$$

where

$$\chi_1 = 2\,(1-\rho)\,\chi_y$$

$$\chi_2 = (1 - 2\rho)\,\chi_y + \chi_h$$

$$\chi_{cr} = \frac{(s-1)\,M_p}{E_r\, I_e} + (1 - 2\,\rho)\,\chi_y + \chi_h$$

CASE for $\rho \leq \dfrac{s-1}{2}$

$$M = \chi\, E\, I_e \qquad \text{for} \quad \chi \leq \chi_1$$

$$M = (1-\rho)M_p \qquad \text{for} \quad \chi_1 \leq \chi \leq \chi_2$$

$$M = (1-\rho)\,M_p + E_r\, I_e\,[\chi - (1 - 2\rho)\,\chi_y - \chi_h] \qquad \text{for} \quad \chi_2 \leq \chi \leq \chi_3$$

$$M = (1+\rho)\,M_p \qquad \text{for} \quad \chi_3 \leq \chi \leq \chi_4$$

$$M = M_p + E_h\, I_e\,(\chi - 2\chi_h) \qquad \text{for} \quad \chi_4 \leq \chi \leq \chi_{cr}$$

where

$$\chi_1 = 2\,(1-\rho)\,\chi_y$$

$$\chi_2 = (1 - 2\rho)\,\chi_y + \chi_h$$

$$\chi_3 = \frac{2\,\rho\, M_p}{E_r\, I_e} + (1 - 2\,\rho)\,\chi_y + \chi_h$$

$$\chi_4 = \frac{2\,\rho\, M_p}{E_r\, I_e} + (1 - 2\,\rho)\,\chi_y + 2\,\chi_h - \chi_y$$

$$\chi_{cr} = \frac{(s - \rho - 1)\,M_p}{E_h\, I_e} + 2\chi_h$$

CASE for $\rho = 0$

$$M = \chi\, E\, I_e \qquad \text{for} \quad \chi \leq \chi_1$$

Table 3.5　Continued

$M = M_p$	for　$\chi_1 \leq \chi \leq \chi_2$
$M = M_p + E_h I_e (\chi - 2\chi_h)$	for　$\chi_2 \leq \chi \leq \chi_{cr}$

where

$$\chi_1 = 2\chi_y$$

$$\chi_2 = 2\chi_h$$

$$\chi_{cr} = \frac{(s-1)M_p}{E_h I_e} + 2\chi_h$$

Table 3.6　Moment versus curvature relationships for an ideal two-flange section with rigid–plastic flow-strain hardening behaviour

CASE for $\rho > \dfrac{s-1}{2}$

$M = (1-\rho)M_p$	for　$\chi \leq \chi_1$
$M = (1-\rho)M_p + 2E_h I_e (\chi - \chi_h)$	for　$\chi_1 \leq \chi \leq \chi_{cr}$

where

$$\chi_1 = \chi_h$$

$$\chi_{cr} = \frac{(s-1)M_p}{2E_h I_e} + \chi_h$$

CASE for $\rho \leq \dfrac{s-1}{2}$

$M = (1-\rho)M_p$	for　$\chi \leq \chi_1$
$M = (1-\rho)M_p + 2E_h I_e (\chi - \chi_h)$	for　$\chi_1 \leq \chi \leq \chi_2$
$M = (1+\rho)M_p$	for　$\chi_2 \leq \chi \leq \chi_3$
$M = M_p + E_h I_e (\chi - 2\chi_h)$	for　$\chi_3 \leq \chi \leq \chi_{cr}$

where

$$\chi_1 = \chi_h$$

$$\chi_2 = \frac{\rho M_p}{E_h I_e} + \chi_h$$

Table 3.6   Continued

$$\chi_3 = \frac{\rho\, M_p}{E_h\, I_e} + 2\,\chi_h$$

$$\chi_{cr} = \frac{(s - \rho - 1)\, M_p}{E_h\, I_e} + 2\,\chi_h$$

CASE $\rho = 0$

$$M = M_p \qquad\qquad\qquad\qquad \text{for} \quad \chi \leq \chi_1$$

$$M = M_p + E_h\, I_e\, (\chi - 2\,\chi_h) \qquad\qquad \text{for} \quad \chi_1 \leq \chi \leq \chi_{cr}$$

where

$$\chi_1 = 2\,\chi_h$$

$$\chi_{cr} = \frac{(s - 1)\, M_p}{E_h\, I_e} + 2\,\chi_h$$

Case for $\rho > (s - 1)/2$

It is easy to show that, for the buckling state, the length of the member which is in plastic range is (Fig. 3.6)

$$\alpha\, L = \frac{s - 1}{s - \rho}\, L \tag{3.46}$$

while the part which remains in elastic range is given by

$$(1 - \alpha)\, L = \frac{1 - \rho}{s - \rho}\, L \tag{3.47}$$

By integrating the curvature diagram, the rotation corresponding to the maximum moment is obtained (Fig. 3.6)

$$\theta_m = \frac{\chi_1\, (1 - \alpha)\, L}{2} + \chi_2\, \alpha\, L + \frac{(\chi_{cr} - \chi_2)\, \alpha L}{2} \tag{3.48}$$

and by introducing (3.11), (3.13), (3.30), (3.46) and (3.47) the following equation is derived

$$\theta_m = \frac{L}{s - \rho} \left\{ (1 - \rho)^2\, \chi_y + (s - 1) \left[(1 - 2\rho)\, \chi_y + \chi_h + \frac{(s - 1)\, M_p}{2 E_r\, I_e} \right] \right\} \tag{3.49}$$

which, for $E \to \infty$, provides

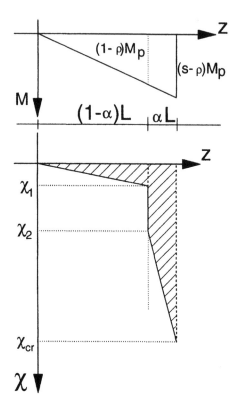

Fig. 3.6   Evaluation of the rotation θ_m for $\rho > (s-1)/2$

$$\theta_m = \frac{s-1}{s-\rho} L \left[\chi_h + \frac{(s-1)\,M_p}{4\,E_h\,I_e} \right] \tag{3.50}$$

This last relation is coincident with the one provided in references [17–19].

Case for $\rho \leq (s-1)/2$

In this case, with reference to Fig. 3.7, the length of the part which remains in elastic range is

$$(1-\alpha-\beta)\,L = \frac{1-\rho}{s-\rho}\,L \tag{3.51}$$

where

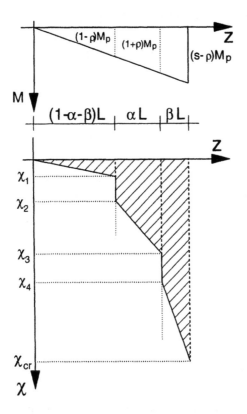

Fig. 3.7   Evaluation of the rotation θ_m for $\rho \le (s-1)/2$

$$\alpha L = \frac{2\rho}{s-\rho} L \qquad (3.52)$$

and

$$\beta L = \frac{s-2\rho-1}{s-\rho} L \qquad (3.53)$$

Therefore, α and β define the part of the member which is in plastic range.

By integrating the curvature diagram, the following equation is obtained (Fig. 3.7)

$$\theta_m = \frac{\chi_1 (1-\alpha-\beta) L}{2} + \chi_2 \alpha L + \frac{(\chi_3-\chi_2) \alpha L}{2} + \chi_4 \beta L + \frac{(\chi_{cr}-\chi_4) \beta L}{2} \qquad (3.54)$$

Combination of equations (3.43), (3.37), (3.27) and (3.34) gives

$$\chi_{cr} - \chi_4 = \frac{M_p \, (s - 2\rho - 1)}{E_h \, I_e} \tag{3.55}$$

From relations (3.13) and (3.35),

$$\chi_3 - \chi_2 = \frac{2 \, \rho \, M_p}{E_r \, I_e} \tag{3.56}$$

By inserting equations (3.11), (3.13), (3.56), (3.37), (3.55), (3.51), (3.52) and (3.53) in relationship (3.54), the final expression of the rotation corresponding to the maximum moment is obtained

$$\theta_m = \frac{L}{s - \rho} \left\{ \, [\, (1 + \rho^2) - 2\rho \, (s - 1) \,] \, \chi_y + 2 \, \chi_h \, (s - \rho - 1) + \frac{M_p}{E_r \, I_e} \, 2 \, \rho \, (s - \rho - 1) \right.$$

$$\left. + \, \frac{(s - 2 \, p - 1)^2 \, M_p}{2 \, E_h \, I_e} \right\} \tag{3.57}$$

which, for $E \to \infty$ provides

$$\theta_m = \frac{L}{s - \rho} \left\{ 2 \, \chi_h \, (s - \rho - 1) + \frac{M_p}{2 \, E_h \, I_e} \, [\, 2 \, \rho^2 + (s - 1) \, (s - 2 \, \rho - 1) \,] \right\} \tag{3.58}$$

which is coincident with the relationship provided in references [17–19].

Case for $\rho = 0$
In the case of simple beams, the relationship for evaluating the rotation θ_m is obtained from equation (3.57) with the condition $\rho = 0$ (absence of axial force), which provides

$$\theta_m = \frac{L}{s} \left\{ \chi_y + 2 \, \chi_h \, (s - 1) + \frac{(s - 1)^2 \, M_p}{2 \, E_h \, I_e} \right\} \tag{3.59}$$

Obviously, for $E \to \infty$, equation (3.59) is simplified as follows

$$\theta_m = \frac{L \, (s - 1)}{s} \left\{ 2 \, \chi_h + (s - 1) \, \frac{M_p}{2 \, E_h \, I_e} \right\} \tag{3.60}$$

3.4.4 Critical stress evaluation

In the previous sections the relationships for evaluating the rotation corresponding to the maximum moment have been provided for beams and beam-columns. The use of such relationships requires the computation of the ratio s between the critical stress leading to buckling and the first yielding stress.

The computation of the buckling stress ratio s can be performed using theoretical relationships or experimental relationships. In the first case the corresponding method for evaluating the stable part of the rotation capacity is a theoretical method, while in the second case it can be defined a semi-empirical method, because experimental relationships for evaluating the buckling stress ratio s are combined with the theoretical expressions obtained in the previous sections.

The two cases of local buckling and lateral–torsional buckling are considered below.

a. Local buckling

The critical stress leading to the inelastic buckling of the compressed flange, neglecting the restraint due to the web, is given [30] by

$$\sigma_{cr} \, b_f \, t_f = \frac{12}{b_f^{\,2}} \, \frac{b_f \, t_f^{\,3}}{3} \, G_h \qquad (3.61)$$

where b_f is the width of the flange, t_f is the thickness of the flange and G_h is the strain-hardening tangential modulus.

Moreover, Lay [2] has proved that

$$G_h = \frac{E}{1 + v + \dfrac{E}{4E_h}} \qquad (3.62)$$

where $v = 0.3$ is Poisson's ratio; therefore

$$s = \frac{\sigma_{cr}}{\sigma_y} = \left(\frac{t_f}{b_f}\right)^2 \frac{16E_h}{5.2\,E_h + E} \frac{1}{\varepsilon_y} \qquad (3.63)$$

A more accurate expression of the critical stress leading to inelastic local buckling can be obtained from the analysis of a plate restrained by a torsional spring, whose stiffness represents the restraining action exerted by the web. From this analysis the following expression of the critical strain is obtained [3, 4, 31, 32]

$$\varepsilon_{cr} = \frac{12}{E_h \, b_f^{\,3} \, t_f} \left[G_h \frac{b_f \, t_f^{\,3}}{3} + E_h \, I_\omega \left(\frac{2n\pi}{L_f}\right)^2 + K\left(\frac{L_f}{2n\pi}\right)^2 \right] \qquad (3.64)$$

where I_ω is the warping constant of the flange, K is the torsional stiffness of the web, L_f/n is the half-wave in which the flange buckles.

The minimum value of the critical strain is obtained by imposing

$$\frac{\partial \, \varepsilon_{cr}}{\partial \, (L_f \, / n)} = 0 \qquad (3.65)$$

which leads to the length of the half-wave corresponding to the minimum critical strain (and, therefore, to the minimum critical stress)

$$\frac{L_f}{n} = 2\pi \left(\frac{E_h \, I_\omega}{K} \right)^{0.25} \qquad (3.66)$$

By substituting equation (3.66) in (3.64) the expression of the critical strain is obtained as a function of the mechanical and geometrical properties of the flange and of the web [4]

$$\varepsilon_{cr} = \frac{12}{E_h \, b_f{}^3 \, t_f} \left[G_h \frac{b_f \, t_f{}^3}{3} + 2 \sqrt{K \, E_h \, I_\omega} \right] \qquad (3.67)$$

The relation (3.67) shows that the critical strain depends upon the material mechanical properties in plastic range E_h and G_h and the restraining condition of the flange. In particular, the stiffness K of the torsional spring has to be related to the restraining action, in elastic or plastic range, exerted by the web and it has also to be able to represent the possibility that, in some conditions, the web is restrained by the flange. The difficulties related to the evaluation of the stiffness K can be surpassed by taking into account that equation (3.67) essentially indicates that the contributions of the second and third term in equation (3.64) have the same weight [32] and, therefore, equation (3.66) can be used in order to eliminate from (3.64) the unknown contribution of the web, leading to a relationship which includes the important influence of the wave length:

$$\varepsilon_{cr} = \frac{12}{E_h \, b_f{}^3 \, t_f} \left[G_h \frac{b_f \, t_f{}^3}{3} + \frac{8 \, \pi^2 \, E_h \, I_\omega}{L_f{}^2} \right] \qquad (3.68)$$

The warping constant can be expressed [32] as

$$I_\omega = \frac{b_f \, t_f{}^3}{144} \qquad (3.69)$$

By combining equations (3.68) and (3.69) the final expression of the critical strain is obtained

$$\varepsilon_{cr} = 4 \frac{G_h}{E_h} \left(\frac{t_f}{b_f} \right)^2 + \frac{2}{3} \, \pi^2 \left(\frac{t_f}{L_f} \right)^2 \qquad (3.70)$$

The above equations enable the theoretical prediction of the critical stress leading to the local buckling of the compressed flange plate.

The main difficulties in the theoretical evaluation of the rotation capacity arise from the necessity to compute the critical stress leading to the local buckling of the compressed flange. The value of this critical stress is influenced by the restraining action exerted by the web, which, depending upon the stress state (elastic or plastic) of the web itself, is very difficult to evaluate. Therefore, as an alternative, the use of empirical relationships based upon experimental evidence is justified. By using data from a great number of stub column tests, Kato [17, 18] has proposed the following relationships which have been derived by means of multiple regression analysis:

- for H-section members

$$\frac{1}{s} = 0.6003 + \frac{1.600}{\alpha_f} + \frac{0.1535}{\alpha_w} \tag{3.71}$$

- for welded square hollow sections (welded SHS)

$$\frac{1}{s} = 0.710 + \frac{0.167}{\alpha} \tag{3.72}$$

- for cold-formed square hollow sections (cold-formed SHS)

$$\frac{1}{s} = 0.778 + \frac{0.13}{\alpha} \tag{3.73}$$

- for cold-formed circular hollow sections (cold-formed CHS)

$$\frac{1}{s} = 0.777 + \frac{1.18}{\alpha} \tag{3.74}$$

where

- $\alpha_f = \dfrac{E}{\sigma_y}\left(\dfrac{t_f}{b_f/2}\right)^2$ is the slenderness parameter of the flange;

- $\alpha_w = \dfrac{E}{\sigma_y}\left(\dfrac{t_w}{d_w}\right)^2$ is the slenderness parameter of the web (d_w is the depth of the web), and

- $\alpha = \dfrac{E}{\sigma_y}\left(\dfrac{t}{B}\right)^2$ and $\alpha = \dfrac{E}{\sigma_y}\left(\dfrac{t}{D}\right)^2$ are the slenderness parameters for SHS

and CHS sections respectively (B is the side of the SHS section, D is the diameter of the CHS section and t is the thickness).

The above relationships can be adopted independently of the grade of steel. Moreover, more accurate relationships, which take into account effects due to the mechanical properties of steel, are given by the same author for H-sections and with reference to steel grades currently adopted in Japan [19].

It must be remembered that the above experimental relationships are based upon stub column tests in which webs are uniformly compressed, while webs in beams and beam-columns have a stress gradient. This difference can be taken into account by introducing the 'effective width' of the web, which represents the compressed part of the web when the section is fully yielded. With reference to Fig. 3.8, as the total axial load ($N = \rho\, A\, \sigma_y$) is sustained by the web ($N = N_w$),

$$\rho\, A\, \sigma_y = (1 - 2\,\beta\,)\, A_w\, \sigma_y \qquad (3.75)$$

therefore

$$\beta = \frac{A_w - \rho\, A}{2\, A_w} \qquad (3.76)$$

and

$$d_{we} = \frac{1}{2}\left(1 + \frac{A}{A_w}\,\rho\right) d_w \leq d_w \qquad (3.77)$$

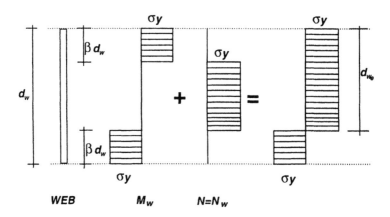

Fig. 3.8   Definition of the 'effective width' of the web

which has to be introduced, instead of d_w, in the definition of α_w.

It must be remembered that theoretical analyses available in the technical literature do not take into account the effects due to the stress gradient in the buckled zone, which derives from the longitudinal variation of the bending moment. In particular, the dependence of the critical stress from the stress gradient, which has been indicated by the experimental results of Kuhlmann [25], is not included in theoretical formulations as well as in empirical relationships such as those proposed by Kato [17–19] (equations 3.71–3.74).

On the basis of experimental evidence it can be stated that the average length of the zone, where local buckling of the compressed flange occurs, is approximately 1.20 b_f [25]. Therefore, the influence of the stress gradient on the critical stress which produces the local buckling of the compressed flange depends upon the b_f/L ratio (where, in general, L is the distance between the section in which the bending moment is zero and the one in which it assumes its maximum value). The analysis of the experimental data collected [16, 18, 25] has led, by means of a multiple regression analysis, to the following empirical relationship [20, 21]

$$\frac{1}{s} = 0.546321 + 1.632533\,\lambda_f^2 + 0.062124\,\lambda_w^2 - 0.602125\,\frac{b_f}{L} + 0.001471\,\frac{E}{E_h}$$

$$+ 0.007766\,\frac{\varepsilon_h}{\varepsilon_y} \tag{3.78}$$

where λ_f and λ_w, which represent the slenderness parameters of the flange and of the web, respectively are given by

$$\lambda_f = \frac{b_f}{2t_f}\,\sqrt{\varepsilon_y} \tag{3.79}$$

$$\lambda_w = \frac{d_{w_e}}{t_w}\,\sqrt{\varepsilon_y} \tag{3.80}$$

For H sections, equation (3.78) is more complete than equation (3.71), because it includes the influence of the slenderness of the flange and of the web as well as the effects due to the stress gradient and the influence of the mechanical properties of the material.

b. Lateral–torsional buckling

The buckling stress ratio s can be computed by means of the relationship presented in the previous section if local buckling occurs before flexural–torsional buckling. This condition is verified when the distance between torsional restraints is sufficiently small in order to prevent the lateral displacement of the compressed flange. In the opposite case, the buckling stress ratio has to be computed as the minimum value between the one leading to local buckling and the one leading to lateral–torsional buckling.

This last value can be computed, taking into account the moment gradient, using the relationships which provide the ultimate equivalent uniform bending moment for lateral–torsional buckling and the relationships correlating end moments to the equivalent uniform bending moment. For these formulations reference can be made to Eurocode 3 [28].

3.4.5 Postbuckling behaviour

It has already been shown that a section does not fail immediately when the point of local buckling has been reached, but an important amount of plastic deformation can develop before the bending resistance of the section falls below the plastic moment. Therefore, the analysis of postbuckling behaviour can represent an important phase in the evaluation of the rotation capacity.

From the theoretical point of view the evaluation of the plastic deformation capacity after the occurrence of local buckling can be performed by means of sophisticated numerical analyses based on the finite element method [25] or on the finite strip method [22–24].

Relationships describing the plate postbuckling behaviour have been derived from yield-line models, which have been calibrated on the basis of the experimental buckling shapes. These buckling models are used in order to determine the stress–strain relationships to be included in the stiffness analysis of the numerical procedure [25]. The stress–strain relationships thus obtained are characterized by a decreasing stress with an increasing strain, which allows the simulation of the reduction of load-carrying capacity of the member. Specific numerical procedures have to be introduced in the finite-element process in order to investigate the softening branch of the moment versus rotation curve [25].

Another approach is based on the evaluation of the yield mechanism curve of the member. This approach, which will be briefly summarized below, has been introduced [6]. This method has been applied by Alexander [33], Pugesley and Macaulay [34] for examining cylindrical shells, and by Climenhaga and Johnson [9] for the analysis both of the buckling of steel members subjected to non-uniform bending and the one due to the negative moment in composite continuous beams. Moreover, a detailed description of its application to double T-shape steel members subjected to uniform bending is provided in reference [11].

In this approach, collapse mechanisms are chosen so that the compati-

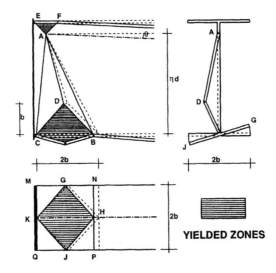

Fig. 3.9 Collapse mechanism suggested by experimental tests

bility requirements, the yield criterion and the associated flow rule are satisfied. These collapse mechanisms are directly suggested by the experimental evidence. The kinematic theorem of plasticity is applied, giving an upper bound.

With reference to the case of in-plane local buckling of an I-shaped cantilever beam, the yield mechanism is shown in Fig. 3.9. The centre line of the flanges remains in the plane of bending. Plastic deformation develops in the shaded area and in the linear plastic hinges required for compatibility with the buckling of the compressed flange. In the tensile zone, plastic deformation develops in the flange section EF, and in zone AEF of the web plate. In the compressed zone, plastic deformation develops in the zone BCD of the web, and plastic hinges AB, AC, AD, BD and CD arise. Moreover, plastic deformations develop in zone GHJK of the compressed flange and plastic hinges MK, QK, NH, HP, GK, GH, JK and JH arise. The yield mechanism in the compressed zone of the web is compatible with that in the compressed flange if points B and H, as well as C and K, coincide. Moreover, a plastic hinge KH between the web and the flange, assumed in the web, arises. By assuming that the wave length of the buckled flange is equal to $b_f = 2b$, the whole mechanism is determined through the position of the point A, which represents the centre of rotation, assumed at a distance ηd from the buckled flange.

The work done by the external load F, applied at the top of the cantilever, is given by

$$W_E = \int_0^\theta F \ [\ L - t_f - (\ 1 - \eta\ )\ d\ ]\ d\theta \qquad (3.81)$$

where L is the length of the cantilever, t_f is the thickness of the beam flange and d is the depth of the beam.

The internal work W_I is obtained through the summation of the single contributions due to the previously specified plastic zones and plastic hinges. These contributions are easy to evaluate by analysis of the distorted geometry of the cross-section, but tedious algebra is involved in the details of the complete solution [9, 11]. This solution leads to simple functions of η and θ.

Equating the work done to the work absorbed and differentiating with respect to θ, the softening branch of the moment versus rotation (i.e. M–θ) curve is provided in the form

$$\frac{M}{M_p} = \frac{L}{[\ L - t_f - (\ 1 - \eta\ )\ d\ ]\ M_p}\ \frac{dW_I}{d\theta} \qquad (3.82)$$

in which **M is the applied moment at the fixed end.**

It is not possible to explicitly minimize the M–θ curve with respect to η, therefore a numerical procedure has to be developed.

After the minimization has been performed, the value of the ultimate rotation θ_u (Fig. 3.1) corresponding to $M/M_p = 1$ can be computed and the rotation capacity can be evaluated.

3.5   THEORETICAL METHODS

3.5.1 General

The methods for evaluating rotation capacity of steel beams and beam-columns, by means of closed form solutions, are defined as 'theoretical methods' when they are based on the theoretical evaluation of the moment versus curvature relationship of a simplified model and on the theoretical evaluation of the critical stress leading to the local buckling of the compressed flange or to the lateral–torsional buckling.

In particular, it has been shown that these methods do not consider the part of the deformation capacity corresponding to the postbuckling behaviour; therefore, they provide a safe-side solution (Fig. 3.1).

3.5.2 Closed-form solution

In the previous section the relationships for evaluating ultimate rotation of beams and beam-columns have been provided for the ideal two-flange section. In order to compute the rotation capacity, the first-yielding rotation is now obtained with reference to the actual section, by means of the relationships

$$\theta_y = (1-\rho) \, L \frac{\varepsilon_y}{h} \tag{3.83}$$

and

$$\theta_y = (1-\rho) \, \frac{M_p \, L}{2 \, E \, I} \tag{3.84}$$

where h and I are the depth and the moment of inertia of the actual section. The stable part of the rotation capacity

$$R_{st} = \frac{\theta_m}{\theta_y} - 1 = \frac{\theta_p}{\theta_y} \tag{3.85}$$

can now be obtained for all behavioural cases.

Case for $\rho > (s-1)/2$
By combining relationships (3.49), (3.83), (3.84), (3.85), (3.12) and (3.14), the following equation is obtained

$$R_{st} = \frac{1}{s-\rho} \left\{ (1-\rho) \frac{h}{h_e} + \frac{s-1}{1-\rho} \left[(1-2\rho) \frac{h}{h_e} + \frac{\varepsilon_h}{\varepsilon_y} \frac{h}{h_e} + (s-1) \frac{E}{E_r} \frac{I}{I_e} \right] \right\} - 1 \tag{3.86}$$

It must be remembered that h and I represent, respectively, the height and the moment of inertia of the actual section, while h_e and I_e are, respectively, the height and the inertia moment of the ideal two-flange section having a plastic section modulus equal to that of the actual section. The limiting case for $E \to \infty$ is obtained by combining equations (3.50), (3.83), (3.84), (3.85) and (3.14). The final expression is, obviously, the same as that provided by Kato [17–19]:

$$R_{st} = \frac{s-1}{2 \, (s-\rho) \, (1-\rho)} \left[2 \frac{\varepsilon_h}{\varepsilon_y} \frac{h}{h_e} + (s-1) \frac{E}{E_h} \frac{I}{I_e} \right] \tag{3.87}$$

Case for $\rho \leq (s-1)/2$
By combining equations (3.57), (3.83), (3.84), (3.85), (3.12) and (3.14), the following relationship is obtained

$$R_{st} = \frac{1}{(s-\rho)(1-\rho)} \left\{ [1+\rho^2 - 2\rho \, (s-1)] \frac{h}{h_e} + 2 \frac{\varepsilon_h}{\varepsilon_y} \frac{h}{h_e} (s-\rho-1) \right.$$

$$+ \frac{E}{E_r} \frac{I}{I_e} 4 \, \rho \, (s - \rho - 1) + (s - 2 \, \rho - 1)^2 \frac{E}{E_h} \frac{I}{I_e} \Bigg\} - 1 \qquad (3.88)$$

The limiting case for $E \to \infty$ is obtained by combining equations (3.58), (3.83), (3.84), (3.85) and (3.14). The final expression,

$$R_{st} = \frac{1}{(s - \rho)(1 - \rho)}$$

$$\left\{ 2 \frac{\varepsilon_h}{\varepsilon_y} \frac{h}{h_e} (s - \rho - 1) + \frac{E}{E_h} \frac{I}{I_e} [2 \, \rho^2 + (s - 1)(s - 2 \, \rho - 1)] \right\} \qquad (3.89)$$

is, obviously, coincident with that given by Kato [13–15].

Case for $\rho = 0$ (pure bending)
The equations for evaluating the rotation capacity of beams are obtained by assuming $\rho = 0$ in relationships (3.88) and (3.89), in the cases of elastic–plastic flow-strain hardening behaviour and rigid–plastic flow-strain hardening behaviour, respectively.

Finally it is useful to point out the general validity of the above equations, from which in particular the case of elastic-strain hardening behaviour can be obtained by assuming $\chi_h = \chi_y$ and the case of rigid-strain-hardening behaviour by assuming $\chi_h = 0$. The relationships based upon the assumption of rigid behaviour (elastic deformability neglected) have been presented above to show the possibility of obtaining Kato's formulations as a particular case; however, from the theoretical point of view, it seems that the contribution of the part of the member which remains in elastic range cannot be neglected, otherwise the definition itself of rotation capacity fails (equations 3.1 and 3.2).

3.5.3 Kemp method

The procedure developed by Kemp [16] is based on the assumption that the material has an elastic–perfectly plastic strain hardening behaviour and the moment versus curvature relationship is evaluated with reference to an ideal two-flange section. The theoretical analysis has been developed by Kemp for beams only ($\rho = 0$), as the relationship

$$R_{st} = \gamma \left(2 \frac{\varepsilon_h}{\varepsilon_y} - 1 + \frac{E}{E_h} \frac{\gamma}{1 - \gamma} \right) \qquad (3.90)$$

where

$$\gamma L = \frac{s - 1}{s} L \qquad (3.91)$$

is the extension of the yielded zone, i.e. the value αL (equation 3.46) for $\rho = 0$ or, also, the value $\alpha L + \beta L$ (equations 3.52 and 3.53) for $\rho = 0$.

Equation (3.90) is a particular case of equation (3.88), as it can be shown by assuming $\rho = 0$ and $h/h_e = I/I_e = 1$ (ideal two-flange section). The use of equation (3.90) requires the definition of the conditions leading to the local buckling of the compressed flange. In this context, the fundamental observation, of Lay and Galambos [4], is that local buckling arises when the extension of the yielded zone is equal to or greater than the wave length over which the first local buckle developed in the flange.

Therefore, the critical condition is given by

$$L_f = \gamma L \tag{3.92}$$

Moreover, Kemp [16] has introduced the hypothesis that the critical strain, provided by the Southward formula (3.70), is attained in the middle of the yielded zone. In this case,

$$\frac{\sigma_{cr}}{\sigma_y} = 1 + \frac{0.5\,\gamma}{1 - \gamma} \tag{3.93}$$

Therefore, by combining equations (3.93) and (3.70) and by assuming $G_h / E_h = 1/3$ following Southward [32], the following is obtained

$$\frac{4}{3}\left(\frac{t_f}{b_f}\right)^2 + \frac{2}{3}\pi^2\left(\frac{t_f}{\gamma L}\right)^2 = \varepsilon_h + \frac{\sigma_y}{E_h}\frac{0.5\,\gamma}{1 - \gamma} \tag{3.94}$$

which is a third-order equation in γ as the unknown.

The value of gamma, obtained from (3.94) by trial and error, provides the value of the available rotation capacity through equation (3.90).

In the case of columns, i.e. $\rho \neq 0$, Kemp assumes, instead of equation (3.90), the approximate relationship

$$R = \frac{h'_w}{2\,h'_c}\,\gamma\left(2\frac{\varepsilon_h}{\varepsilon_y} - 1 + \frac{E}{E_h}\frac{\gamma}{1 - \gamma}\right) \tag{3.95}$$

in which h'_w is the depth between the centres of the two flanges and h'_c the height from the centre of the compressed flange to the plastic neutral axis.

Therefore, the ratio $h'_w/2h'_c$ is a function of the non-dimensional axial load ρ.

3.6   SEMI-EMPIRICAL METHODS

3.6.1 General

As shown above, theoretical methods are based upon the approximate theoretical evaluation of the moment versus curvature relationship, by neglecting shape effects and mechanical imperfections; they also include the theoretical analysis of the stability phenomena such as the local buckling of the compressed flange.

Alternatively, in the semi-empirical methods the theoretical evaluation of the moment versus curvature relationship is integrated by the empirical determination of the conditions leading to the local buckling of the compressed flange on the basis of the experimental evidence. Moreover, in the absence of torsional restraints, control of the conditions leading to the lateral torsional buckling of the member is necessary.

3.6.2 Kato method

The procedure proposed by Kato [17–19] is based upon a simplified model represented by an ideal two-flange section statically equivalent to the actual section and on the assumption that the material has a rigid, perfectly plastic-strain hardening behaviour. The computation of the rotation capacity is, therefore, performed through relationships (3.87) and (3.89). In these the parameter s, which defines the conditions leading to the local buckling of the compressed flange, has to be evaluated by means of the experimental relationships (3.71–3.74) and by using the effective width of the web given by equation (3.77).

3.6.3 Mazzolani–Piluso method

This method is also based on the simplified model represented by the ideal two-flange section, but an elastic–plastic flow-strain hardening behaviour has been assumed [20, 21].

Moreover, the simplified model (ideal two-flange section) has been retained intentionally in the computation of the first yielding rotation. As a consequence, this method is based upon the use of equations (3.86) and (3.88) with $h/h_e = I/I_e = 1$ for evaluating rotation capacity. The computation of the critical stress leading to the local buckling of the compressed flange is performed by equation (3.78).

This method starts with the observation that the rotation capacity of steel beams and beam-columns is influenced not only by the width-to-thickness ratios of the compressed parts of the cross-section, but also by the stress gradient along the cross-section (which, as shown in Fig. 3.8, can be taken into account by means of the concept of effective width of the web), by the longitudinal stress gradient due to the bending moment variation and by the distance between torsional restraints which plays a fundamental role in preventing flexural–torsional buckling.

By considering these new important parameters, the authors have shown that the concept of cross-section behavioural classes should be substituted by the concept of member behavioural classes [35], which gives a more general meaning to the classification.

The ductile behaviour corresponds to the capability of the member to provide high values of the rotation capacity, while the compact behaviour is associated with a moderate value of the rotation capacity. Therefore, a classification criterion based directly on the rotation capacity value is proposed, rather than on the width-to-thickness ratios of the member section. As an example, Fig. 3.10 provides the values of the non-dimensional buckling stress s which are necessary to obtain a given value of the rotation capacity ($R = 2, 4, 6, 8, 10$) for any given value of the non-dimensional axial load. The value of s obtained provides, through equation (3.78), the limit values of the slenderness parameters of the flange and of the web. Equations (3.79) and (3.80) therefore yield the corresponding limit values of the width-to-thickness ratios for any given value of the distance L between the zero moment point and the

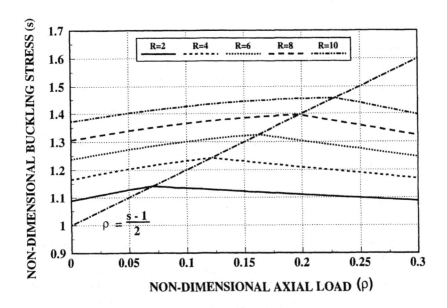

Fig. 3.10   Values of the non-dimensional buckling stress requested in order to obtain a given value of the rotation capacity (for Fe430 steel)

plastic hinge section. In other words, it is possible to define the width-to-thickness ratios that the member section has to possess in order to guarantee a given behavioural class, i.e. a given value of the rotation capacity.

The curves of Fig. 3.10 have been plotted by assuming $h/h_e = l/l_e = 1$, as previously stated, and by taking into account that the values of the ratios E/E_h and $\varepsilon_h/\varepsilon_y$ of current structural steels are approximately given by

$$\frac{E}{E_h} = 37.5 \quad \text{and} \quad \frac{\varepsilon_h}{\varepsilon_y} = 12.3 \quad \text{for Fe360 steel}$$

$$\frac{E}{E_h} = 42.8 \quad \text{and} \quad \frac{\varepsilon_h}{\varepsilon_y} = 11.0 \quad \text{for Fe430 steel}$$

$$\frac{E}{E_h} = 48.2 \quad \text{and} \quad \frac{\varepsilon_h}{\varepsilon_y} = 9.8 \quad \text{for Fe510 steel}$$

The limit values of the width-to-thickness ratios obtained by the above procedure guarantee the assigned value of the rotation capacity, provided that torsional restraints are located so that lateral–torsional buckling is prevented.

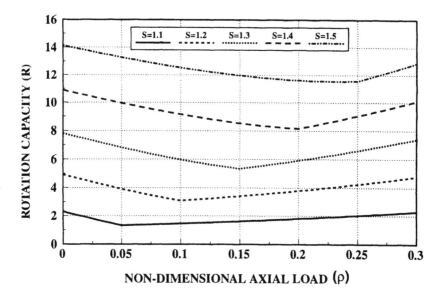

Fig. 3.11   Rotation capacity as a function of the non-dimensional axial load and of the non-dimensional buckling stress (for Fe430 steel)

Furthermore, Fig. 3.11 can be used for calculating the rotation capacity of members when the cross-section, non-dimensional axial load and bending moment distribution are given. In fact, if the geometrical properties of the section are assigned, equation (3.78) provides the corresponding non-dimensional local buckling stress which has to be used in Fig. 3.11 in order to obtain the value of the available rotation capacity.

3.7 EMPIRICAL METHODS

3.7.1 General

Empirical relationships, based upon test results, for evaluating the deformation capacity of beams and beam-columns have been proposed by Japanese researchers [13–15, 36]. A different approach has been adopted by Sedlacek and Spangemacher [26, 27], who have derived an empirical relationship on the basis of both experimental and numerical simulation data. The formulations derived from both these approaches are presented. In the following section a comparison among theoretical, semi-empirical and empirical relationships and experimental data is provided.

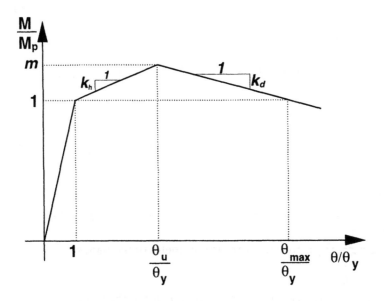

Fig. 3.12   Assumed simplification for the moment versus rotation curve

3.7.2 Kato–Akiyama method

For members subjected to double curvature bending, Kato and Akiyama [15, 36] have proposed the use of a simplified moment versus rotation curve which is composed of three linear branches; the non-dimensional moment versus rotation curve is shown in Fig. 3.12.

The first increasing branch represents the phase of elastic behaviour. The second branch is due to plastic behaviour and it is characterized by means of the parameter k_h representing the ratio of the stiffness of the strain-hardening range to the elastic one. As soon as the maximum strength is reached, due to buckling, a gradual load reduction follows without any sudden failure. The softening branch is characterized by k_d, which is defined as the ratio between the slope of the degrading segment and the elastic one. Finally, the ratio of the maximum strength to the yield strength m is defined as the 'stress-increasing ratio'.

The following equations are applicable to H-section members, by satisfying the limitations

$$\frac{b_f}{2t_f} \leq \frac{0.52}{\sqrt{\varepsilon_y}} \quad \text{and} \quad \frac{d_w}{t_w} \leq \frac{2.4}{\sqrt{\varepsilon_y}} \tag{3.96}$$

In the case of beams ($\rho = 0$), the stress-increasing ratio can be computed as the larger value provided by the relationship

$$m_o = 1 + [\,(\,0.043 - 0.0744\ \lambda_f\,)^2 - (\,0.00024\ \lambda_w - 0.00025\,)\,]\frac{1}{\varepsilon_y} \tag{3.97}$$

For $\lambda_y \leq 200$,

$$m_o = 1.46 - [0.63\ \lambda_f + 0.053\ \lambda_w + 0.02\ (\,\lambda_y - 50)\ \sqrt{\varepsilon_y}\,] \tag{3.98}$$

or, for $\lambda_y > 200$,

$$m_o = 1.46 - (\,0.63\ \lambda_f + 0.053\ \lambda_w + 3\ \sqrt{\varepsilon_y}\,) \tag{3.99}$$

where:

- m_o is the value of the stress-increasing ratio m for $\rho = 0$ (beams);
- $\lambda_f = \dfrac{b_f}{2t_f}\ \sqrt{\varepsilon_y}$ is the slenderness parameter of the flange;
- $\lambda_w = \dfrac{d_w}{t_w}\ \sqrt{\varepsilon_y}$ is the slenderness parameter of the web;

- λ_y is the slenderness of the member with respect to the weak axis of the section.

In all cases the limitation $1 \leq m_o \leq \sigma_u/\sigma_y$ has to be satisfied, where σ_u is the tensile strength of the material.

In the case of beam-columns ($\rho \neq 0$) the stress-increasing ratio is given by

$$m = \frac{Z'_\rho}{Z_\rho} m_o \tag{3.100}$$

where Z'_ρ is the plastic section modulus corresponding to an axial stress ratio ρ/m_o and Z_ρ the plastic section modulus corresponding to an axial stress ratio ρ.

Where $\rho > 2 A_w/3A$, the slenderness parameter λ_w in equations (3.97) and (3.98) has to be doubled and where $\lambda_y > 100$ and $\rho < 2 A_w/3A$, the term λ_w in equation (3.97) has to be factored by 1.5.

The above relationships are valid only when local buckling occurs. In the case when flexural–torsional buckling takes place, the stress-increasing ratio has to be computed using the condition

$$m = \left(\frac{6.82}{\lambda_e^{1.7}} + 0.9 \right)^{1/2} \tag{3.101}$$

where $\lambda_e = \lambda_y \sqrt{\varepsilon_y}$ for $\rho = 0$ (beams) and $\lambda_e = 1.28 \lambda_y \sqrt{\varepsilon_y}$ for $\rho \neq 0$ (beam-columns).

The maximum strength is deemed to be scarcely affected by the coupling of flexural–torsional buckling and local buckling, and therefore m can be assumed as the smaller value provided by equations (3.97) – (3.99) or (3.100) and (3.101).

The stiffness ratio of the strain-hardening branch can be evaluated by the equation

$$k_h = 0.03 + 0.04 \, \rho \tag{3.102}$$

When only local buckling occurs, the stiffness ratio of the softening branch can be calculated as the smaller value, provided by

$$k_d = - \, 0.355 \, \lambda_w \sqrt{\varepsilon_y} \tag{3.103}$$

and

$$k_d = - \left[- 1.33 + (10.6 \, \lambda_f - 2) \, (0.63 + 0.33 \, \lambda_w) \right] \sqrt{\varepsilon_y} \qquad (3.104)$$

In equations (3.103) and (3.104) λ_w has to be doubled if $\rho > 2 \, A_w / 3A$.

The influence of the coupling of flexural–torsional buckling and local buckling is considered to be significant in the post-buckling range, after the maximum load is reached. In order to judge the possibility of coupling of the buckling modes, another parameter has to be computed

$$m^* = \left(\frac{2.36}{\lambda_e^{1.7}} + 0.9 \right)^{\!\! \frac{1}{2}} \qquad (3.105)$$

If the value of the stress-increasing ratio m is greater than m^*, coupling of the buckling modes occurs leading to a higher slope of the softening branch, so that the value provided by equations (3.103) or (3.104) has to be factored by 3.0.

The above relationships have led to the complete definition of the simplified moment versus rotation curve. Now, the stable part of the rotation capacity can be computed as

$$R_{st} = \frac{m - 1}{k_h} \qquad (3.106)$$

The deformation capacity due to the softening branch can also be included, so that the total rotation capacity becomes

$$R = \frac{\theta_u - \theta_y}{\theta_y} = (m - 1) \left(\frac{1}{k_h} + \frac{1}{| \, k_d \, |} \right) \qquad (3.107)$$

The above derivations are for strong-axis bending. However, as members subjected to weak-axis bending are less sensitive to local buckling, the above formulae can be used also for weak-axis bending without significant error [36]. Furthermore, in the case of weak-axis bending, flexural–torsional buckling does not occur and, therefore, local buckling only has to be considered.

The authors have also derived relationships for defining moment versus rotation curves for cases of SHS and CHS sections [15, 36].

3.7.3 Mitani–Makino method

Empirical formulae for estimating the rotation capacity of beam-columns have been established by Mitani and Makino [14] on the basis of experimental results.

The authors provide the relationships

$$R_{st} = \left(\frac{500}{k \, \dfrac{L_x}{i_x} \dfrac{L_b}{i_y}} \, \frac{235}{\sigma_y} \right)^{1/2} [\, 80 \, (\lambda_f - 0.65)^2 - 4.0 \, \lambda_w + 6 \,] \qquad (3.108)$$

which is valid for $\rho < A_w / 2A$ and

$$R_{st} = \left(\frac{500}{k \, \dfrac{L_x}{i_x} \dfrac{L_b}{i_y}} \, \frac{235}{\sigma_y} \right)^{1/2} [\, 50 \, (\lambda_f - 0.65)^2 - 5.5 \, \lambda_w + 7 \,] \qquad (3.109)$$

which is valid in the opposite case. Moreover, both relations have been derived for $\lambda_f < 0.65$.

In equations (3.108) and (3.109) the following symbols have been used:

- L_x is the distance between the plastic hinge and the zero-moment point;
- L_b is the distance between lateral bracings;
- i_x is the radius of gyration around the strong axis;
- i_y is the radius of gyration around the weak axis;
- k is a numerical coefficient, which depends on the moment distribution, defining the effective length for lateral–torsional buckling as $k \, L_b$.

The above relationships have been derived from experimental results on cantilever beams. The value of $k = 0.7$ has been assumed with reference to test conditions [14]. In cases of beams and beam-columns subjected to double curvature bending, with end moments $M_1 > M_2$ (both positive for double curvature), equations (3.108) and (3.109) can be used by assuming

$$L_x = \frac{L_b}{1 + \dfrac{M_2}{M_1}} \qquad (3.110)$$

and

$$k = 0.7 \left(\frac{1.75}{c_b} \right)^{1/2} \qquad (3.111)$$

where

$$c_b = 1.75 + 1.05 \frac{M_2}{M_1} + 0.3 \left(\frac{M_2}{M_1} \right)^2 \leq 2.3 \qquad (3.112)$$

3.7.4 Nakamura method

By analysing 121 experimental measurements, Nakamura [13] has related the rotation capacity of beams to the non-dimensional lateral–torsional slenderness by means of the condition

$$R = \frac{\beta}{\bar{\lambda}_{LT}} - 1 \qquad (3.113)$$

where $\bar{\lambda}_{LT} = \sqrt{M_p / c \, M_{cr}}$ is the non-dimensional lateral–torsional slenderness. Here, M_p is the plastic moment, M_{cr} is the elastic critical moment of the simply supported beam under uniform moment, and c is the bending coefficient to evaluate the lateral buckling moment ($c \, M_{cr}$) of a beam under arbitrary moment distributions, lateral bracing and end conditions. The term β is a coefficient, which represents the effect of the moment gradient on the rotation capacity. It is given by

$$\beta = 1 - \frac{1}{2}\alpha \qquad \text{for} \qquad -1 \le \alpha \le 0 \qquad (3.114)$$

$$\beta = 1 - \alpha \qquad \text{for} \qquad 0 \le \alpha \le \frac{2}{3} \qquad (3.115)$$

$$\beta = \frac{1}{3} \qquad \text{for} \qquad \frac{2}{3} \le \alpha \le 1 \qquad (3.116)$$

where α is the ratio between the end moments, assumed positive when the curvature does not change its sign along the member. It should be mentioned that the experimental data examined by the author mainly relate to the cases $\alpha = 0$ and $\alpha = 1$.

3.7.5 Sedlacek–Spangemacher method

This method is mainly based on results obtained by numerical simulations using the finite-element method. In particular, these simulations have been divided into different series, in which a single parameter affecting rotation capacity has been varied, keeping all the others constant [26, 27]. Analysis of the results has led to the empirical relationship

$$R = R_o\left(\frac{f_u}{f_y}\right) + \Delta R(t_f) + \Delta R\left(\frac{b_f}{t_f}\right) + \Delta R\left(\frac{L}{b_f}\right) - \Delta R(K_\theta) \qquad (3.117)$$

where

$$R_o\left(\frac{f_u}{f_y}\right) = 0.75 \left(\frac{f_u}{f_y}\right)^{6.5} \qquad (3.118)$$

$$\Delta R(t_f) = \left[2.53 \left(\frac{f_u}{f_y} \right) - 2.63 \right] \alpha \ (15 - t_f) \tag{3.119}$$

$$\alpha = 1.0 \qquad \text{for} \qquad t_f < 15 \text{ mm} \tag{3.120}$$

$$\alpha = 0.5 \qquad \text{for} \qquad t_f \geq 15 \text{ mm} \tag{3.121}$$

$$\Delta R \left(\frac{b_f}{t_f} \right) = \left[2.81 \left(\frac{f_u}{f_y} \right) - 2.74 \right] \left(20 - \frac{b_f}{t_f} \right) \tag{3.122}$$

$$\Delta R \left(\frac{L}{b_f} \right) = \left[2.70 \left(\frac{f_u}{f_y} \right) - 2.70 \right] \left(5 - \frac{L}{b_f} \right) \tag{3.123}$$

and

$$\Delta R(K_\theta) = S_{K_\theta} \ \Delta K_\theta \tag{3.124}$$

$$\Delta K_\theta = 9.31 - 0.035 \left(\frac{f_u}{f_y} \right)^{6.5} - \frac{G_h \ t_w^3}{3 \ d_w} \tag{3.125}$$

$$S_{K_\theta} = 0.35 \left(\frac{f_u}{f_y} \right)^4 \qquad \text{for} \qquad \Delta K_\theta > 0 \tag{3.126}$$

$$S_{K_\theta} = 0 \qquad \text{for} \qquad \Delta K_\theta \leq 0 \tag{3.127}$$

In equation (3.125) G_h is the strain hardening tangential modulus of the web evaluated according to Lay's equation (3.62).

In addition, it is useful to be aware that the axial load is not considered; therefore, Sedlacek and Spangemacher formulations can be applied only to beams.

3.8   COMPARISON WITH EXPERIMENTAL DATA

In the previous sections different procedures for evaluating the rotation capacity of beams and beam-columns have been introduced. In addition, it has been shown that both theoretical and semi-empirical methods provide an estimate of just the stable part of the rotation capacity. However, empirical methods can also take into account the deformation capacity corresponding to the postbuckling behaviour. Only the empirical method of Kato and Akiyama allows a distinction between the stable and unstable parts of the rotation capacity.

The reliability of the different methods can be assessed by comparison with experimental results available in the technical literature [16, 18, 25, 27].

Unfortunately this comparison cannot be carried out for all the previously described methods, because the geometrical and mechanical description of the specimens is often not completely exhaustive.

For this reason, in what follows, comparison with experimental data will be limited to those cases in which a correct application of the described methods has been possible.

As a consequence, the empirical method of Nakamura is excluded from this comparison, while the empirical method of Kato and Akiyama is applied with reference to the stable part only (R_{st}) of the total rotation capacity.

Comparison between experimental results and the rotation capacity values predicted by means of the methods previously described is given in Figs. 3.13–3.18. Figures 3.13–3.17 refer to the stable part (R_{st}) of the rotation capacity, while Fig. 3.18 refers to the total value (R). For each procedure the predicted versus experimental value relationship is given, with reference to beams and beam-columns under the moment gradient. The same figures also give the 'scatter areas', characterized by a ±20% or a ±30% difference between predicted and experimental values (dotted lines).

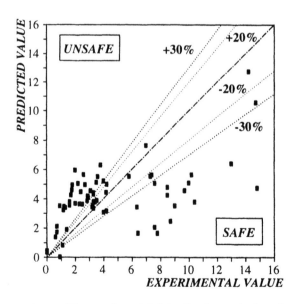

Fig. 3.13   Comparison with experimental data: the theoretical method of Kemp

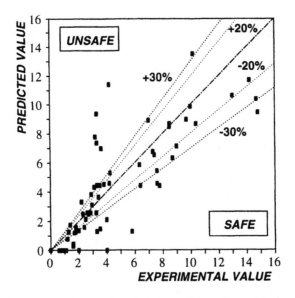

Fig. 3.14 Comparison with experimental data: the semi-empirical method of Kato

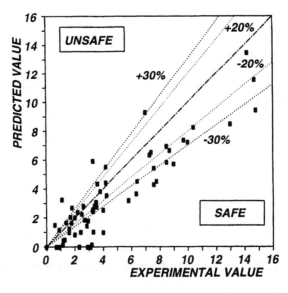

Fig. 3.15 Comparison with experimental data: the semi-empirical method of Mazzolani and Piluso

These representations have pointed out that the theoretical method of Kemp (Fig. 3.13) is not conservative for low values of rotation capacity, while it is too conservative in the opposite case. In particular, in most cases the scatters between predicted values and experimental values are very high. This considerable scattering is due to difficulties arising in the theoretical evaluation of the critical stress, which produces the local buckling of the compressed flange.

A strong reduction of the scatters between predicted and experimental values is obtained by the semi-empirical method of Kato (Fig. 3.14). In this case, the greatest part of the experimental data is such that the corresponding points are included in the 'scatter areas' and only in a few cases they are situated in the unsafe zone.

The semi-empirical method of Mazzolani and Piluso (Fig. 3.15), including both the influence of that part of the member which remains in elastic range and the influence of the stress gradient and of the mechanical properties of the material, seems able to provide a considerable reduction of scatter with respect to the previous methods, together with sufficiently conservative results.

Finally, the empirical methods of Kato and Akiyama (Fig. 3.16) and of Mitani and Makino (Fig. 3.17) allow predictions which are, without a doubt, better than those provided by the theoretical method of Kemp, but the scatters are more significant than for the semi-empirical methods. Moreover, the number of cases in which predictions are not conservative is increased for both methods.

Taking into account the complexity of the phenomenon and the advantages due to the use of an equation in 'closed form', the approximation obtained by semi-empirical methods can be considered satisfactory.

The empirical method of Sedlacek and Spangemacher has to be considered apart, because it is the only one for which the comparison with experimental data includes the deformation capacity due to postbuckling behaviour. Figure 3.18 shows that the predicted values are in good agreement with the experimental ones [27], and the magnitude of the scatters is similar to those obtained by the semi-empirical methods.

In conclusion, the use of semi-empirical methods for evaluating the stable part of the rotation capacity can be recommended both for beams and beam-columns, while prediction of the total deformation capacity, for beams only, can be performed by the empirical method of Sedlacek and Spangemacher.

The reliability connected to the degree of approximation of the above recommendations can be emphasized by observing that some researchers [37] have performed a numerical simulation of the experimental tests of Lukey and Adams [8] by using the finite-element method and adopting nine-node Lagrangian shell finite elements with 2×2 Gauss integration

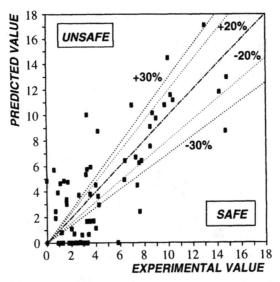

Fig. 3.16   Comparison with experimental data: the empirical method of Kato and Akiyama

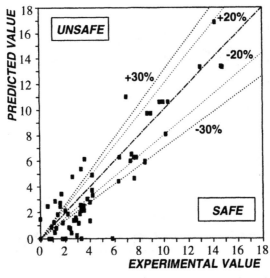

Fig. 3.17   Comparison with experimental data: the empirical method of Mitani and Makino

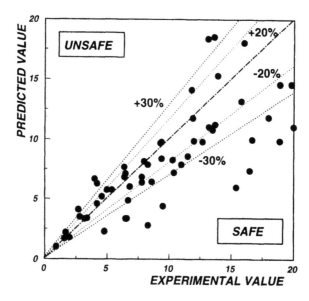

Fig. 3.18   Comparison with experimental data: the empirical method of
Sedlacek and Spangemacher

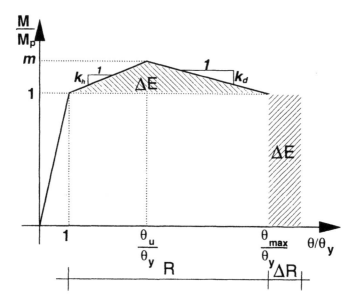

Fig. 3.19   Definition of the equivalent elastic–perfectly plastic rotation capacity

and a bounding surface model based on the linear flow theory. However, the scatters obtained between predicted and experimental values of the rotation capacity were very rough, varying from –40% to +60%, despite the rigour of this approach.

Finally, it should be remembered that none of the described procedures nor the definition itself of the rotation capacity are able to take into account that, in the actual case of members belonging to a real structure, the distance between the inflection point and the location of the maximum bending moment can change during the loading process.

3.9   STOREY DUCTILITY

In the seismic design of steel structures by means of the energy approach, which will be described in detail in Chapter 4, the possibility of establishing a relationship between the rotation capacity of members and the 'storey ductility' is particularly important. Kato and Akiyama [15, 36] have introduced the concept of 'equivalent rotation capacity', which represents the rotation capacity of a member with ideal elastic–perfectly plastic behaviour dissipating the same amount of energy as the actual member. With reference to Fig. 3.19, due to the strain hardening and to the degradation of the flexural capacity, which follows the attainment of the maximum bending moment, an increase of dissipated energy is obtained with respect to the elastic–perfectly plastic behaviour. This increase of dissipated energy, in non-dimensional form, is given by

$$\Delta E = \frac{(m-1)R}{2} \tag{3.128}$$

Therefore, the rotation capacity of the equivalent elastic–perfectly plastic member is given by $R_{eq} = R + \Delta R$, being $\Delta R = \Delta E$. Taking into account that the rotation capacity is provided by equation (3.107),

$$R_{eq} = \frac{m^2 - 1}{2} \left(\frac{1}{k_h} + \frac{1}{|k_d|} \right) \tag{3.129}$$

For shear-type frames, only columns are deformed; therefore, the 'storey ductility' $\bar{\mu}$ is coincident with the rotation capacity of the columns

$$\bar{\mu} = R_{eq} \tag{3.130}$$

In the case of real frames, as all members are involved in the storey sway, a relationship of the following type can be derived [15]

$$\bar{\mu} = u\, R_{eq} \tag{3.131}$$

where, according to the authors [15], the value 2/3 can be assumed for the coefficient u, in order to account for the beam flexibility.

The authors assume that collapse is attained when the energy dissipated at a storey under cyclic loads is coincident, for each direction (positive and negative), with the one dissipated under monotonic loading. Under this hypothesis, the available cumulated ductility η in one direction (positive or negative) is given by the one available under monotonic loading $\eta = \bar{\mu}$. This assumption plays a fundamental role in seismic design based upon the energy approach [38].

Finally, to account for the dissipation capacity that members are still able to provide when they are deformed in the postcritical range, the same authors, on the basis of a great number of numerical simulations of the inelastic response of steel frames [15], have proposed the relationship

$$\eta = \frac{2}{3} R_{eq} + 2 \tag{3.132}$$

where the additional '2' with respect to (3.131) takes into account energy dissipation capacity during the post-critical behaviour. This formula has been introduced into the Japanese seismic code for evaluating the 'storey ductility'.

3.10 CYCLIC BEHAVIOUR

3.10.1 General

In the previous sections, the problem of evaluating rotation capacity of steel beams and beam-columns has been described with specific reference to monotonic loading conditions. The methods examined are of particular use when defining the limiting values of the width-to-thickness ratios assuring a given value of the plastic deformation capacity.

However, the above treatment is not completely exhaustive, because steel members are cyclically loaded during seismic motions.

Knowledge of monotonic behaviour can be sufficient to estimate the collapse condition, when the criterion of maximum plastic deformation can be adopted. According to this criterion, collapse is attained when the required local ductility is greater than that available. This assumption represents a good approximation in cases of members being subjected to a deformation history characterized by only one cycle with a big plastic engagement. In the opposite case, of deformation histories characterized by many significant plastic excursions, two needs arise. On one hand, a sufficiently accurate modelling of the cyclic behaviour is requested and, on the other hand, the occurrence of failure should be characterized by taking into account all plastic excursions.

A very simple criterion for establishing the failure condition of cyclically loaded members is the energy criterion. It can be assumed [36, 38] that collapse occurs when the cumulated plastic ductility in one direction, positive or negative, is greater than that available under monotonic loading conditions.

In the following, the cyclic behaviour of bent steel members will be briefly examined on the basis of the experimental evidence [39–41] and the problems arising in its modelling will be discussed [42–44]. In addition, some recent proposals for collapse interpretation through a low-cycle fatigue approach will be presented [45–47].

3.10.2 Experimental research

The first experimental activity in the field of cyclic behaviour of steel members goes back to 1965, thanks to Bertero and Popov, who investigated the effects of large alternate strains on steel beams, up to their collapse due to inelastic buckling [48]. This was followed by other experimental tests on the cyclic inelastic behaviour of thin-walled tubular columns [49]. The study of the cyclic behaviour of thin-walled box beams has been faced [50, 51], while experimental investigations devoted to thin-web welded H-columns have been carried out [52].

An extensive testing program dealing with the cyclic behaviour of both welded and rolled steel members has been carried out by Ballio at the Politecnico di Milano. In particular [39], the behaviour of welded box and H shapes under variable amplitude loading tests has been examined. The geometrical properties of various sections have been selected in order to examine the influence of the width-to-thickness ratios on the cyclic behaviour. The test apparatus has been described in [53]. The loading program, under displacement control, has been developed according to ECCS Recommendations [54]. Some load F versus displacement v hysteretic curves are given in Figs. 3.20–3.23.

The first observation is that, once a specimen has buckled, the maximum load cannot be reached in the following cycles. The magnitude of the deterioration increases as the width-to-thickness ratio of the flanges increases. It can be noted that, for a given value of the b_f/t_f ratio, the strength deterioration of H sections is higher than for box-shaped beams. Independently of the section shape, the extension of the buckled zone increases as the number of cycles increases. In some cases, some cracks arose in the web-to-flange connection zone.

The effects of local buckling and fatigue, as well as of the loading history on the cyclic behaviour of rolled steel shapes, such as the HEA, HEB and IPE commercial series, have been investigated [40]. The ex-

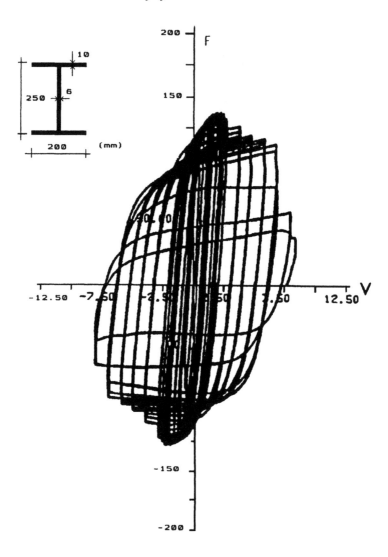

Fig. 3.20   Cyclic behaviour of an H-shaped section with $b/t = 20$ [39]

perimental results have confirmed the influence of the width-to-thickness ratios on the magnitude of the strength deterioration after the occurrence of local buckling. In all cases collapse was reached with fracture of the flange which first buckled. Moreover, for a given value of the b_f/t_f ratio of the flanges, the behaviour is strongly influenced by the d_w/t_w ratio of

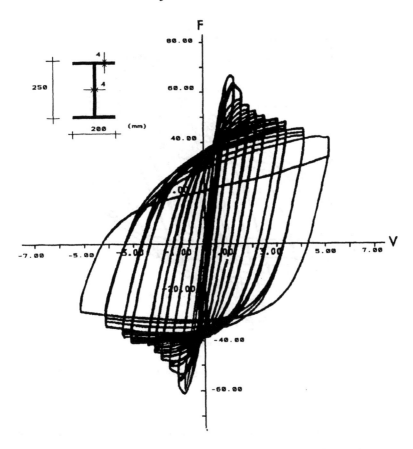

Fig. 3.21 Cyclic behaviour of an H-shaped section with $b/t = 50$ [39]

the web, because deterioration of strength and energy dissipation capacity increases as the d_w/t_w ratio increases. This result agrees with the previously described formulations for evaluating rotation capacity under monotonic loading, which have shown the need for a cross-section classification in which the width-to-thickness ratios both of flanges and web are simultaneously considered (Sections 3.5–3.7).

The experimental investigation of the cyclic behaviour of rolled members under both constant and variable amplitude loading has been performed [41, 47] with two objectives: to develop a cumulative damage model based on low-cycle fatigue, and the better calibration of a cyclic model previously proposed [43].

For seismic design purposes, the idea to interpret the collapse of steel members under cyclic loads as a low-cycle fatigue phenomenon goes

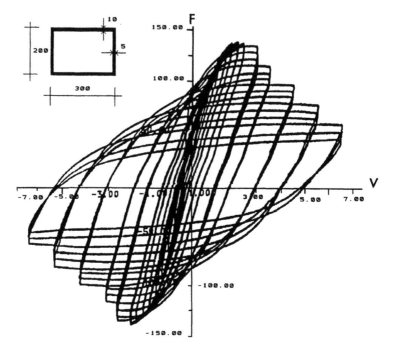

Fig. 3.22   Cyclic behaviour of a box-shaped section with $b/t = 30$ [39]

back to Krawinkler and Zohrei [45]. Their experimental work has shown that deterioration is usually of one of the two types shown in Figs. 3.24 and 3.25. In the first case (Fig. 3.24), the number of cycles without significant deterioration (deterioration threshold) is small and the rate of deterioration is low, so that some deterioration should be accepted in a conventional collapse definition. In the second case (Fig. 3.25), once deterioration arises, it occurs at a very high rate soon leading to collapse. In this case, the occurrence of significant deterioration has to be assumed as collapse.

The first behaviour type is characteristic of local buckling failure modes; the second is characteristic of crack propagation and fracture at welds. In fact, localized crack growth does not cause significant deterioration until the crack has grown considerably.

The authors have found that the rate of deterioration per reversal, (Δd), for constant amplitude cycling, can be accurately expressed as

$$\Delta d = a \, (\Delta \delta_p)^b \tag{3.133}$$

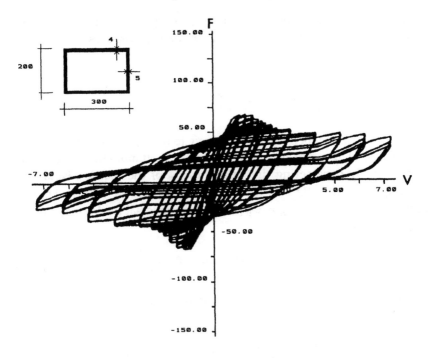

Fig. 3.23 Cyclic behaviour of a box-shaped section with $b/t = 75$ [39]

where a and b are parameters which depend on the properties of the structural component and $\Delta\delta_p$ is the plastic (generalized) displacement range. In equation (3.133) d usually represents the deterioration of strength, although generally, it can be referred to also as stiffness or the energy dissipation capacity. Under the assumption of linear damage accumulation, both for variable and constant amplitude reversals, the total accumulated deterioration can be expressed as

$$d = \sum_{i=1}^{n} \Delta d_i = a \sum_{i=1}^{n} (\Delta\delta_p)^b \qquad (3.134)$$

where n is the number of reversals or plastic excursions.

Failure can be defined as the attainment of a limiting value of deterioration $d_{\lim}$, therefore the number of reversals to failure for constant amplitude cycles is given by

$$N_{rf} = \frac{d_{\lim}}{a} (\Delta\delta_p)^{-b} = \frac{1}{B} (\Delta\delta_p)^{-b} \qquad (3.135)$$

148 *Local ductility of beams and beam-columns*

As a consequence, under the Miner assumption of linear damage accumulation [55], the total damage D can be expressed as

$$D = \sum_{i=1}^{n} \frac{1}{N_{rfi}} = \frac{a}{d_{lim}} \sum_{i=1}^{n} (\Delta\delta_{pi})^b = B \sum_{i=1}^{n} (\Delta\delta_{pi})^b \qquad (3.136)$$

where the parameters B (or a) and b have to be experimentally evaluated.

A value of D equal to one represents the attainment of the collapse condition. By considering the value $\Delta\delta_{p,mon}$ of the plastic displacement leading to collapse ($D = 1$) under monotonic loading ($n = 1$), i.e. under only one reversal ($\Delta\delta_{p,mon} = \Delta\delta_{p,1}$ amplitude reversal for $N_{rf} = 1$), equation (3.136) yields

$$1 = B\ (\Delta\delta_{p1})^b \quad \Rightarrow \quad B = \frac{1}{(\Delta\delta_{p1})^b} \qquad (3.137)$$

Therefore, equation (3.136) can be written in the form

$$D = \sum_{i=1}^{n} \left(\frac{\Delta\delta_{pi}}{\Delta\delta_{p1}}\right)^b = \sum_{i=1}^{n} \left(\frac{\Delta\delta_{pi}}{l}\Delta\delta_{p,mon}\right)^b \qquad (3.138)$$

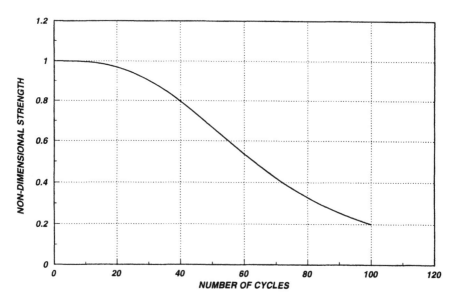

Fig. 3.24   Behaviour for local buckling failure modes

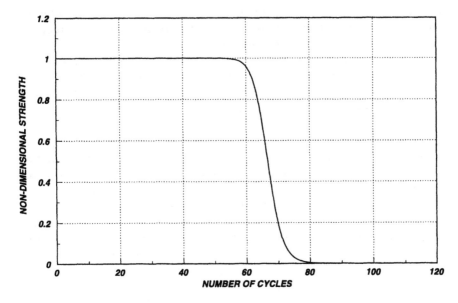

Fig. 3.25   Behaviour for crack propagation and fracture failure modes

It is important to stress that the parameter b and, therefore, $\Delta\delta_{p,1} = \Delta\delta_{p,mon}$ are strictly dependent on the collapse definition through the limit value of deterioration d_{lim}, as shown by equation (3.135).

In addition, equation (3.135) can be written as

$$\ln(\Delta\delta_{pi}) = -\frac{1}{b}\,\ln N_{rf} - \ln B \qquad (3.139)$$

which, in the plane $\ln\Delta\delta_{p,i} - \ln N_{rf}$ represents a straight line with a slope given by $-1/b$.

Parameter b therefore provides a measure of the gradient of deterioration trend. In fact, section sensitivity to the strength deterioration due to local buckling increases as the b decreases.

The authors have pointed out that the low-cycle fatigue approach, expressed in the form (3.136), can be adopted both for local buckling failure modes and for crack propagation and fracture failure modes.

In particular, two rolled H sections have been tested [41]. The first, having a flange b_f/t_f ratio equal to 18.9, failed due to progressive deterioration caused by local buckling. The second, having a b_f/t_f ratio equal to 11.5 failed due to crack propagation phenomena. The parameter assumed values of 1.69 and 2.12 in the two cases respectively, in agreement with the physical meaning of parameter b.

The problem of the collapse interpretation through a low-cycle fatigue approach has been also faced [47] where, according to customary practice, the number of cycles rather than reversals is considered. Moreover, these authors evaluated the cycle amplitude rather than the plastic range of the cycle.

The associated experimental data [41, 47] are presented in Figs. 3.26 and 3.27, where reference is made to the number of reversals to failure and the corresponding plastic amplitude. The values 1.68 and 2.53 are obtained for HE220A and HE220B shapes, respectively (these shapes are characterized by $b_f/t_f = 20$ and $b_f/t_f = 13.75$ respectively).

Significant experimental research in the field of collapse characterization through the low-cycle fatigue approach has been developed by Yamada [46] who has investigated the cyclic behaviour both of H-shaped and box-shaped sections. In both cases, the parameter b increases as the flange width-to-thickness ratio decreases, but for a given value of the b_f/t_f ratio the parameter b for H sections is less than that of box-shaped

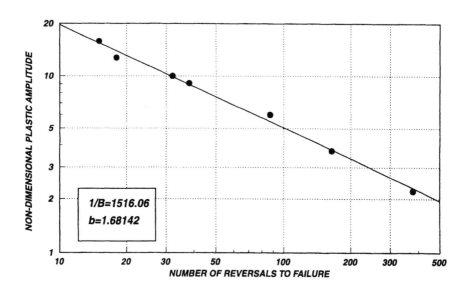

Fig. 3.26   Experimental data provided by Ballio and Castiglioni for HE220A

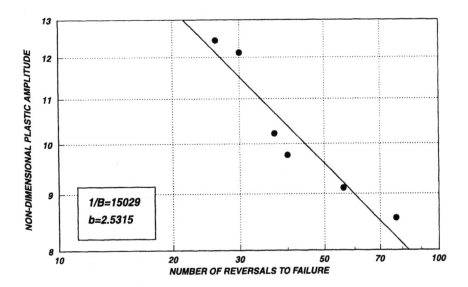

Fig. 3.27 Experimental data provided by Ballio and Castiglioni for HE220B (local buckling failure mode)

sections. This confirms that box-shaped sections are less sensitive to the strength deterioration arising from the occurrence of local buckling.

3.10.3 Modelling of cyclic behaviour

On the basis of experimental evidence, it has been recognized that an exhaustive modelling of cyclic behaviour of bent steel members has to take into account the strain hardening of the material, the local buckling of the flanges, and the Bauschinger effect, as well as the progressive damage due to fatigue and fracture.

The modelling of the cyclic behaviour of bent steel members can be carried out in two ways.

The first approach is based on the subdivision of the member cross-section into elementary areas. It is possible to associate at each elementary area the given values of mechanical properties and of residual stress. All phenomena are modelled with reference to the elementary area, while the behaviour of the whole section is derived through the equilibrium equations

$$\sum \sigma_i \, A_i = N \qquad (3.140)$$

$$\sum \sigma_i \, A_i \, y_i = M \qquad (3.141)$$

where σ_i, A_i and y_i are, respectively, the stress, the area and the distance from centre of gravity.

The equations governing the behaviour of the section have to be integrated by those describing the behaviour of the member as a whole.

This approach has been widely used for simulating the buckling curves of steel columns [56] and, more recently, for evaluating the moment–curvature relationship of cold-formed steel sections [57].

In this section, attention is focused only on those aspects which have to be considered in the simulation of the cyclic behaviour.

The first problem is represented by the modelling of the cyclic constitutive law of the material. By comparison of the numerical results obtained when adopting different constitutive laws suggested in the technical literature, it has been shown [42] that the best numerical–experimental agreement is obtained when the cyclic behaviour of the material is described by the Giuffré–Menegotto-Pinto model Fig. 3.28 [58, 59]. According to this model,

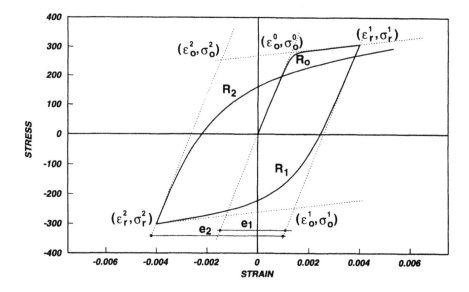

Fig. 3.28   Cyclic model proposed by Giuffré, Menegotto and Pinto [58,59]

$$\sigma^* = \rho_h \, \varepsilon^* + (1 - \rho_h) \, \frac{\varepsilon^*}{\left(1 + |\varepsilon^*|^{R(e)}\right)^{1/R(e)}} \qquad (3.142)$$

where σ^* and ε^* are the non-dimensional stress and strain, respectively, while ρ_h is the ratio between the hardening modulus and the elastic modulus.

The exponent R is a decreasing function of a parameter e measuring the plastic flow during the previous cycle. It is given by

$$R(e) = R_o - \frac{A_1 \, e}{A_2 + e} \qquad (3.143)$$

where R_o, A_1 and A_2 have to be experimentally determined.

The parameter e represents the difference between the strain corresponding to the point of load reversal and that corresponding to the intersection point of the two asymptotes of the previous branch.

The non-dimensional stress and strain are given by

$$\sigma^* = \frac{\sigma - \sigma_o}{\sigma_o - \sigma_r} \qquad \varepsilon^* = \frac{\varepsilon - \varepsilon_o}{\varepsilon_o - \varepsilon_r} \qquad (3.144)$$

where $(\sigma_o, \varepsilon_o)$ are the coordinates of the intersection point of the two asymptotes of the current loading curve and $(\sigma_r, \varepsilon_r)$ are the coordinates of the previous point of load reversal.

As has been shown [42] and extensively discussed [43], this model allows a correct simulation of displacement histories symmetric with respect to the undeformed condition. In the case of random variations of the imposed displacements, the model should possess a memory of all previous branches of the loading history so that the hysteresis loop follows the previous loading branch, as soon as the new reloading curve has reached it. As this is impractical from a computational point of view, Filippou, Bertero and Popov [60] have proposed to limit the memory of the model to only four fundamental curves:

- the initial monotonic envelope;
- the descending lower branch curve, originating at the reversal point with the largest strain value;
- the ascending upper branch curve, originating at the reversal point with the smaller strain value;
- the current curve originating at the most recent reversal point.

Moreover, in order to take into account isotropic strain hardening, the

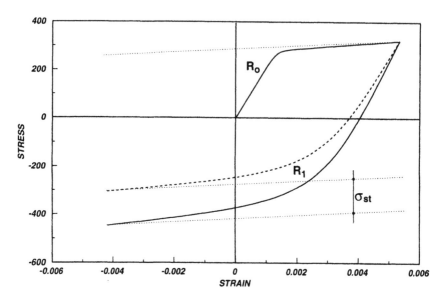

Fig. 3.29   Shifting of the asymptote according to Filippou, Bertero and Popov [60]

authors propose shifting of the asymptote of the hardening branch according to the value given by (Fig. 3.29)

$$\sigma_{st} = A_3 \; \sigma_y \left(\frac{\varepsilon_{p,m}}{\varepsilon_y} - A_4 \right) \tag{3.145}$$

where $\varepsilon_{p,m}$ is the maximum plastic strain before the occurrence of the load reversal, while A_3 and A_4 are parameters to be experimentally evaluated.

Due to local buckling, fatigue and fracture in the web-to-flange connection zone, damage can occur in each elementary area during cyclic loading.

It is assumed [41] that the collapse of an elementary area due to one of the above phenomena is attained for a limit value of the strain. The authors [41] describe damage phenomenon through three partial damage indexes I_{LB}, I_{LCF} and I_F for local buckling, low cycle fatigue and fracture respectively. The actual damage is due to the combination of the above phenomena and is evaluated by means of an effective damage index I_{ED} defined as

$$I_{ED} = \sqrt{ C_{LB} \; I_{LB}^2 + C_{LCF} \; I_{LCF}^2 + C_F \; I_F^2 } \tag{3.146}$$

where the combination coefficients C_{LB}, C_{LCF} and C_F (all less than or equal to one) have to be calibrated on the basis of experimental results. The authors have pointed out [41] that for a given section shape the values of the combination coefficients are not affected by the loading history.

The damage process is simulated by reducing the initial elementary area A_{oi} according to the following relationship

$$A_i = A_{oi} \ (1 - I_{ED}) \tag{3.147}$$

so that $I_{ED} = 1$ corresponds to the collapse of the elementary area, which is definitely cancelled.

It is assumed that the limiting value of the strain leading to local buckling of the elementary area is the elastic one ε_{cr} expressed through the relationship

$$\frac{\varepsilon_{cr}}{\varepsilon_y} = \frac{\pi^2 \ E}{12 \ (1 - v^2) \ \sigma_y} \frac{k}{(x/t \ )^2} \tag{3.148}$$

where x is the distance between the centroid of the elementary area and the web-to-flange connection, k is a buckling coefficient which has a unit value under the assumption that the restraining part buckles simultaneously with the restrained part, and t is the plate thickness.

According to Ballio and Castiglioni [41] the partial damage index for local buckling is provided by

$$I_{LB} = \sum_{i=1}^{N_c} \left(\frac{\Delta\varepsilon_{max}}{\varepsilon_{cr}} \right)_i \left(\frac{1}{N_L} \right)^{C_1} \tag{3.149}$$

The term $(\Delta\varepsilon_{max} / \varepsilon_{cr})_i$ represents the ratio between the maximum absolute strain range experienced by the elementary area during the loading history before the ith cycle and the Euler critical strain ε_{cr}.

Moreover, N_c is the current number of cycles and N_L is the number of cycles having an amplitude not greater than $\Delta\varepsilon_{max}$. If a new cycle, having an amplitude greater than $\Delta\varepsilon_{max}$ occurs, the value of N_L is reset to 1 and the value of $\Delta\varepsilon_{max}$ is updated.

Finally, the exponent C_1, which has to be experimentally calibrated, defines a progressive reduction of the rate of strength deterioration corresponding to cycles having an amplitude not greater than the maximum one.

With respect to the original formulation of Castiglioni and Di Palma [43], equation (3.149) takes into account that, under constant amplitude loading history, the strength deterioration due to local buckling develops in the early cycles after which a stage of stabilization is reached (Fig. 3.30). A further increase of the out-of-plane buckling and, therefore, of

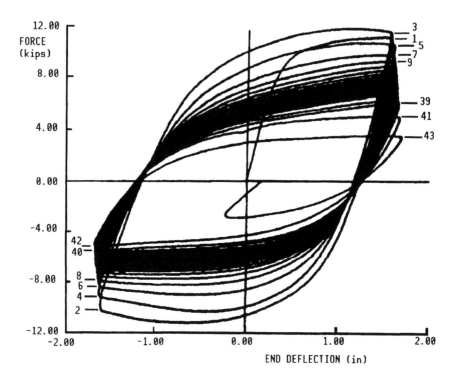

Fig. 3.30   Cyclic behaviour for local buckling failure mode

the strength deterioration is verified only by imposing new cycles having an amplitude greater than the previous ones.

During experimental tests cracking arises in the web-to-flange connection zone, both in welded [39] and in rolled [40] shapes. Crack opening and its propagation develop mainly in the flange due to the local buckling and the stiffening effect of the web, which produce a stress concentration. Therefore, the authors [41, 43] define the partial damage index for fracture as

$$I_F = \frac{\Delta\varepsilon_{max}\, C_2}{\varepsilon_u} \tag{3.150}$$

where ε_u is the ultimate strain of the material and C_2 is a strain concentration coefficient, which assumes a value of 4 at a distance from the web-to-flange connection less than the plate thickness and linearly reduces to 1 until the above distance reaches three times the plate thickness.

Fatigue effects play a very important role in the damage process of

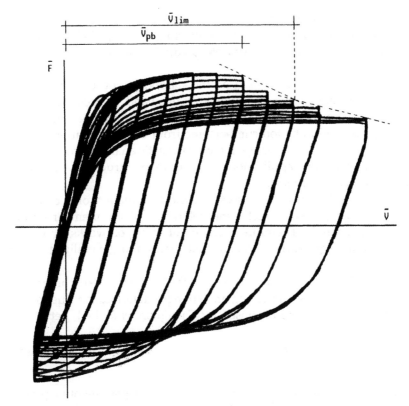

Fig. 3.31 Basic definitions [44]

steel members under cyclic loading. This type of damage and its location within the cross-section are very similar to those occurring for fracture effects, but in this case the damage magnitude depends both on the strain amplitude $\Delta\varepsilon$ and on the number of cycles N_c.

According to Ballio and Castiglioni [41], the partial damage index for low-cycle fatigue is given by

$$I_{LCF} = \sum_{i=1}^{N_c} \frac{\Delta\varepsilon_i \; C_3}{\varepsilon_u} \left[1 - \left(1 - \frac{\Delta\varepsilon_i}{\varepsilon_u} \right)^i \right] \qquad (3.151)$$

where C_3 is a concentration coefficient, which for rolled sections can be assumed equal to C_2.

With respect to the original formulation of Castiglioni and Di Palma [43], the term in square brackets has been added in order to take into account the increasing importance of fatigue effects as the number of imposed cycles increases.

Very good agreement between numerical simulations and experimental behaviour has been shown [41], so that it can be concluded that the above modelling of the phenomena affecting the cyclic behaviour of bent steel members can give an accurate prediction of the seismic behaviour of steel structures.

The above modelling, based on the subdivision of the member cross-section into elementary areas, allows, on one hand, an accurate prediction of the cyclic behaviour under seismic loading, but, on the other hand, it has been proved to be too cumbersome from the computational point of view, for use in a step-by-step dynamic inelastic analysis program. Therefore, for practical applications on complex structures, a different approach has to be used.

This second approach is based on the formulation of a constitutive law able to represent directly the behaviour of the member or of the section as a whole. Therefore, a model of this type works by means of global parameters such as generalized forces and generalized displacements.

A very simple model of this kind, which includes Bauschinger effect, strain hardening, local buckling and strength deterioration due to fatigue and fracture is the one proposed by Castiglioni *et al.* [44]. Let

$$\bar{F} = \frac{F}{F_y} \qquad \text{and} \qquad \bar{v} = \frac{v}{v_y} \qquad (3.152)$$

be the non-dimensional generalized force and displacement respectively. With reference to Fig. 3.31, the following quantities are defined

- $\bar{v}_i$ is the non-dimensional conventional amplitude of the *i*th semi-cycle, representing the displacement range between the point with zero force and the following reversal point.
- $\bar{v}_{pb}$ is the value of $\bar{v}_i$ corresponding to the attainment of the maximum force and, therefore, preceding the occurrence of local buckling.
- $\bar{v}_{\lim}$ is the value of $\bar{v}_i$ corresponding to the stabilization of the cycles.

The parameter $\bar{v}_{\lim}$ is also used as a conventional upper bound for the semi-cycle amplitude. In other words, semi-cycles having an amplitude $\bar{v}_i > \bar{v}_{\lim}$ are considered as having an amplitude $\bar{v}_{\lim}$. Therefore, in the following, $\bar{v}_i \le \bar{v}_{\lim}$ is assumed.

The cyclic behaviour is modelled by means of linear branches (Fig. 3.32). The non-dimensional stiffness of the hardening portion of the curve is expressed at the *i*th cycle as

$$\bar{K}_{h_i} = \bar{K}_{h_o}(1 - H_1\,\bar{v}_{\max}) \qquad (3.153)$$

where $\bar{K}_{h_o}$ is the non-dimensional stiffness of the hardening portion of the

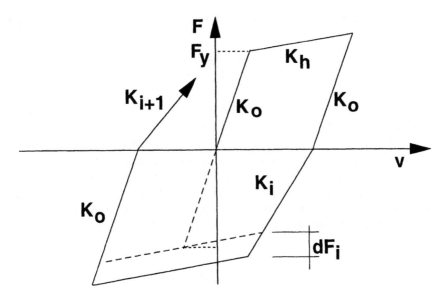

Fig. 3.32   Trilinear model proposed by Castiglioni, Di Palma and Moretta [44]

virgin curve, $\bar{v}_{max}$ is the maximum conventional non-dimensional amplitude of the semi-cycles previously completed during the loading history (with $\bar{v}_{max} \leq \bar{v}_{lim}$), and H_1 is a numerical coefficient describing the progressive reduction of the hardening stiffness during the loading history.

The numerical coefficient H_1 can be determined by identifying on the experimental test results the non-dimensional amplitude $\bar{v}_{k_h}$ of the semi-cycle corresponding to the attainment of the minimum value ($\overline{K}_{h_{min}}$) of $\overline{K}_h$.

From equation (3.153),

$$H_1 = \left(1 - \frac{\overline{K}_{h,min}}{\overline{K}_{h_o}}\right) \frac{1}{\bar{v}_{k_h}} \qquad (3.154)$$

Analogously, the non-dimensional stiffness of the reloading branches of the hysteresis loops, with reference to the ith cycle, is given by

$$\overline{K}_i = \overline{K}_o \left(1 - H_2 \, \bar{v}_{max}\right) \qquad (3.155)$$

where $\overline{K}_o$ is the non-dimensional value of the initial stiffness of the virgin curve and H_2 is a numerical coefficient describing the progressive reduction of the above stiffness.

On the basis of experimental test results, the coefficient H_2 can be evaluated as

$$H_2 = \left(\frac{1 - \overline{K}_{min}}{\overline{K}_o}\right) \frac{1}{\overline{v}_k} \qquad (3.156)$$

where $\overline{v}_k$ is the non-dimensional conventional amplitude of the semi-cycle corresponding to the attainment of the minimum value ($\overline{K}_{min}$) of $\overline{K}$.

Moreover, the model is based on the shifting of the hardening branch in order to simulate both the isotropic hardening and the structural damage due to local buckling and fatigue. The first effect, isotropic hardening, tends to expand the hysteresis curve in the force direction, while the second, structural damage, provides an opposite contribution.

The shifting of the hardening branch of the ith semi-cycle is given by

$$dF_i = H_3\, \overline{v}_{max} - H_4\, \overline{v}_{lb} - \sum_{j=1}^{i-1} H_5\, \overline{v}_j \qquad (3.157)$$

where H_3, H_4 and H_5 are numerical coefficients, $\overline{v}_{lb}$ is a conventional value representing the part of the non-dimensional displacement range exceeding the one ($\overline{v}_{pb}$) preceding the occurrence of local buckling

$$\overline{v}_{lb} = \overline{v}_{max} - \overline{v}_{pb} \geq 0 \qquad (3.158)$$

The first term of equation (3.157) describes the effects of isotropic hardening; the second term represents the effects of local buckling; the third one simulates, according to the Miner rule [55], the cumulated damage due to low-cycle fatigue.

The numerical coefficient H_5 is evaluated on the basis of experimental test results by analysing semi-cycles having an amplitude greater than $\overline{v}_{lim}$, so that $\overline{v}_{max} = \overline{v}_{lim}$ and $\overline{v}_{lb} = \overline{v}_{lim} - \overline{v}_{pb}$ have to be assumed in equation (3.157). By considering two cycles greater than $\overline{v}_{lim}$,

$$dF_i = H_3\, \overline{v}_{lim} - H_4\, (\overline{v}_{lim} - \overline{v}_{pb}) - \sum_{j=1}^{i-1} H_5\, \overline{v}_j \qquad (3.159)$$

and

$$dF_{i+1} = H_3\, \overline{v}_{lim} - H_4\, (\overline{v}_{lim} - \overline{v}_{pb}) - \sum_{j=1}^{i} H_5\, \overline{v}_j \qquad (3.160)$$

Therefore,

$$dF_{i+1} - dF_i = -H_5 \, \bar{v}_i \qquad (3.161)$$

where $\bar{v}_i$ exceeds $\bar{v}_{\lim}$ and so the limit condition $\bar{v}_i = \bar{v}_{\lim}$ has to be applied. As a consequence

$$H_5 = \frac{dF_i - dF_{i+1}}{\bar{v}_{\lim}} \qquad (3.162)$$

The numerical coefficient H_3 can be derived by analysing semi-cycles having an amplitude less than $\bar{v}_{pb}$. For such semi-cycles, where $\bar{v}_{\max} < \bar{v}_{pb}$, then $\bar{v}_{lb} = 0$. When $\bar{v}_{\max} = \bar{v}_i$, equation (3.157) becomes

$$dF_i = H_3 \, \bar{v}_i - H_5 \sum_{j=1}^{i-1} \bar{v}_j \qquad (3.163)$$

By considering a second semi-cycle with $\bar{v}_{i+1} > \bar{v}_i$, but less than $\bar{v}_{pb}$, $\bar{v}_{\max} = \bar{v}_{i+1}$ and $\bar{v}_{lb} = 0$, therefore

$$dF_{i+1} = H_3 \, \bar{v}_{i+1} - H_5 \sum_{j=1}^{i} \bar{v}_j \qquad (3.164)$$

As a consequence, by combining equations (3.163) and (3.164)

$$H_3 = \frac{(F_{i+1} - dF_i) + H_5 \, \bar{v}_i}{\bar{v}_{i+1} - \bar{v}_i} \qquad (3.165)$$

Finally, by denoting as $i + 1$ the first semi-cycle with an amplitude greater than $\bar{v}_{pb}$ so that $\bar{v}_i = \bar{v}_{pb}$,

$$dF_i = H_3 \, \bar{v}_{pb} - H_5 \sum_{j=1}^{i-1} \bar{v}_j \qquad (3.166)$$

and

$$dF_{i+1} = H_3 \, \bar{v}_{i+1} - H_4 \, (\bar{v}_{i+1} - \bar{v}_{pb}) - H_5 \sum_{j=1}^{i} \bar{v}_j \qquad (3.167)$$

which, when combined, yields

$$H_4 = \frac{H_3 \, (\bar{v}_{i+1} - \bar{v}_{pb}) - H_5 \, \bar{v}_{pb} - (F_{i+1} - dF_i)}{\bar{v}_{i+1} - \bar{v}_{pb}} \qquad (3.168)$$

Comparison between this simplified model and experimental test results [44] has shown the model is able to simulate the absorbed energy in the hysteresis loops with an average error smaller than 5%. The model therefore allows an accurate prediction of the cycle response of steel members and at the same time is sufficiently simple to be used in a step-by-step dynamic inelastic analysis program.

3.11 REFERENCES

[1] *G. Haaijer, B. Thürlimann:* 'On Inelastic Buckling in Steel', Journal of Engineering Mechanics Division, ASCE, April, 1958.

[2] *M.G. Lay:* 'Flange Local Buckling in Wide Flange Shapes', Journal of Structural Division, ASCE, vol. 91, December, 1965.

[3] *M.G. Lay, T.V. Galambos:* 'Inelastic Beams Under Moment Gradient', Journal of Structural Division, ASCE, Vol. 93, February, 1967.

[4] *M.G. Lay, T.V. Galambos:* 'Inelastic Beams Under Uniform Moment', Journal of Structural Division, ASCE, Vol. 91, December, 1965.

[5] *G. Haaijer:* 'Plastic Buckling in the Strain-Hardening Range', Journal ASCE, Vol. 83, EM2, April 1957.

[6] *B. Kato:* 'Buckling Strength of Plates in the Plastic Range', Publications of IABSE, Vol. 25, 1965.

[7] *A.F. Lukey, P.F. Adams:* 'Rotation Capacity of Wide-Flange Beams under Moment Gradient', Journal of the Structural Division, ASCE, Vol. 95, ST6, pp. 1173–1188, June 1969

[8] *A.F. Lukey, P.F. Adams:* 'Rotation capacity of beams under moment gradient', Welding Research Council Bulletin, N.142, pp. 1–19, July 1969.

[9] *J.J. Climenhaga, R.P. Johnson:* 'Moment-Rotation Curves for Locally Buckling Beams', Journal of the Structural Division, ASCE, Vol. 98, ST6, June, 1972.

[10] *M. Ivanyi:* 'Yield Mechanism Curves for Local Buckling of Axially Compressed Members', Periodica Polytechnica, Civil Engineering, Vol. 23, N.3–4, pp. 203–216, 1979.

[11] *M. Ivanyi:* 'Moment-Rotation Characteristics for Locally Buckling Beams', Periodica Polytechnica, Civil Engineering, Vol. 23, N.3–4, pp. 217–230, 1979.

[12] *V. Gioncu, G. Mateescu, S. Orasteanu:* 'Theoretical and Experimental Research regarding the Ductility of Welded I-sections subjected to Bending', Stability of Metal Structures, Proceedings of the Fourth International Colloquium on Structural Stability, Asian Session, Beijing, China, pp. 289–298, 1989.

[13] *T. Nakamura:* 'Strength and Deformability of H-Shaped Steel Beams and Lateral Bracing Requirements', Journal of Constructional Steel Research, pp. 217–228, N.9, 1988.

[14] *I. Mitani, M. Makino:* 'Post Local Buckling Behaviour and Plastic Rotation Capacity of Steel Beam-Columns', 7th World Conference on Earthquake Engineering, Istanbul, 1980.

[15] *B. Kato, H. Akiyama:* 'Ductility of Members and Frames subject to Buckling', ASCE Convention, May 11–15, 1981.

[16] *A.R. Kemp:* 'Interaction of Plastic Local and Lateral Buckling', Journal of Structural Engineering, ASCE, Vol. 111, October, 1985.

[17] *B. Kato:* 'Rotation Capacity of Steel Members subject to Local Buckling', 9th World Conference on Earthquake Engineering, Vol. IV, paper 6-2-3, August 2–9, Tokyo-Kyoto, 1988.

[18] *B. Kato:* 'Rotation Capacity of H-Section Members as Determined by Local Buckling', Journal of Constructional Steel Research, pp. 95–109, N.13, 1989.

[19] *B. Kato:* 'Deformation Capacity of Steel Structures', Journal of Constructional Steel Research, pp. 33–94, N.17, 1990.

[20] *V. Piluso:* 'Il Comportamento Inelastico dei Telai Sismo-Resistenti in Acciaio', Tesi di Dottorato (PhD Thesis) in Ingegneria delle Strutture, IV Ciclo, Università di Napoli, Febbraio 1992.

[21] *F.M. Mazzolani, V. Piluso:* 'Evaluation of the Rotation Capacity of Steel Beams and Beam-Columns', 1st Cost C1 Workshop, Strasbourg, 28–30 October, 1992.

[22] *M.A. Bradford:* 'Inelastic Local Buckling of Fabricated I-Beams', Journal of Constructional Steel Research, N.7, 317–334, 1987.

[23] *J.L. Dawe, G.L. Kulak:* 'Plate Instability of W Shapes', Journal of Structural Engineering, ASCE, Vol. 110, N.6, June, 1984.

[24] *G.J. Hancock:* 'Local, Distortional and Lateral Buckling of I-Beams', Journal of Structural Division, ASCE, Vol. 104, ST11, November, 1978.

[25] *U. Kuhlmann:* 'Definition of Flange Slenderness Limits on the Basis of Rotation Capacity Values', Journal of Constructional Steel Research, pp. 21–40, 1989.

[26] *R. Spangemacher, G. Sedlacek:* 'On the Development of a Computer Simulator for Tests of Steel Structures', Proceedings of the First World Conference on Constructional Steel Design, Acapulco, Mexico, 6–9 December 1992.

[27] *R. Spangemacher:* 'Zum Rotationsnachweis von Stahlkonstruktionen, die nach dem Traglastverfahren berechnet werden', Dissertation RWTH Aachen 1992.

[28] *Commission of the European Communities:* 'Eurocode 3: Design of Steel Structures', 1992.

[29] *ECCS (European Convention for Constructional Steelwork):* 'European Recommendations for Steel Structures in Seismic Zones', 1988.

[30] *Ch. Massonnet, M. Save:* 'Calcolo Plastico a Rottura delle Costruzioni', CLUP Milano, 1980.

[31] *E.Z. Stowell:* 'Compressive Strength of Flanges', Technical Note No. 2020, National Advisory Committee for Aeronautics, 1950.

[32] *R.E. Southward:* 'Local Buckling in Universal Sections', Internal Report, University of Cambridge, Engineering Department, 1969.

[33] *J.M. Alexander:* 'An Approximate Analysis of the Collapse of Thin Cylindrical Shells Under Axial Loading', Quarterly Journal of Mechanical and Applied Mathematics, Vol. 13, 1960

[34] *A. Pugesley, M. Macaulay:* 'The Large-Scale Crumpling of Thin Cylindrical Columns', Quarterly Journal of Mechanical and Applied Mathematics, Vol. 13, 1960

[35] *F.M. Mazzolani, V. Piluso:* 'Member Behavioural Classes for Steel Beams and Beam-Columns', XXVI C.T.A, Giornate Italiane della Costruzione in Acciaio, Viareggio, Ottobre 1993.

[36] *H. Akiyama:* 'Earthquake Resistant Limit State Design for Buildings', University of Tokyo Press, 1985.

[37] *G. Greshick, D.W. White, W. McGuire, J.F. Abel:* 'Toward the prediction of flexural ductility of wide-flange beams for seismic design', Proceedings of Fourth U.S. National Conference on Earthquake Engineering, Palm Springs, California, Vol. 2, pp. 107–115, May 20–24, 1990.

[38] *H. Akiyama:* 'Earthquake Resistant Design Based on the Energy Concept', Proceedings of the 9th World Conference on Earthquake Engineering, Tokyo, Kyoto, paper 8-1-2, Vol.V, August 2–9, 1988.

[39] *G. Ballio, L. Calado:* 'Steel Bent Sections under Cyclic Loads: Experimental and Numerical Approaches', Costruzioni Metalliche, N.1, 1986

[40] *C.A. Castiglioni, N. Di Palma:* 'Experimental Behaviour of Steel Members under Cyclic Bending', Costruzioni Metalliche, N.2/3, 1989.

[41] *G. Ballio, C.A. Castiglioni:* 'Seismic Behaviour of Steel Sections', paper accepted for publication in Journal of Constructional Steel Research, 1993.

[42] *C.A. Castiglioni:* 'Numerical Simulation of Steel Shapes under Cyclic Bending: Effect of the Constitutive Law of the Material', Costruzioni Metalliche, N.3, 1987

[43] *C.A. Castiglioni, N. Di Palma:* 'Steel Members under Cyclic Loads: Numerical Modelling and Experimental Verifications', Costruzioni Metalliche, N.6, 1988.

[44] *C.A. Castiglioni, N. Di Palma, E. Moretta:* 'A Trilinear Constitutive Model for the Seismic Analysis of Steel Structures', Costruzioni Metalliche, N.2, 1990

[45] *H. Krawinkler, M. Zohrei:* 'Cumulative Damage in Steel Structures Subjected to Earthquake Ground Motion', Computers & Structures, Vol. 16, N.1–4, pp. 531–541, 1983.

[46] *M. Yamada:* 'Low Cycle Fatigue Fracture Limits of Structural Materials

and Structural Elements', published in 'Testing of Metal Structures', edited by F.M. Mazzolani, E & FN Spon, RILEM, 1992.

[47] *G. Ballio, C.A. Castiglioni:* 'A Unified Approach for the Design of Steel Structures under Low and/or High Cycle Fatigue', Paper accepted for publication on Journal of Constructional Steel Research, 1993.

[48] *V. Bertero, E. Popov:* 'Effect of Large Alternating Strains of Steel Beams', Journal of Structural Division, ASCE, Vol. 91, ST1, February, 1965.

[49] *E. Popov, V. Zayas, S. Mahin:* 'Cyclic Inelastic Buckling of Thin Tubular Columns', Journal of Structural Division, ASCE, Vol. 105, ST11, November 1979

[50] *Y. Fukumoto, H. Kusama:* 'Cyclic Tests of Thin-Walled Box Beams', Proc. of JSCE Structural Engineering, Earthquake Engineering, Vol. 2, N.1, April, 1985.

[51] *Y. Fukumoto, H. Kusama:* 'Cyclic Behaviour of Thin-Walled Box Stub Columns and Beams', Stability of Metal Structures, Paris, 16–17 November 1983.

[52] *R. Avent, S. Wells:* 'Experimental Study of Thin-Web Welded H Columns', Journal of Structural Division, Vol. 108, ST7, July, 1982.

[53] *G. Ballio, R. Zandonini*: 'An Experimental Equipment to Test Steel Structural Members and Subassemblages Subjected to Cyclic Loads', Ingegneria Sismica, N.2, 1985.

[54] *ECCS-CECM-EKS*: 'Recommended Testing Procedure for Assessing the Behaviour of Structural Steel Elements Under Cyclic Loads', Doc. ECCS TWG 1.3, N.45/86, 1986.

[55] *M.A. Miner*: 'Cumulative Damage in Fatigue', Journal of Applied Mechanics, September, 1945.

[54] *C. Faella, F.M. Mazzolani:* 'Simulazione del Comportamento di "Aste Industriali" Inelastiche sotto Carico Assiale', Costruzioni Metalliche, N.4, 1974

[57] *A. De Martino, F.P. De Martino, A. Ghersi, F.M. Mazzolani:* 'Il Comportamento Flessionale dei Profili Sottili Sagomati a Freddo: Impostazione della Ricerca', Acciaio, Settembre, 1989.

[58] *A. Giuffré, P.E. Pinto:* 'Il Comportamento del Cemento Armato per Sollecitazioni Cicliche di Forte Intensità', Giornale del Genio Civile, N.5, 1970

[59] *M. Menegotto, P.E. Pinto:* 'Method of Analysis for Cyclically Loaded Reinforced Concrete Plane Frames including Changes in Geometry and Nonelastic Behaviour of Elements under Combined Normal Force and Bending', Proceedings IABSE Symposium, Lisbona, 1973.

[60] *F.C. Filippou, V.V. Bertero, E.P. Popov:* 'Effects of Bound Deterioration on Hysteretic Behaviour of Reinforced Concrete Joints', Earthquake Engineering Research Center, Report 83/19, August, University of California, Berkeley, 1983.

4

Evaluation of the q-factor

4.1 HISTORICAL REVIEW

Structures are usually designed so that some of the energy input during a severe earthquake is dissipated through inelastic deformations. In order to prevent collapse, the magnitude of these plastic deformations has to be limited to values which are compatible with the available local and global ductility and with the energy dissipation capacity of the structure.

With reference to ultimate limit state, the design of seismic-resistant structures against destructive earthquakes can be performed nowadays by means of two main methods of structural analysis. The first approach is represented by the use of dynamic non-linear analyses which are able to provide the time-history of the response of the structure subjected to generated or historical earthquakes characterizing the seismic zone with a sufficient degree of accuracy. The second approach is based on the use of modal analysis in the linear range by assuming a design spectrum, which provides, as a function of period T, the normalized pseudo-acceleration, required for a specified level of inelastic response. These inelastic spectra are obtained in seismic codes by modifying the linear elastic design response spectrum by means of a factor, namely the q-factor, which takes into account the dissipative capacity of the structure up to failure (Chapter 1).

The first approach requires an accurate and sophisticated modelling of the cyclic behaviour of the material and of the structural details, which is often provided only by object-oriented computer programs. In addition, from the computational point of view, numerical analyses are particularly cumbersome. For these reasons, the dynamic non-linear analysis is limited to research purposes or to the design of structures presenting an increased risk for the population, such as nuclear power plants.

Therefore, from the practical point of view, modal analysis in the linear range provides an efficient method for office routine. In this second case, it is assumed that the fulfilment of the design and detailing rules provided by the seismic codes are able to assure the specified level of ductility and energy dissipation capacity, so that the designer performs in a single 'shot' both the check against the serviceability limit state and the one against the ultimate limit state, while the task to provide reliable values of the q-factor is reserved for the seismic codes.

The use of simplified methods, which have been proposed by many authors, is justified by the fact that the correct evaluation of the q-factor, defined as the ratio between the acceleration leading to collapse and the one corresponding to the occurrence of first yielding, requires several dynamic analyses for different ground motions and, therefore, is particularly cumbersome to use in the case of real structures.

In order to evaluate the degree of safety of a structure against destructive earthquakes, the primary problem to be faced is represented by the definition of the collapse conditions or of an 'admissible level' of structural damage.

The prediction of the amount of seismic damage that a structure is likely to sustain during its lifetime is a probabilistic matter. In fact, there are many uncertainties in damage computation, particularly in the modelling assumptions for structural analysis and in the choice of earthquake ground motions. On one hand, the true strengths of structural materials and members may be very different from the ones assumed for the analysis and, on the other hand, the ground motion itself can be the biggest source of uncertainty.

In spite of the above difficulties, deterministic analysis can represent a valuable tool.

For a long time, the problem of damage evaluation has been faced by identifying structural damage with the required global ductility and, as a consequence, by assuming that collapse is attained when the required global ductility is greater than that available, which is usually evaluated under monotonic loading conditions. Successively, it has been shown that the ductility concept does not take into account either the number of plastic excursions or the plastic engage required by each of them. As a consequence, the energy approach has been recognized as a very promising tool for evaluating the seismic reliability of structures, due to the relative ease of its computation and to the possibility of sketching out a general design method based on energy balance [1]. In this approach, collapse conditions are attained when a given amount of energy dissipation is reached.

In more recent times, it has been shown that the maximum plastic deformation approach, i.e. the ductility approach, can be useful in cases of deformation histories characterized by few cycles having a large plastic excursion, while the energy approach represents a good approximation in cases of many cycles of small plastic amplitude. Since the deformation history is unlikely to consist of regular cycles, different situations can arise in which both the maximum plastic deformation and the energy dissipation play an important role in the damage mechanism. As a consequence, new damage parameters have been proposed in order to improve the damage characterization [2–5] (Chapter 1).

The simplest way to evaluate the cumulative damage is to sum the inelastic deformation, which corresponds to the energy approach. This ignores the fact that the damage caused by a large number of small plastic deformations is likely to be less than the one due to a smaller number of large plastic deformations. For this reason, the most modern approach to the damage characterization is represented by the application of the fatigue theory in the low-cycle range and by the contemporary assumption of linear damage cumulation according to the Miner rule [4].

It has been shown [5] that damage parameters can be evaluated at three different levels, namely 'structure', 'substructure' and individual 'member'. Damage parameters at the structure level, such as global ductility, are overall response quantities which are particularly useful for a quick damage assessment.

Damage parameters at the substructure level are calculated for subassemblages, such as individual frames or storeys, providing information on the damage distribution within the structure. Damage parameters at the member level are computed for individual members, giving the most detailed information.

A very important problem is how to relate the value of a damage parameter at the structure level to the value at the member level. This problem has been faced for plane frames in Chapter 2 with reference to the ductility criterion. It is clear that, as far as the degree of sophistication of the damage characterization increases, the difficulties arising in the attempt to provide the above relation increase. In fact, experimental studies have shown that the strength and stiffness properties of elements and structures deteriorate during cycle loading, because materials, and therefore elements and structures, have a memory of past loading history. However, these experimental studies, which have focused the damage process as a low-cycle fatigue phenomenon, have been developed at the 'member' level. As a consequence, there is no a simple rule with which to derive damage at 'member' level from damage quantified at 'structure' level and vice versa. Up to now, therefore, the use of sophisticated damage parameters, such as low-cycle fatigue, has not led to simplified methods for evaluating the q-factor, but is limited to procedures applicable in research activities only. A further discussion of this aspect will be provided in Section 4.4.

As a consequence of the difficulties arising in the use of sophisticated damage parameters able to consider the whole deformation histories of the dissipative zones of the members constituting the structure, the simplified methods for evaluating the q-factor are mainly based on the damage characterization through the maximum plastic deformation criterion or through the energy criterion.

A first attempt to take into account the effects of the low-cycle fatigue in a simplified method is represented by the introduction of the concept of equivalent ductility factors, which corresponds to the definition of a reduced ductility supply reflecting the influence of cyclic response and which should be used instead of the conventional monotonic ductility supply in design procedures [6–10].

Inelastic response spectra of single degree of freedom (SDOF) systems for earthquake motions have attracted considerable attention since the early 1960s. In the early work of Veletsos and Newmark [11] response analyses have been used to grasp the ratio between the maximum response deformations of the one-mass elastic–perfectly plastic system and the maximum response deformation in the elastic system. On the base of their studies, Veletsos and Newmark proposed two approaches for relating the elasto-plastic spectrum to that of the corresponding elastic system.

The first suggested approach, considering the maximum deformations to be the same in the two systems, leads to a shear force coefficient for the elasto-plastic system which is obtained by dividing the corresponding value of the elastic system by the value of the ductility factor. This method, which is nowadays considered suitable for small magnitude of damping and for $T > 0.5$ s, is often remembered as the theory of ductility factor [11].

The second approach, useful for structures having natural period $T < 0.5$ s, makes an estimate of the value of q assuming an equivalence in strain energy between the elastic and the elastic–perfectly plastic system. The inelastic spectrum is therefore reduced, with respect to the elastic one, by the ratio $1/(2\mu-1)^{1/2}$ [11].

Penzien derived a similar conclusion through the response analysis of the multi-mass elastic plastic system [12].

The above studies are the starting point of the well-known Newmark and Hall method (1973) [13] (Section 4.2.3).

In recent times Mahin and Bertero [14] have shown that the scatter in peak response of systems designed according to the Newmark and Hall method is so large that it cannot be considered reliable in order to limit the ductility demands to specified values, even for elastic–perfectly plastic simple degree of freedom systems.

In more recent times, the study of the seismic inelastic response of SDOF systems has been faced by many researchers with different purposes [6, 7, 15–22]. Some have provided more detailed formulations relating the q-factor of the SDOF system to the available ductility and to the period of vibration [15, 17, 18, 21]. Also, the influence of second-order effects has been investigated [18, 19, 20, 22]. Other researchers [23, 24] have investigated the influence of the parameters adopted for damage characterization pointing out that the low-cycle fatigue parameter provides intermediate results between the kinematic ductility criterion and the energy criterion based on the total hysteretic ductility, which leads to the most severe results.

In many cases the Park and Ang parameter and the low-cycle fatigue parameter lead to similar conclusions.

The influence of different hysteresis models has also been investigated [6, 7]. Another interesting result is represented by the study, from the probabilistic point of view, of the distribution of the plastic excursions which provides some indications on the features of the loading histories to which specimen should be subjected in experimental tests in order to simulate the actual conditions under earthquake action [25, 26].

The purpose of this chapter is to critically review and compare some recent methods for evaluating the q-factor in steel-framed structures. As the use of the different damage criteria in evaluating the q-factor does not yet have a consolidated background, reference will be made in the following mainly to the kinematic ductility criterion and to the energy approach.

4.2   SIMPLIFIED METHODS

4.2.1 General

The existing methods for evaluating the q-factor can be grouped into three categories:

• methods based on ductility factor theory;
• methods based on extension of the results concerning the dynamic inelastic response of simple degree of freedom systems;
• methods based on the energy approach.

In methods of the first group, the theory of ductility factor (reliable for steel structures being generally $T > 0.5$ s) is used to establish relations between the q-factor and some parameters characterizing the post-elastic behaviour of steel frames which are easier to determine [27–30]. In other cases it is used to interpret the results of inelastic dynamic analyses [31, 32]. In the first case, i.e. for relations obtained from simplified models, the hypothesis of the so-called 'structural regularity' is required (Chapter 8), excluding closely spaced modes, torsional effects and cases in which the structure cannot be modelled two-dimensionally. These relations are also deduced for a global-type mechanism, leaving to a second phase the analysis of effects due to undesiderable collapse mechanisms.

Methods belonging to the second group, starting from the dynamic behaviour of inelastic SDOF systems, require hypotheses of structural regularity and global collapse mechanisms.

Although the research on inelastic spectra is strictly applicable to SDOF systems only, its use has been extended to multidegree of freedom systems. The critical point of this extension derives from the number of parameters which are necessary to characterize the pattern of yieldings of multistorey frames. In fact, while a single parameter, such as the required global ductility, can be sufficient to characterize SDOF response, in case of MDOF systems different patterns of yieldings could correspond to the same maximum inelastic displacement. Furthermore, the period of vibration and participation factor of the 'fundamental mode' are increased due to inelastic deformations, as shown by Veletsos and Vann [33]. Moreover, when the structure becomes inelastic, axial forces in exterior columns can exceed the values predicted by modal analysis and inelastic spectra leading to a reduction of plastic moment capacities and an increase of required ductility [34]. Inelastic spectra of SDOF systems cannot take into account such effects.

It is useful to note that the validity range of the theory of ductility factor has been obtained from the analysis of SDOF systems [15]. The limitations explained above generally apply also to the methods of the first group.

The third group of methods are the most general ones. These methods, which are an extension of Housner's methodology (1956) [35, 36], do not require structural regularity and global mechanism hypotheses.

It is useful to remember that, in the methodology used by Veletsos and Newmark, the equivalence of energy is only an aid to accessing the maximum deformation response; the quantitative evaluation of the input energy has been beyond the scope of the method.

Many researchers [1–4, 25, 37] have pointed out that a design based only on displacement ductility neglects the number of yield excursions and reversals, which may be useful in understanding the amount of damage undergone by structures during an earthquake motion.

4.2.2 Structural characterization

Under the assumption that the first mode of vibration is the most significant, and therefore it is possible to assume a pattern of lateral forces depending on a single parameter, the behaviour of framed structures can be characterized by the α–δ curve, where δ is the displacement of the upper storey (Fig. 4.1).

As discussed in Chapter 2, the parameters affecting the structural behaviour can be summarized as follows:

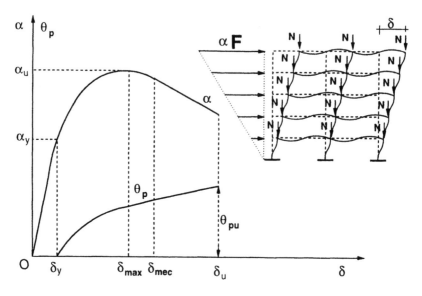

Fig. 4.1   Multiplier of horizontal forces versus displacement relationship

- the plastic redistribution capacity of the structure expressed by means of the ratio α_u/α_y;
- the global ductility $\mu = \delta_u/\delta_y$;
- the slope γ of the softening branch of the α–δ curve;
- the collapse mechanism type;
- the local ductility.

Characterization of dynamic behaviour also requires the period of vibration T, while damping is usually considered constant and equal to 5% of critical value.

4.2.3 Methods based on the ductility factor theory

The method proposed in references [27–30] derives from the need to take into account, in steel framed structures, the P–Δ effect due to vertical loads.

It is based on the global-type mechanism hypothesis. Assuming the theory of ductility factor as valid, the following relationship applies

$$q = \frac{\delta_u}{\delta_y} = \alpha_c \left(\frac{\alpha_u}{\alpha_y} - \beta \right) + \beta \qquad (4.1)$$

obtained from the analysis of the simplified model shown in Fig. 4.2, where α_c is the critical elastic multiplier for vertical loads.

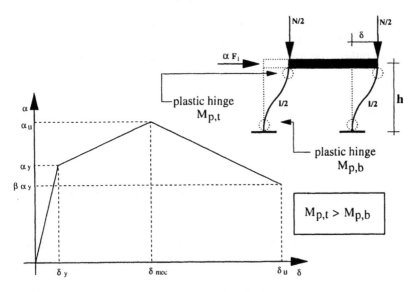

Fig. 4.2 Simplified model with plastic redistribution capacity

In equation (4.1) β depends upon the period of vibration. It is a dynamic equivalence coefficient which defines, by means of the fraction of first-yielding static load, the equivalence between the static elastic–plastic model and the dynamic model. Let

$$\beta = \beta' \, \frac{\alpha_u}{\alpha_y} \qquad (4.2)$$

equation (4.1) can be rewritten as

$$q = \frac{\alpha_u}{\alpha_y} \, [\, (1 - \beta') \, \alpha_c + \beta' \,] \qquad (4.3)$$

The coefficient β' can be derived using results, available in the technical literature, on the dynamic behaviour of SDOF systems [38].

Starting from the analysis of these results the authors suggest the following relationship

$$\beta' = 1 - T \qquad \text{with} \qquad \beta' > 0.5 \qquad (4.4)$$

Therefore, the q-factor assumes a constant value for $T > 0.5$ s, while it is a linear function of T in the opposite case.

Equation (4.3) has been derived by neglecting any limitation which can arise from local ductility. Therefore the value resulting from equation

(4.3) represents only an upper bound. The actual value must also take into account the rotational capacity which can be provided by members. In this context the authors divide members into three classes by means of their local slenderness coefficient $(b/t)(f_y/235)^{1/2}$ (f_y is expressed in N/mm^2), in analogy with Eurocode 3 [39].

The limitations derived from local ductility are fixed as follows:

- plastic or ductile sections – $q \leq 6$
- compact sections – $q \leq 4$
- semicompact sections – $q \leq 2$

This approach has been used in ECCS Recommendations and in Eurocode 8.

A different approach is the one proposed by Ballio and Setti [31, 32]. Their method is based on the use of ductility factor theory as a tool to interpret the results of dynamic inelastic analyses.

The main feature of the method is that a knowledge of the rotational capacity of members is not requested, because a collapse definition at 'member' level is not required for the application of the method.

The theory of ductility factor states that the q-factor is coincident with the global ductility, provided that there are no limitations due to local ductility.

The shear design seismic force is given by

$$F_d = \frac{a_u R(T) M}{q} \tag{4.5}$$

so that the q-factor is the ratio between the ground acceleration leading to collapse a_u and the design value $a_d = a_u/q$.

If the design force F_d corresponds to the first yielding ($F_d = F_y$) then the structure will be able to resist an acceleration q times greater than the design value, if and only if the available global ductility is greater than q.

If the inelastic response is lower than the indefinitely elastic one, the design based on an elastic spectrum will be on the safe side. In the opposite case the design will be on the unsafe side. The greatest value which can be assigned to the q-factor can be derived by means of a series of dynamic inelastic analyses, in which the peak ground acceleration is increased step by step. At each step the ratios a/a_d and δ/δ_d must be computed, where a_d and δ_d are, respectively, the design acceleration and the corresponding maximum displacement (evaluated by means of a first-order elastic analysis).

The bisectrix of the axes δ/δ_d and a/a_d represents the indefinitely elastic response. The design based on the elastic spectrum is valid until δ/δ_d is less than a/a_d; therefore the maximum value which can be assigned to the q-factor is given by the intersection between the curve δ/δ_d–a/a_d coming from the dynamic inelastic analyses and the bisectrix (Fig. 4.3).

Sedlacek and Kuck [40] observe that if the gradient of the curve δ/δ_d–a/a_d in the range of dynamic instability is high, the procedure results in realistic values of the q-factor. In other cases, where the influence of the second-order effects is lower and hence the gradient of the curve is not so high in the range of dynamic instability, the resulting values of the q-factor are conservative and a modified definition of the q-factor seems to be useful. Hereby the q-factor can be defined through the intersection point of the k-dependent linear curves with the non-linear curve (Fig. 4.4). Different values for the q-factor can therefore be determined according to the chosen k-value.

If high values for k are selected, the point of dynamic instability is a good approximation. Conversely, the deflections depending on it are high ($\delta_{max} = q \ k \ \delta_d$).

4.2.4 Methods based on the response of SDOF systems

Among the methods based on the response of inelastic SDOF systems, that of Newmark and Hall deserves pride of place because it is the first and best known method of this group and its background has already

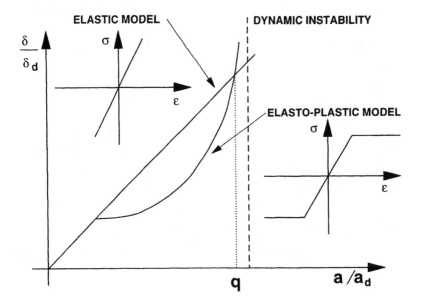

Fig. 4.3 Ballio and Setti method

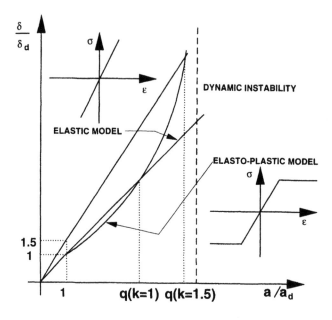

Fig. 4.4 Modification proposed by Sedlacek to Ballio's method

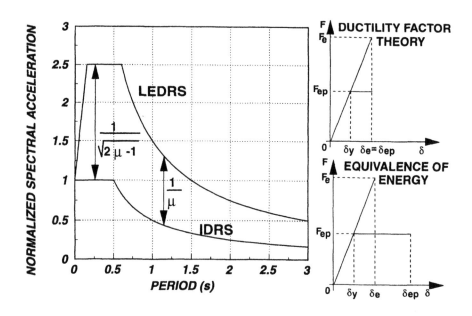

Fig. 4.5 Newmark and Hall method

been recalled in the introduction. Figure 4.5 shows how to obtain the inelastic design response spectrum (IDRS) starting from a linear elastic design response spectrum (LEDRS) [13]. Due to the fact that steel structures very often have a period T higher than 0.5 s, their q-factor coincides with the global ductility.

On the basis of the analysis of the seismic inelastic response of SDOF systems subjected to different ground motions, other researchers have proposed more complete relations expressing the q-factor of the SDOF model as a function of the period and of the available ductility.

By using ductility response spectra of stiffness degrading SDOF systems under several ground motions, Giuffré and Giannini [15] obtained the following approximated formula relating the q-factor to the global ductility μ and to the period of vibration T:

$$q = 1 + (\mu - 1)^{0.87 - 0.05\,T} \qquad \text{for } T > T_o \qquad (4.6)$$

$$q = 1 + 2\,(q_o - 1)\frac{T}{T_o}\left(1 - \frac{T}{2\,T_o}\right) \qquad \text{for } T < T_o \qquad (4.7)$$

where T_o is the value of the period corresponding to the beginning of the softening branch of the elastic response spectrum and q_o the value provided by equation (4.6) for $T = T_o$.

By means of a similar procedure, Palazzo and Fraternali [17, 18] have obtained an approximated formula, which considers also the P–Δ effect due to vertical loads (Chapter 5).

The analysed model is a stiffness degrading SDOF system. Stiffness degradation is a function of the displacement and of the energy dissipated in the previous cycles. The P–Δ effect is introduced by using the γ parameter which represents the slope of the softening branch of the force versus displacement law (Fig. 4.6).

On the basis of the results of numerical analyses, the authors propose values of the q-factor linearly varying from 1 to q_1 in the period range between 0 and T_1, being

$$q_1 = 1 + \left[0.5 + \left(\frac{T_1}{T_o} - 0.5\right)(2\,\mu^*\gamma - \mu^{*2}\,\gamma^2)\right](q_o - 1) \qquad (4.8)$$

where T_1 is the period value corresponding to the beginning of the plateau of the elastic design response spectrum and T_o the period corresponding to the beginning of the softening branch.

In the range between T_o and T_1, the q-factor varies from q_1 to q_o, where

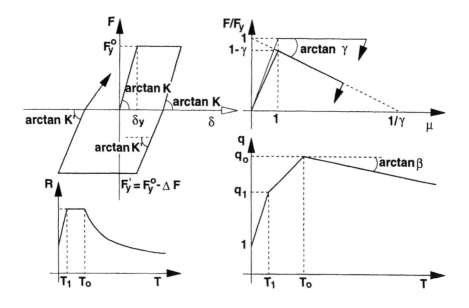

Fig. 4.6   Restoring forces versus displacement and q-factor versus period
relationships according to Palazzo and Fraternali

$$q_0 = \mu^{*\,[\,1 - (\mu\, +\, 100\,\gamma\,)/40\,]} \qquad (4.9)$$

Finally for $T \geq T_o$, the q-factor is constantly equal to q_0 if $\gamma > 0.10$. For $\gamma <$
0.10 its value linearly drops with a slope given by

$$\beta = [(\mu^* - 1)/20]\,(1 - 10\gamma) \qquad (4.10)$$

The parameter μ^* of equations (4.8), (4.9) and (4.10) is given by

$$\mu^* = \mu \quad \text{if} \quad \mu \leq \frac{1}{\gamma} \quad \text{or} \quad \mu^* = \frac{1}{\gamma} \quad \text{if} \quad \mu > \frac{1}{\gamma} \qquad (4.11)$$

A different relation, which is characterized by a mathematical structure having some interesting properties, is that proposed by Krawinkler
and Nassar [21] for the elastic strain hardening SDOF model

$$q = [\,c\,(\mu - 1\,) + 1\,]^{1/c} \qquad (4.12)$$

where

$$c\,(T,\bar{K}_h) = \frac{T^a}{1 + T^a} + \frac{b}{T} \tag{4.13}$$

Here $\bar{K}_h$ is the ratio between the stiffness of the hardening branch and the initial stiffness, while a and b are two numerical parameters which have been derived by non-linear regression analysis. Their values are given in Table 4.1 for different values of $\bar{K}_h$.

Table 4.1 Parameters for the Krawinkler and Nassar formulation

$\bar{K}_h$	a	b
0.00	1.00	0.42
0.02	1.00	0.37
0.10	0.80	0.29

The mathematical structure of equations (4.12) and (4.13) has some fundamental properties. First of all, for $\mu = 1$ equation (4.12) provides $q = 1$ for any given value of T according to the elastic behaviour. Moreover, for $T = 0$ equation (4.13) states that $c \to \infty$ and equation (4.12) again provides $q = 1$ according to the behaviour of infinitely rigid structures. Finally, for $T \to \infty$ equation (4.13) provides $c = 1$ and equation (4.12) states that $q = \mu$, representing the behaviour of infinitely flexible structures.

The values of q given by the above relationships are mean values. Moreover, they have been obtained from the analysis of the SDOF bilinear system subjected to ground motion records from sites corresponding to rock or stiff soils. According to the authors [21], as the average q-factor was found to be insensitive to relatively small variations in average response spectra shapes, the above relationships can probably be used for different soil conditions as well, excluding the case of motion in soft soils, which are affected by the site-soil column.

Figure 4.7 shows the comparison among the above formulations. The representation of the formulation of Krawinkler and Nassar is referred to the elastic–perfectly plastic model ($\bar{K}_h = 0$). The values of the q-factor corresponding to this case are greater than those provided by the other authors due to the fact that the energy dissipation capacity of the bilinear model is, for a given value of the available ductility μ, greater than for the stiffness degrading model.

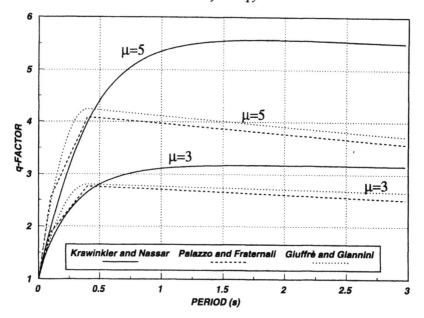

Fig. 4.7 Comparison between the different fromulations for SDOF systems

Moreover, it can be seen from Fig. 4.7 that the theory of the ductility factor ($q \approx \mu$) represents, for $T \geq 0.8$ s, a cautelative assumption in the case of bilinear SDOF systems.

The extension of the previous results, obtained from the inelastic response of SDOF systems to multistorey structures, requires the use of an equivalent SDOF system, which can be derived starting from the α–δ curve of the frame. The hypotheses of structural regularity and global collapse mechanism are also required. Torsional effects and, therefore, cases in which the structure cannot be two-dimensionally modelled are excluded.

An attempt to rationally combine the background regarding the seismic inelastic response of SDOF systems with the knowledge on the static inelastic response of steel framed structures has been made in [41]. The method is based on the identification of the structure through an SDOF model, whose parameters are defined by means of the behavioural curve, $\alpha - \delta$, of the structure. Regularity conditions are still required, so that the structure can be two-dimensionally modelled.

Considering a constant value for viscous damping, usually 5%, the q-factor is given by

$$q = q\left(\mu, T, \gamma, \frac{\alpha_u}{\alpha_y}\right) \qquad (4.14)$$

The method is based on the following steps.

a. Evaluation of rotational capacity R of members: this can be achieved through existing results [42–46] (Chapter 3).
b. Evaluation of global ductility by means of the approximated relationship

$$\mu = 1 + \theta_1 \left[R - 2 \left(\frac{\alpha_u}{\alpha_y} - 1 \right) \right] \qquad (4.15)$$

where θ_1 is a coefficient depending on the ratio between column and beam stiffness and on the number of stories; its average value is 2/3 [30, 41] (Chapter 2).
c. Computation of the structural coefficient q' of the SDOF system having the same period and the same ductility of the actual structure. This value can be provided, neglecting geometrical non-linearity, by the formulations previously presented.
d. Evaluation of the reduction factor φ which takes into account the P–Δ effect due to vertical loads; this coefficient can be obtained by means of known results [19], based on a large statistical sample (discussed in Chapter 5).

From the above steps, the q-factor is finally given by

$$q = \frac{\alpha_u}{\alpha_y} q' \frac{\left(\mu, T, \gamma = 0, \frac{\alpha_u}{\alpha_y} = 1 \right)}{\varphi(\gamma, \mu, T)} \qquad (4.16)$$

Equation (4.16) states that the value of q for $\gamma = 0$ and $\alpha_u/\alpha_y = 1$ must be amplified by the overstrength coefficient α_u/α_y, which takes into account the redistribution capacity of the structure, and has to be reduced through the coefficient $\varphi \geq 1$ due to P–Δ effects.

The values of γ and R must be modified in order to take into account the type of collapse mechanism, which is not usually of the gobal type [30, 41] (Chapter 2).

The evaluation of μ and γ can be conveniently done by using the results of static inelastic analyses, which are in any case required to compute α_u/α_y. In this way the value of the available global ductility is more accurate than that given by equation (4.15). As an alternative, the 'mechanism curve method' described in Chapter 2 can be adopted for evaluating the available ductility.

4.2.5 Energy methods

A simplified energy approach has been proposed by Como and Lanni [47, 48]. The method is based on a simplified model of the energy exchanges occurring during the earthquake. The complex evolution of the seismic motion of a structure is divided into a series of simplified cycles of energy exchanges. Each cycle is made up of a first phase of kinetic energy storing, during which there is a gradual increasing of the elastic oscillations and a second phase in which the energy cumulated in the first phase is transformed into elastic–plastic work. In this energy dissipation phase the kinetic energy which the ground continues to transmit to the structure is neglected.

The kinetic energy given by the design earthquake, with maximum ground acceleration a and response spectrum S, is called E_k. The accelerogram of the severest design earthquake is characterized by the multiplier m of the design accelerogram. The kinetic energy E_{km} given by this destructive earthquake, which can be obtained using the maximum pseudo-velocity evaluated on the spectrum mS, is related to E_k by the relationship

$$E_{km} = m^2 \, E_k \qquad (4.17)$$

The equation of energy balance gives

$$E_{ku} = W_o + D_u - E_{2u} \qquad (4.18)$$

where

- W_o is the elastic energy stored in the first stage;
- D_u is the energy dissipated through inelastic deformations;
- E_{2u} is the work done by vertical loads during the whole process of deformation;
- E_{ku} is the maximum kinetic energy which the structure is able to withstand and to dissipate.

If E_{ku} is greater than E_{km} the motion of the structure is inverted by means of the elastic restoring forces and other cycles of deformation can occur.

The maximum level m_{lim} of intensity of the earthquake which the structure can sustain is

$$E_{k,m_{\text{lim}}} = E_{ku} \qquad (4.19)$$

which gives

$$m_{\text{lim}}^2 T = W_o + D_u - E_{2u} \qquad (4.20)$$

If E_k is the kinetic energy related to the earthquake under which the structure has to remain in elastic range, the following is true

$$E_k = W \qquad (4.21)$$

where W is the elastic energy stored during this kind of earthquake.

Under destructive earthquakes, the inertia forces increase from zero up to the limit value $\alpha_o F$, which produces the first yielding and, under the assumption of contemporary formation of all the plastic hinges, the attainment of the collapse mechanism. From this hypothesis,

$$W_o = \alpha_o W \qquad (4.22)$$

Therefore, from equations (4.20), (4.21) and (4.22),

$$m_{\text{lim}} = \left[\alpha_o \left(1 + \frac{D_u}{W_o} - \frac{E_{2u}}{W_o} \right) \right]^{1/2} \qquad (4.23)$$

The authors define 'seismic ductility' and 'seismic slenderness', respectively, by the following ratios:

$$D_s = \frac{D_u}{W_o} \qquad (4.24)$$

$$\Lambda_s = \frac{E_{2u}}{W_o} \qquad (4.25)$$

Their difference,

$$D_{sr} = D_s - \Lambda_s \qquad (4.26)$$

is defined as the 'reduced seismic ductility'.

Therefore, the capability of a structure to resist strong earthquakes depends on a property which is a combination of resistance, by means of α_o, and ductility, by means of D_{sr}. This property, defined by the authors as 'seismic toughness' τ_s, is

$$\tau_s = \sqrt{\alpha_0 \, (1 + D_{sr})} \tag{4.27}$$

This parameter has the same meaning as the q-factor ($q = \tau_s$), because it represents a reduction factor of the accelerogram of the severest design earthquake due to the seismic toughness of the structure.

It is easy to see that, by removing the assumption of contemporary formation of all the plastic hinges and by assuming conservatively that equation (4.17) is still valid, equation (4.27) can be written as

$$q = \left(\frac{W_u}{W_y}\right)^{1/2} \tag{4.28}$$

Here W_y is the elastic strain energy stored by the system in the yield state and W_u is the total energy stored and dissipated up to failure.

From the practical point of view, W_y and W_u can be approximately evaluated by considering the work done by a system of equivalent horizontal forces, statically applied and distributed according to a combination of a selected number of vibration modes. Therefore, from the computational point of view, a static inelastic analysis providing the behavioural curve α–δ is required when evaluating the q-factor given by equation (4.28).

A simplified method, which attempts to account for the damage distribution, is the energy method proposed by Kato and Akiyama [1, 49–53].

This ultimate strength design method for steel-framed buildings, subjected to severe earthquakes, represents the background of the Japanese seismic code. It is based on the energy concept; in fact the safety of a structure against the design major earthquake is assessed by comparing the structure energy dissipating capacity with the earthquake input energy into the structure.

The procedure for evaluating the earthquake-resisting capacity of a steel structure must be applied at each storey. The method is valid for a shear-type system, such as 'weak-column strong-beam frames'.

Although strong-column weak-beam frames are seismically more advantageous, because the damage concentration into a storey can be avoided, it is important to note that most buildings are classified as weak-column structures in which large amounts of damage arise from the input energy concentration into a relatively weak storey. However, it is easy to find sound reasons with which to justify the presence of weak-column structures.

• As the concrete floor slab contributes to the resistance of beams, the effective strength of the beams is higher than the design value.

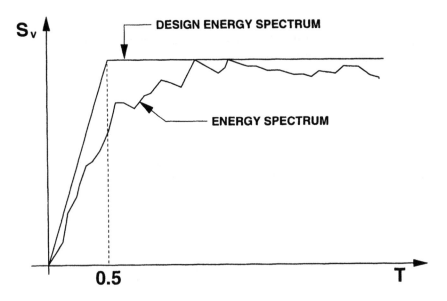

Fig. 4.8 Design energy spectrum for the Kato and Akiyama method

- Long-span girders are sized on the basis of vertical loads rather than of earthquake actions.
- Simple design rules which exclude the possibility of column yielding have not yet been developed.

According to the Japanese approach, weak-column structures can be therefore identified as the most common, and the shear-type model can be generally accepted for a safe design.

A quantitative evaluation of the total energy input from an earthquake was made by Housner [35, 36] using the velocity response spectrum and assuming that the energy input responsible for the damage of an elastic–plastic system is the same as that producing damage in the elastic system. This energy can be obtained as

$$E = \frac{1}{2} M S_v^2 \tag{4.29}$$

where M is the total mass of the structure and S_v is the spectral pseudo-velocity response for a damped system.

The average pseudo-velocity spectrum can be approximated as shown in Fig. 4.8. For values of the fundamental period $T \geq 0.5$ s, S_v is independent of T. This property is very important in the evaluation of the earthquake energy input into a structure considering that the fundamental period changes during the evolution of plastic deformations.

The energy balance provides

$$\frac{1}{2} M S_v^2 = W_e + W_p \tag{4.30}$$

where W_e is the elastic strain energy and W_p is the cumulative plastic strain energy.

The survival of a structure against the design earthquake requires that the structure capacity of cumulative plastic energy dissipation W_{ps} is not less than cumulative plastic energy response W_p

$$W_{ps} \geq W_p = (\frac{1}{2})MS_v^2 - W_e \tag{4.31}$$

The elastic strain energy W_e is given approximately by

$$W_e = \frac{1}{2} M \left(\frac{T}{2\pi} \alpha_1 g \right)^2 \tag{4.32}$$

where α_1 is the yield base shear coefficient defined as

$$\alpha_1 = \frac{Q_y}{M g} \tag{4.33}$$

and Q_y is the yield base shear force.

The first step of the check procedure requires equation (4.31) to be satisfied with respect to the first storey; then, the safety check can be extended to each storey using the plastic work distribution law obtained by the same authors.

The hysteretic energy dissipation of a steel frame in one direction can be assumed equal to the plastic energy absorption under monotonic loading with a displacement amplitude equal to the cumulative plastic deformation. Therefore, the plastic work of the ith storey is given by (Fig. 4.9)

$$W_{pi} = 2 Q_{yi} (\delta_{ui} - \delta_{yi}) \tag{4.34}$$

and, by accounting for all plastic excursions in both positive and negative ranges by means of the cumulated ductility ratio, i.e.

$$\eta_i^c = \sum \frac{\delta_{p_i}}{\delta_{y_i}} \tag{4.35}$$

equation (4.34) gives

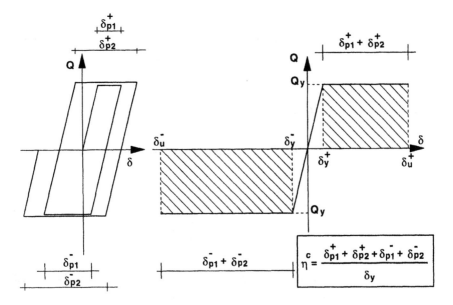

Fig. 4.9 Cumulated ductility ratio

$$W_{pi} = Q_{yi}\,\delta_{yi}\,\eta_i^c = \frac{Q_{yi}^2}{K_i}\,\eta_i^c \tag{4.36}$$

Let K be the equivalent stiffness of the frame

$$K = M\,\frac{4\,\pi^2}{T^2}$$

$$\tag{4.37}$$

Then, by defining $\chi_i = K_i/K$,

$$K_i = \chi_i\,M\,\frac{4\,\pi^2}{T^2} \tag{4.38}$$

and equation (4.36) becomes

$$W_{pi} = \frac{\alpha_i^2 g^2\,\eta_i^c\,T^2}{\chi_i\,M\,4\,\pi^2}\left(\sum_{j=i}^{n_s} m_j\right)^2 \tag{4.39}$$

where

$$\alpha_i = \frac{Q_{yi}}{\displaystyle\sum_{j=i}^{n_s} m_j \, g} \tag{4.40}$$

is the yield shear coefficient of the ith storey, with n_s being the number of storeys.

Equation (4.39) can also be rewritten in non-dimensional form

$$A_{pi} = \frac{W_{pi}}{\dfrac{M g^2 T^2}{4\pi^2}} = \frac{\alpha_i^2 \, \eta_i^c}{\chi_i} \left(\sum_{j=i}^{n_s} \frac{m_j}{M} \right)^2 \tag{4.41}$$

The distribution of plastic work W_{pi} is affected by the distribution of the yield shear coefficients α_i. The distribution of α_i, which produces a uniform distribution of cumulated plastic ductility $\eta_i = \eta_0$ (common response), is defined as optimum distribution. The optimum lateral shear coefficient distribution was found from of a series of parametric studies [1, 50, 52] which gave

$$\frac{\alpha_{opt,i}}{\alpha_1} = 1 + 1.593X - 11.85X^2 + 42.6X^3 - 59.5X^4 + 30X^5 \tag{4.42}$$

in the case of shear-type structures, and

$$\alpha'_{opt,\,i} / \alpha_1 = \alpha_{opt,\,i} / \alpha_1 + 1.3\, X^4 \tag{4.43}$$

in the case of flexural shear-type structures, where $X = x_i/H$, x_i being the height of the floor of relevant storey and H the total height of the building. This gives $X = (i-1)/n_s$ if all storeys have the same height.

If η_0 is the common response at each storey, the optimum distribution of plastic work is given, in non-dimensional form, by

$$A_{p,opt\,i} = \left(\sum_{j=i}^{n_s} \frac{m_j}{M} \right)^2 \frac{\alpha_{opt,i}^2 \, \eta_0}{\chi_i} \tag{4.44}$$

therefore, at the first storey,

$$A_{p,opt\,1} = \frac{\alpha_1^2 \, \eta_0}{\chi_1} \tag{4.45}$$

When the shear coefficient distribution differs from the optimal value, the structure is poorly sized and the input energy will concentrate into a particular storey.

The unavoidable scatter between the optimum and the actual yield shear coefficient distribution can be represented by the coefficient

$$p_i = \frac{\alpha_i}{\alpha_{opt,\,i}} \tag{4.46}$$

The following empirical relation for damage distribution has been found [1, 49, 51]:

$$\frac{A_{p,i}}{A_{p,1}} = \frac{W_{p,i}}{W_{p,1}} = \frac{A_{p,opt\,i}}{A_{p,opt\,1}}\, p_i^{-\xi} \tag{4.47}$$

Using equations (4.44) and (4.45), this gives

$$s_i = \frac{A_{p,opti}}{A_{p,opt\,1}} = \left(\sum_{j=i}^{n_s} \frac{m_j}{M} \right)^2 \left(\frac{\alpha_{opt,\,i}}{\alpha_1} \right)^2 \frac{K_1}{K_i} \tag{4.48}$$

Equation (4.47) provides, in non-dimensional form, the plastic work at the *i*th storey as a function of mass, stiffness and yield shear coefficient distributions starting from the plastic work at the first storey

$$A_{p,i} = s_i\, p_i^{-\xi}\, A_{p,1} \tag{4.49}$$

The non-dimensional plastic work due to the whole structure can be evaluated by

$$A_p = A_{p,1} \sum_{i=1}^{n_s} s_i\, p_i^{-\xi} \tag{4.50}$$

From equation (4.50)

$$a_1 = \frac{A_p}{A_{p,1}} = \frac{W_p}{W_{p,1}} = \sum_{i=1}^{n_s} s_i\, p_i^{-\xi} \tag{4.51}$$

where a_1 is the ratio of the plastic work done by the whole structure to that done by the first storey.

The exponent ξ of equations (4.47) and (4.51) is called the 'damage concentration index'. It assumes the value $\xi = 12$ in the case of 'shear-type' frames (i.e. weak-column strong-beam) and the value $\xi = 6$ for strong-column weak-beam frames (Chapter 2).

The plastic energy dissipation capacity of the first storey is given by

$$W_{pls} = 2Q_{y1}\delta_{y1}\eta_1 = 4W_{el}\eta_1 \tag{4.52}$$

where W_{e1} is the elastic strain energy of the first storey and

$$\eta_1 = \frac{(\delta_{u1} - \delta_{y1})}{\delta_{y1}} \tag{4.53}$$

is the cumulated plastic ductility ratio in one direction which can be related [1] to the available rotation capacity through the relationship $\eta_i = (2/3) R_i + 2$ (the index i refers to the ith storey and R_i the rotation capacity of the ith storey members).

It is easy to show that

$$\frac{W_{e1}}{W_e} = \frac{\delta_{y1}}{\delta_{eq}} = \frac{K}{K_1} = \frac{1}{\chi_1} = c_1 \tag{4.54}$$

and, therefore,

$$W_{pls} = 4W_e\, c_1\eta_1 \tag{4.55}$$

Using equations (4.32), (4.51), (4.54), (4.55), the condition (4.31) for the survival of the structure becomes

$$\left(\frac{T}{2\pi}\right)^2 (\alpha_1 g)^2 (1 + 4\, c_1\, a_1\, \eta_1) \geq S_v^2 \tag{4.56}$$

The spectral pseudo-velocity response S_v is related to the spectral acceleration response S_a by

$$S_v = \frac{T}{2\pi} S_a \tag{4.57}$$

thus, equation (4.56) becomes

$$\alpha_1 g \geq \frac{S_a}{\sqrt{1 + 4\, c_1\, a_1\, \eta_1}} \tag{4.58}$$

or, more conveniently,

$$Q_{y1} \geq \frac{1}{\sqrt{1 + 4\, c_1\, a_1\, \eta_1}}\, S_a M \qquad (4.59)$$

in which $Q_{y1} = \alpha_1 g M$ is the required yield base shear strength and $S_a M$ is the elastic maximum shear force corresponding to the spectral acceleration response S_a.

The term

$$D_s = \frac{1}{\sqrt{1 + 4\, c_1\, a_1\, \eta_1}} \qquad (4.60)$$

represents the structural ductility characteristic of the frame and is independent of the earthquake intensity.

D_s corresponds to the structural factor q used in European seismic codes by means of the relationship

$$q = \frac{1}{D_s} = \sqrt{1 + 4\, c_1\, a_1\, \eta_1} \qquad (4.61)$$

The value of q given by equation (4.61) can be interpreted as a measure of the safety of the structure considered as a whole. It can be adopted if and only if the required cumulated plastic ductility ratio at each storey is less than the available one. In other words, it has to be verified that

$$W_{pi} \leq W_{pis} \qquad (4.62)$$

where W_{pi} is the plastic work at the ith storey and W_{pis} is the plastic dissipation capacity of the structure at the ith storey.

It is, therefore, important to show how this approach can be used to evaluate the design q-factor, so that it can be adopted to resize a structure having a weak storey other than the first storey.

By combining equations (4.47), (4.48) and (4.50), the **damage partition coefficients** are obtained as

$$\rho_i = \frac{1}{a_i} = \frac{W_{p,i}}{W_p} = \frac{s_i\, \bar{p_i}^{-\xi}}{\displaystyle\sum_{i=1}^{n_s} s_i\, \bar{p_i}^{-\xi}} \qquad (4.63)$$

and, from equation (4.39),

$$W_p = a_i \; \frac{g^2 T^2}{4 \pi^2} \; \frac{\alpha_i^2 \; \eta_i^c}{\chi_i} \; \frac{\left(\displaystyle\sum_{j=i}^{n_s} m_j \right)^2}{M} \tag{4.64}$$

The following position is deduced:

$$c_i = \frac{1}{\chi_i} \left(\sum_{j=i}^{n_s} \frac{m_j}{M} \right)^2 \tag{4.65}$$

The introduction of the cumulated plastic ductility in one direction (positive and negative) $\eta_i = \eta_i^c / 2$, allows transformation of equation (4.64) to

$$W_p = a_i \; \frac{M g^2 T^2}{4 \pi^2} \; 2 c_i \; \alpha_i^2 \; \eta_i \tag{4.66}$$

Therefore, taking into account equations (4.66) and (4.32), the energy balance equation (4.30) can be written as:

$$\frac{M g^2 T^2}{4 \pi^2} \; \alpha_1^2 \left[\frac{1}{2} + 2 a_i c_i \eta_i \left(\frac{\alpha_i}{\alpha_1} \right)^2 \right] = \frac{1}{2} M \; S_v^2 \tag{4.67}$$

and, from equation (4.57),

$$\alpha_1^2 = \left[1 + 4 a_i c_i \eta_i \left(\frac{\alpha_i}{\alpha_1} \right)^2 \right] = \frac{S_a^2}{g^2} \tag{4.68}$$

therefore, taking in account equation (4.33), (4.68) becomes

$$Q_y = \frac{M S_a}{\sqrt{1 + 4 a_i c_i \eta_i \left(\dfrac{\alpha_i}{\alpha_1} \right)^2}} \tag{4.69}$$

Equation (4.69) shows that the lower bound of the q-factor arises from the limitation of the ith storey, as expressed by

$$q_i = \sqrt{1 + 4\,a_i\,c_i\,\eta_i \left(\frac{\alpha_i}{\alpha_1}\right)^2} \tag{4.70}$$

As a consequence, the value of the q-factor that satisfies for each storey the limitation provided by the previous condition is given by

$$q = \min(q_1, q_2, \ldots, q_N) \tag{4.71}$$

From the practical point of view it is important to emphasize that the energy method of Kato and Akiyama is the only one that does not require any elasto-plastic analysis. Other methods require static elasto-plastic analyses for evaluating the α–δ curve or complex dynamic elasto-plastic analyses. Contrary to other methods, it is also applicable to irregular frames.

The last, but not least, merit of the Kato and Akiyama method is the large amount of information provided. The importance of this feature is shown in a worked example [54].

4.3 LOW-CYCLE FATIGUE APPROACH

The simplified methods for evaluating the q-factor which have been examined so far are all based on an approximate characterization of the structural damage; the collapse conditions have been defined mainly through the maximum displacement criterion or, sometimes, through the energy criterion. It has already been shown that the first criterion gives a good approximation for deformation histories characterized by only one excursion with a large plastic engage. In this case the total damage is accumulated quite exclusively during the single cycle with large ductility demand and, therefore, it can be directly related with good approximation to the available kinematic ductility.

On the contrary, in cases of deformation histories characterized by many cycles with large plastic engage, the accumulated damage mainly depends on the dissipated energy. As a consequence, the energy criterion is preferred.

Since the deformation histories are not composed of regular cycles, the damage mechanism is generally governed both by the maximum plastic displacement and by the dissipated energy. A very promising approach to account for the actual process of damage cumulation is the fatigue approach [4], which is based on an extension of the classical fatigue theory in the low-cycle range and on the use of the linear damage cumulation assumption according to the Miner rule.

A first proposal for evaluating the q-factor of steel structures on the basis of the low-cycle fatigue has been developed by Ballio and Castiglioni [55].

These authors use the extension of the fatigue theory in low-cycle range, evaluating the cycle amplitude by means of the total displacement ·range. Therefore, the fatigue curve is expressed as

$$N_{rf} = \frac{1}{B} \left(\frac{\Delta v}{v_y} \right)^{-b} \tag{4.72}$$

According to the assumption of linear damage cumulation, the damage index I_D is defined as

$$I_D = \sum_{i=1}^{n_g} \frac{n_i}{N_{rf_i}} \tag{4.73}$$

where n_g is the number of groups of semi-cycles having the same amplitude and n_i is the number of semi-cycles of the ith group, while N_{rf_i} represents for the ith group the number of semi-cycles leading to failure evaluated through equation (4.72). Collapse conditions are attained when the damage index is $I_D \geq 1$.

The procedure for evaluating the q-factor, according to the low-cycle fatigue approach proposed by Ballio and Castiglioni [55], is based on the definition of the q-factor as the one that corresponds to equality of the damage index I_D, estimated with linear elastic and non-linear methods.

The procedure has been applied by the same authors to cantilever columns and the following description of the procedure is referred to this case.

From the computational point of view, the following steps have to be performed.

a. The acceleration records of the seismic motions to be used in the dynamic analyses have to be selected.

b. Dynamic linear and dynamic non-linear analyses of the cantilever column subjected to the ground motions previously specified have to be carried out. The time histories of the end displacements of the cantilever have to be evaluated.

c. Since the time histories do not consist of regular cycles, they have to be reprocessed in order to obtain an equivalent, from the damage point of view, time history made of different groups of semicycles having the same amplitude. This reprocessing of the time histories can be performed by the 'Rain Flow Counting Method' [56, 57].

d. For each time history the damage index has to be computed. Therefore, with reference to the kth acceleration record, the damage index is given by

$$I_{D_k} = \sum_{i=1}^{n_{g_k}} \frac{n_i}{N_{rf_i}} \qquad (4.74)$$

where n_{g_k} is the number of groups of semi-cycles, each having the same amplitude, obtained from the reprocessing of the kth time history.

Successively, the damage index of the cantilever column is obtained as the mean value

$$I_D = \sum_{k=1}^{n_a} \frac{I_{D_k}}{n_a} \qquad (4.75)$$

where n_a is the number of acceleration records considered in the dynamic analyses.

As this step, together with step c, has to be performed both for the time histories obtained from the linear dynamic analyses and the ones obtained from the non-linear dynamic analyses, two values of I_D are computed. The first, $I_D(DL)$, refers to results obtained from linear dynamic analyses and, the second, $I_D(DNL)$, refers to the non-linear analyses.

e. The steps b, c, and d have to be repeated for progressively increasing values of the peak ground acceleration of the ground motion records selected for the dynamic analyses. The analyses are repeated until the value of the peak ground acceleration a^* is obtained for which the equality $I_D(DL) = I_D(DNL)$ holds true.

The q-factor is evaluated as

$$q = \frac{a^*}{a_y} \qquad (4.76)$$

where a_y is the design value of the peak ground acceleration.

The authors [55] have applied the above procedure to a cantilever column, of HE220A shape, for which equation (4.72) has been experimentally evaluated. The corresponding results are given in Fig. 4.10. It can be noted that the above procedure has led to a value of q approxi-

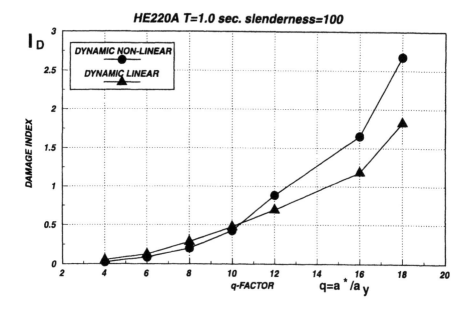

Fig. 4.10 Application of the Ballio and Castiglioni method to a cantilever column

mately equal to 10. As the failure condition $I_D(DNL) = 1$ is attained for a value of q approximately equal to 12.5, the described procedure has led to a conservative result.

The procedure proposed by Ballio and Castiglioni [55] represents a very interesting attempt to introduce the modern damage characterization through the low-cycle fatigue approach in the evaluation of the ultimate limit state of steel-framed structures. The method in its original formulation can be applied to single storey buildings as well as in the case of shear-type structures where the columns can be modelled as cantilevers under cyclic bending due to interstorey drifts.

A similar approach has been applied to cantilever columns by Calado and Azevedo [57].

Some general observations can be made with reference to the practical application of the low-cycle fatigue approach. First of all it requires great computational effort – in cases of actual complex structural schemes there are many dissipative zones where plastic hinges are likely to occur. For these generalized plastic hinges the time histories of the generalized plastic displacements should be evaluated, through a dynamic non-linear analysis, and reprocessed in order to compute, for each plastic hinge, the damage index $I_D(DNL)$. This requires that equation (4.72) should be ex-

pressed by means of the plastic amplitude of the semi-cycles, as previously suggested [4].

It is interesting to observe that, as demonstrated by numerical simulations [55], the fatigue curve is strongly influenced by the presence of
an axial load (Fig. 4.11). This means that for cases in which the critical
plastic hinge, i.e. the plastic hinge where the failure condition
$I_D(DNL) = 1$ is attained, is located into a column, the accuracy of the
q-factor can be quite illusory due to the continued variation of the axial
load in the columns, expecially the exterior ones, during the seismic motion. However, this limit is common to other methodologies. On the
contrary, very accurate predictions of the q-factor are obtained in cases
of frames failing in the global mode, in which the dissipative zones are
limited to the beam ends where axial load can be neglected.

4.4   APPLICATIONS

In order to investigate the reliability of the simplified methods for
evaluating the q-factor, a first comparison among the results obtained
from the application of some of the previously described methods has
been developed [58]. The analysed frames are shown in Fig. 4.12. All have
a uniform vertical load of 52 kN/m at each storey. All members are in

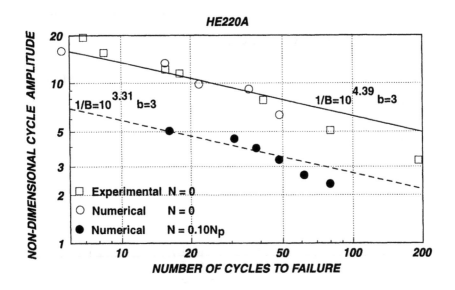

Fig. 4.11   Influence of the axial load on the fatigue curve

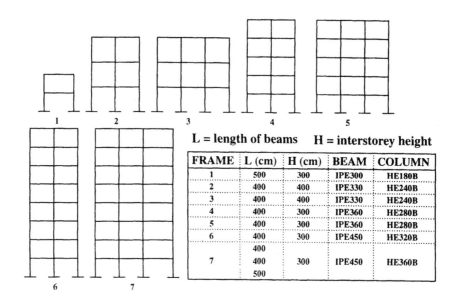

L = length of beams   H = interstorey height

FRAME	L (cm)	H (cm)	BEAM	COLUMN
1	500	300	IPE300	HE180B
2	400	400	IPE330	HE240B
3	400	400	IPE330	HE240B
4	400	300	IPE360	HE280B
5	400	300	IPE360	HE280B
6	400	300	IPE450	HE320B
	400			
7	400	300	IPE450	HE360B
	500			

Fig. 4.12   Analysed frames

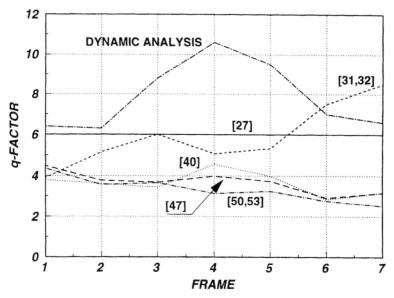

Fig. 4.13   Comparison between *q*-factor values computed by different simplified methods

Fe430 steel. In Fig. 4.13 values of the q-factor, computed by different simplified methods, are given and are identified by the corresponding reference. The highest curve corresponds to q-factor values evaluated by dynamic inelastic analyses, as the ratio between the peak ground acceleration leading to collapse a_u and the design one a_d. The occurrence of collapse has been assumed to correspond to the attainment of the ultimate plastic rotation at the end of a member.

Even if dynamic inelastic analyses were performed for a limited number of ground motions, an encouraging fact is that the use of simplified methods is generally cautelative. However, depending on the method of computation, a considerable scatter of q-factor values can be observed. It is clear that more analyses are necessary to judge the full reliability of all approximated methods.

Further investigations have been carried out with reference to a great number of frames, but focusing attention on only two methods: the energy method of Como and Lanni [47] (Section 4.2.5) and the method proposed by Cosenza *et al.* [41], which is based on the identification of the structure by means of an equivalent SDOF system (Section 4.2.4).

The number of bays in the analysed frames has been varied from 1 to 6, while the number of storeys has been varied from 2 to 8; thus, 42 different structural schemes have been obtained. In all cases, the span of the bays is equal to 4.50 m and the interstorey height is equal to 3.00 m. The uniform load acting on the beams is equal 32.6 kN/m. In addition,

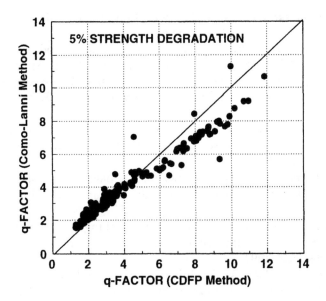

Fig. 4.14   Comparison between the Como-Lanni method and the CDFP method

five different design criteria have been applied. Four of these correspond to application of the Italian seismic code, ECCS Recommendations, Euro-code 8 and UBC88. The fifth design procedure corresponds to the method described in Chapter 2, which has assured in all cases a collapse mecha-nism of the global type. Therefore, 210 different frames have been designed.

For these frames, the value of the q-factor has been computed accord-ing to the two methods to be compared. The attainment of a top sway displacement corresponding to a 5% degradation of the load-carrying capacity has been assumed as the ultimate condition in both methods. The comparison between the q-factor values obtained through the two methods is presented in Fig. 4.14.

This comparison shows very good agreement between the analysed methods. This result is particularly interesting and encouraging, because it has been obtained with two methods starting from completely different theoretical bases.

Despite these satisfactory results, the problem of evaluating the q-fac-tor by means of simplified methods deserves further investigations. In particular, the correspondence with the results of the dynamic inelastic analysis has to be further verified.

4.5   REFERENCES

[1] *H. Akiyama:* 'Earthquake-Resistant Limit State Design for Buildings', University of Tokyo Press, 1985.
[2] *Y.J. Park, A.H.S. Ang:* 'Mechanic Seismic Damage Model for Reinforced Concrete', Journal of Structural Engineering, ASCE, April 1985.
[3] *J.E. Stephens, J.T.P. Yao:* 'Damage Assessment Using Response Meas-urements', Journal of Structural Engineering, ASCE, Vol. 113, April, 1987.
[4] *H. Krawinkler, M. Zohrei:* 'Cumulative Damage in Steel Structures Subjected to Earthquake Ground Motion', Computer & Structures, Vol. 16, N.1–4, 1983.
[5] *G.H. Powell, R. Allahabadi:* 'Seismic Damage Prediction by Determin-istic Methods: Concepts and Procedures', Earthquake Engineering and Structural Dynamics, Vol. 16, pp. 719–734, 1988.
[6] *P. Fajfar:* 'Equivalent Ductility Factors, taking into account Low-Cycle Fatigue', Earthquake Engineering and Structural Dynamics, Vol. 21, pp. 837–848, 1992.
[7] *P. Fajfar, T. Vidic, M. Fischinger:* 'On Energy Demand and Supply in SDOF Systems', in 'Nonlinear Analysis and Design of Reinforced Con-crete Buildings', Eds. P. Fajfar and H. Krawinkler, Elsevier, London, 1992.

[8] *B. Lashari:* 'Seismic Risk Evaluation of Steel Structures based on Low-Cycle Fatigue', Reliability Eng. System Safety, Vol. 20, pp. 297–302, 1988.

[9] *S.L. McCabe, W.J. Hall:* 'Assessment of Seismic Structural Damage', Journal of Structural Engineering, ASCE, Vol. 115, pp. 2166–2183, 1989.

[10] *P. Fajfar, M. Fischinger:* 'A Seismic Design Procedure including Energy Concept', 9th European Conference on Earthquake Engineering, Vol. 2, pp. 312–321, Moscow, September, 1990

[11] *A.S. Veletsos, N.M. Newmark:* 'Effect of Inelastic Behaviour on the Response of Simple Systems to Earthquake Motions', 2nd World Conference on Earthquake Engineering, Japan, pp. 895–912, 1960.

[12] *J. Penzien:* 'Elasto-Plastic Response of Idealized Multi-Storey Structures Subjected to a Strong Motion Earthquake', 2nd World Conference on Earthquake Engineering, Japan, pp. 739–760, 1960.

[13] *N.M. Newmark, J.W. Hall:* 'Procedures and Criteria for Earthquake Resistant Design', Building Practice for Disaster Mitigation, Building Science Series 45, National Bureau of Standards, Washington, pp. 94–103, Feb. 1973.

[14] *A. Mahin, V.V. Bertero:* 'An Evaluation of Inelastic Design Spectra', Journal of Structural Division, ASCE, Vol. 107, pp. 1777–1795, September 1981.

[15] *A. Giuffré, R. Giannini:* 'La Duttilit delle Strutture in Cemento Armato', ANCE-AIDIS, Roma, 1982.

[16] *A. De Luca, G. Serino:* 'L'Approccio Energetico nella Progettazione Sismica', Ingegneria Sismica, Anno V, N.3, 1988.

[17] *B. Palazzo, F. Fraternali:* 'L'Uso degli Spettri di Collasso nell'Analisi Sismica: Proposta per una Diversa Formulazione del Coefficiente di Struttura', 3 Convegno Nazionale l'Ingegneria Sismica in Italia, Roma, 1987.

[18] *B. Palazzo, F. Fraternali:* 'L'Influenza dell'Effetto P-Δ sulla Risposta Sismica di Sistemi a Comportamento Elasto-Plastico: Proposta di una Diversa Formulazione del Coefficiente di Struttura', Giornate Italiane delle Costruzione in Acciaio, C.T.A., Trieste 1987.

[19] *E. Cosenza, C. Faella, V. Piluso:* 'L'Effetto del Degrado Geometrico sul Coefficiente di Struttura', 4 Convegno Nazionale l'Ingegneria Sismica in Italia, Milano, Ottobre 1989.

[20] *C. Faella, O. Mazzarella, V. Piluso:* 'L'influenza della non-linearit geometrica sul danneggiamento strutturale', VI Convegno Nazionale, L'Ingegneria Sismica in Italia, Perugia, 13–15 Ottobre 1993.

[21] *H. Krawinkler, A.A. Nassar:* 'Seismic Design based on Ductility and Cumulative Damage Demands and Capacities', in 'Nonlinear Analysis and Design of Reinforced Concrete Buildings', Eds. P. Fajfar and H. Krawinkler, Elsevier, London, 1992.

[22] *D. Bernal:* 'Amplification Factors for the Inelastic Dynamic P-Δ Effects in Earthquake Analysis', Earthquake Engineering and Structural Dynamics, Vol. 15, 1987.

[23] *E. Cosenza, G. Manfredi, R. Ramasco:* 'An Evaluation of the Use of Damage Functionals in Earthquake Engineering Design', 9th European Conference on Earthquake Engineering, Moscow, September, 1990.

[24] *E. Cosenza, G. Manfredi, R. Ramasco:* 'The Use of Damage Functionals in Earthquake Engineering: A Comparison between Different Methods', Earthquake Engineering and Structural Dynamics, Vol. 22, 1993.

[25] *E. Cosenza, G. Manfredi, R. Ramasco:* 'La Caratterizzazione della Risposta Sismica dell' Oscillatore Elasto-Plastico', Ingegneria Sismica, N.3, 1989.

[26] *E. Cosenza, G. Manfredi:* 'Low Cycle Fatigue: Characterization of the Plastic Cycles due to Earthquake Ground Motion' in 'Testing of Metals for Structures', edited by F.M. Mazzolani, E & FN Spon, London, 1992.

[27] *E. Cosenza, A. De Luca, C. Faella, F.M. Mazzolani:* 'On a Simple Evaluation of Structural Coefficients in Steel Structures', 8th European Conference on Earthquake Engineering, Lisbon, Portugal, September 1986.

[28] *E. Cosenza, A. De Luca, C. Faella:* 'Criteri di Valutazione della Duttilità Globale nelle Strutture Metalliche Intelaiate', Costruzioni Metalliche, N. 5, 1986.

[29] *E. Cosenza, A. De Luca, C. Faella, F.M. Mazzolani:* 'Una Proposta per la Definizione del Coefficiente di Struttura nei Telai Metallici', 3 Convegno Nazionale l'Ingegneria Sismica in Italia, Roma, 1987.

[30] *E. Cosenza:* 'Duttilità Globale delle Strutture Sismo-Resistenti in Acciàio', Tesi di Dottorato di Ricerca, Universit di Napoli, Aprile 1987.

[31] *P. Setti:* 'Un Metodo per la Determinazione del Coefficiente di Struttura per le Costruzioni Metalliche in Zona Sismica', Costruzioni Metalliche, N.3, 1985

[32] *G. Ballio:* 'ECCS Approach for the Design of Steel Structures against Earthquakes', Symposium on Steel in Buildings, Luxembourg, 1985, IABSE-AIPC-IVBH Report, Vol. 48, pp. 373–380, 1985.

[33] *A.S. Veletsos, W.P. Vann:* 'Response of Ground-Excited Elasto-Plastic Systems', Journal of Structural Division, ASCE, Vol. 97, pp. 1257–1281, April 1971.

[34] *S.A. Anagnostopoulos, R.H. Haviland, J.M. Biggs:* 'Use of Inelastic Spectra in Aseismic Design', Journal of Structural Division, Vol. 104, pp. 95–109, January 1978.

[35] *G. Housner:* 'Limit Design of Structures to Resist Earthquakes', 1st World Conference on Earthquake Engineering, 1956.

[36] *G. Housner:* 'Behaviour of Structures During Earthquakes', ASCE, EM4, 1959.

[37] *T.F. Zahrah, J.W. Hall:* 'Earthquake Energy Absorption in SDOF Structures', Journal of Structural Division, Vol. 110, N.8, ASCE, pp. 1757–1772, August 1984.

[38] *G. Al-Sulaimani, J.M. Roesset:* 'Design Spectra for Degrading Systems', Journal of Structural Engineering, ASCE, December 1985.

[39] *Commission of the European Communities:* Eurocode 3: Design of Steel Structures, 1992.

[40] *G. Sedlacek, J. Kuck:* 'Determination of q-factors for Eurocode 8', Aachen den 31.8.1993.

[41] *E. Cosenza, A. De Luca, C. Faella, V. Piluso:* 'A Rational Formulation for the q-Factor in Steel Structures', 9th World Conference on Earthquake Engineering, Tokyo, August, 1988.

[42] *B. Kato, H. Akiyama:* 'Ductility of Members and Frames Subjected to Buckling', ASCE, May, New York, 1981.

[43] *I. Mitani, M. Makino:* 'Post Local Buckling Behaviour and Plastic Rotation Capacity of Steel Beam-Columns', 7th World Conference on Earthquake Engineering, Istanbul, 1980.

[44] *F.M. Mazzolani, V. Piluso:* 'Evaluation of the Rotation Capacity of Steel Beams and Beam-Columns', 1st COST C1 Workshop, Strasbourg, 28–30 October, 1992

[45] *B. Kato:* 'Deformation Capacity of Steel Structures', Journal of Constructional Steel Research, pp. 33–94, N.17, 1990.

[46] *R. Spangemacher, G. Sedlacek:* 'On the Development of a Computer Simulator for Tests of Steel Structures', Proceedings of the First World Conference on Constructional Steel Design, Acapulco, Mexico, 6–9 December 1992.

[47] *M. Como, G. Lanni:* 'Aseismic Toughness of Structures', Meccanica, 18, pp. 107–114, 1983.

[48] *M. Como, G. Lanni:* 'Duttilità e Calcolo allo Stato Limite delle Strutture Antisismiche', Quaderni di Teoria e Tecnica delle Strutture, N.487, Universit di Napoli, Istituto di Tecnica delle Costruzioni, 1984.

[49] *B. Kato, H. Akiyama:* 'Earthquake Resistant Design for Steel Buildings', 6th World Conference on Earthquake Engineering, 1977.

[50] *B. Kato, H. Akiyama:* 'Seismic Design of Steel Buildings', Journal of Structural Division, ASCE, August, 1982.

[51] *B. Kato, H. Akiyama:* 'Energy Concentration of Multi-Storey Buildings', 7th World Conference on Earthquake Engineering, Istanbul, 1980.

[52] *B. Kato:* 'Seismic Design Criteria for Steel Buildings', personal communication to F.M. Mazzolani, November, 1989.

[53] *H. Akiyama:* 'Earthquake Resistant Design Based on Energy Concept', 9th World Conference on Earthquake Engineering, Tokyo-Kyoto, Japan, Vol. V, paper 8-1-2, August 2–9, 1988.

[54] **F.M. Mazzolani, V. Piluso:** 'ECCS Manual on Design of Steel Structures in Seismic Zones', ECCS, European Convention for Constructional Steelwork, 1993

[55] **G. Ballio, C.A. Castiglioni:** 'An Approach to the Seismic Design of Steel Structures based on Cumulative Damage Criteria', Paper accepted for publication on Earthquake Engineering and Structural Dynamics, 1993.

[56] **M.T. Suidan, R.A. Eubanks:** 'Cumulative Fatigue Damage in Seismic Structures', Journal of Structural Division, ASCE, Vol. 99, pp. 923–943, 1973.

[57] **L. Calado, J. Azevedo:** 'A Model for Predicting the Failure of Structural Steel Elements', Journal of Constructional Steel Research, Vol. 14, pp. 41–64, 1989.

[58] **C.A. Guerra, F.M. Mazzolani, V. Piluso:** 'Evaluation of the q-factor in Steel Framed Structures: State-of-art', Ingegneria Sismica, N.2, 1990.

5

Overall stability effects

5.1 INTRODUCTION

The term P–Δ effect refers to the influence of gravity loads acting on a structure in its deformed configuration. In the case of structures stressed in the elastic range and subjected to static loads, this effect produces an amplification of the internal actions with respect to those computed by means of a first-order analysis.

Despite the fact that second-order analyses can be easily performed nowadays, the effects which arise from geometric non-linearity are usually taken into account by amplification formulae in which both the P–δ effect at member level and P–Δ effect at structure level are considered [1].

In the case of structures subjected to dynamic loads, geometric non-linearity causes an increase of the natural period of vibration so that, depending on the loading features, an improvement or a worsening of the structural response can arise. This is a fundamental difference with respect to the case of static loading conditions, where the P–Δ effect always determines an increase of the internal actions. Within the elastic range, P–Δ effects are not always significant; on the contrary they cannot be neglected in the seismic design of structures [2–8] and for steel structures in particular [9, 10].

As discussed in Chapter 3, the ultimate conditions of steel frames under seismic loads are mainly related to the rotational capacity of members, which value is limited by the overall stability phenomena of the member as a whole and by the local buckling of the compressed parts of its cross-section [11–13]. In addition, from the global point of view, vertical loads influence the shape of the horizontal forces versus deflection curve. In fact the slopes of both the increasing elastic branch and of the softening inelastic one are affected by the magnitude of the ratio between vertical and critical load of columns (Chapter 2). Both aspects, local and global, play an important role in the dynamic inelastic behaviour of structures.

Geometrical non-linearity interacts with the mechanical non-linearity due to the yielding of dissipative zones, leading to an increase of the required ductility in order to prevent collapse [6–8, 10, 14].

In addition, it is very important to forecast the collapse mechanism in order to correctly foresee not only the damage concentration, but also the amplification of the ductility demand due to the P–Δ effect. In order to obtain a collapse mechanism of the global type, modern seismic codes provide simple design rules based on the amplification of the stresses for which columns must be dimensioned. As emphasized in Chapter 2, the use of these simple 'column overstrength factors' (COF) seems to be unsatisfactory [15, 16]. It has been seen that the control of the failure mode is most important in order to limit the values of the slope of the softening branch of the multiplier of horizontal forces versus top displacement curve, which is related to the influence of P–Δ effects.

Despite the fact that the problem of evaluating the overall stability effects in steel structures under seismic loads has attracted significant attention, and that research effort has provided much information, a very small part of these results has been codified. In fact, in seismic codes the provisions regarding the influence of geometric non-linearity are insufficient, because they are mainly based on the extension of static–elastic results.

This chapter is devoted to a more complete and exhaustive analysis of the influence of the overall stability phenomena on the inelastic response of steel frames under seismic loads.

In particular, after examination of the provisions suggested in the seismic codes, attention is now focused on the possibility of including the increase in structural damage under seismic loads due to second-order effects, by reducing the q-factor of the examined structural type by a coefficient φ_Δ, which depends on the level of geometric non-linearity. This approach has been initiated [10] and a statistical evaluation of the reduction factor φ_Δ has been performed [14, 17] for the SDOF system with reference to both historical [17] and simulated earthquakes [14]. The influence of the period of vibration, the available ductility and, obviously, the level of vertical loads has been considered.

The last part of the chapter is devoted to the comparison between values of the reduction factor φ_Δ obtained from the SDOF analysis and values computed by numerical simulation of the seismic response of MDOF systems. This comparison shows that the above approach can be usefully adopted for estimating second-order effects in real steel frames [18].

5.2   CODE PROVISIONS FOR SECOND-ORDER EFFECTS

It has been shown that existing seismic codes exploit just a small part of the research results which have been obtained in analysing the overall stability effects in seismic-resistant steel frames [19]. The most rational way for taking into account the influence of second-order effects is their inclusion in the definition of the q-factor design value; however, this process is not directly provided in seismic codes and no mention is made for its rational evaluation.

At present, the provisions are mainly limited to the extension of rules for non-seismic zones. As an example, UBC91 [20] states that the P–Δ effect can be neglected when, for each floor, the following condition is satisfied

$$\Delta_e < \frac{V\,h}{10\,N} \tag{5.1}$$

where Δ_e is the elastic interstorey drift, h is the distance between the floors, V is the total shear force acting above the considered floor, and N is the total axial force acting above the considered floor. This provision is coincident with that given in Eurocode 3 [21].

A different approach is provided in the Mexican Seismic Code [22] where, in order to account for the overall stability effects, the design forces have to be amplified by a coefficient depending on the magnitude of vertical loads. The amplification coefficient is given by

$$\varphi_\Delta = \frac{1}{1 - \mu\,\gamma} \tag{5.2}$$

where μ is the available ductility and γ is the stability coefficient (Chapter 2). The amplification of the design horizontal forces is practically equivalent to the reduction of the design value of the q-factor.

In both Eurocode 8 and the ECCS Recommendations [23, 24], equation (5.1) is applied with reference to the inelastic displacements. The inelastic displacements can be estimated from elastic displacements by

$$\Delta_p \approx q\,\Delta_e \tag{5.3}$$

where Δ_p and Δ_e are the inelastic and elastic drifts, respectively, and q is the behaviour factor. This relationship is based on the so-called ductility factor theory (Chapter 4), stating that the q-factor q is approximately equal to the structural ductility ($q \approx \mu$).

Consequently, according to Eurocode 8 and ECCS, P–Δ effects can be neglected when, for each floor, the following condition is true:

$$\Delta_e < \frac{V\,h}{10\,q\,N} \tag{5.4}$$

In particular, in Eurocode 8 [23] different cases are defined on the basis of the value of the so-called interstorey drift sensitivity coefficient θ, defined as

$$\theta = \frac{N q \Delta_e}{V h} \tag{5.5}$$

The following three cases are identified:

a. θ ≤ 0.10, second-order effects can be neglected as the condition (5.4) is satisfied;
b. 0.10 < θ ≤ 0.30, second-order effects have to be taken into account;
c. θ > 0.30, the structure has to be stiffened due to its excessive sensitivity to second-order effects.

In the second case, a simplified method to account for second-order effects is suggested. It consists of the amplification of the seismic action effects according to the coefficient

$$\varphi_\Delta = \frac{1}{1 - \theta} \tag{5.6}$$

An approximate evaluation of the critical multiplier of vertical loads is given by [25]

$$\alpha_{cr} \approx \frac{V h}{N \Delta_e} \tag{5.7}$$

In the case of frames failing in global mode (Chapter 2),

$$\gamma \approx \frac{1}{\alpha_{cr}} \tag{5.8}$$

so that equation (5.6) can be written as (θ ≈ γq)

$$\varphi_\Delta \approx \frac{1}{1 - \gamma q} \tag{5.9}$$

Therefore, under the assumption of validity of the ductility factor theory (q ≈ μ), it can be recognized that the provision (5.6) is substantially equivalent to that (5.2) suggested by the Mexican Seismic Code. In addition, in Eurocode 8 [23] it is believed that the above simplified approach can be adopted in the case 0.10 < θ < 0.20.

It is useful to note that, where θ ≈ (1/α_{cr}) q, case a arises for $\alpha_{cr} \geq 10 q$, case b for $10 q/3 \leq \alpha_{cr} < 10 q$, and case c for $\alpha_{cr} < 10 q/3$.

Finally, it can be recognized that equation (5.1) corresponds to the least severe provision, because it is substantially equivalent to state that P–Δ effects can be neglected when the critical elastic multiplier of the vertical loads α_{cr} is greater than 10.

5.3 OVERALL STABILITY EFFECTS IN SDOF SYSTEMS

5.3.1 Mechanical model

The simplified model adopted for numerical analyses consists of the SDOF system with geometrical degradation (Fig. 5.1).

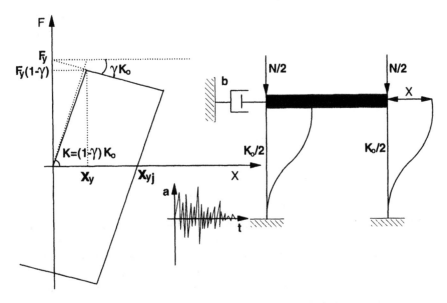

Fig. 5.1   The mechanical model adopted for numerical analyses

The hysteretic model is the elastic–perfectly plastic one. Due to geometrical and mechanical non-linearities, the lateral strength depends upon the horizontal displacement x and is influenced by the vertical load N. The stiffness corresponding to the elastic increasing branch is given by

$$K = K_0 \, (1 - \gamma) \qquad\qquad (5.10)$$

where K_0 is the value of the elastic stiffness in the absence of vertical loads. The parameter γ, known as the 'stability coefficient', takes into account the level of vertical load. In particular, for frames failing in global mode, the stability coefficient γ can be approximately evaluated by the simplified relationship (5.8). The slope of the softening plastic branch is $-K_o\gamma$.

Considering the SDOF system, the 'stability coefficient' γ is given by

$$\gamma = \frac{\pi^2}{12} \frac{N}{N_{cr}} = \frac{\pi^2}{12} \frac{1}{\alpha_{cr}} \approx \frac{1}{\alpha_{cr}} \qquad (5.11)$$

The study of the influence of the P–Δ effect on the seismic inelastic response of seismic-resistant structures is reduced, through the above mechanical model, to the analysis of a simple degree of freedom system subjected to a given ground motion.

Denoting with ω and v, respectively, the frequency and the viscous damping of the system, the motion equation for the elastic branch is

$$\ddot{x} + 2v\omega\dot{x} + \omega^2(1 - \gamma)(x - x_{y,j}) = -a(t) \qquad (5.12)$$

while for the plastic branch it becomes

$$\ddot{x} + 2v\omega\dot{x} + \omega^2(x_y - \gamma x) = -a(t) \qquad (5.13)$$

Here $a(t)$ is the considered accelerogram, x_y the first yielding displacement ($x_y = F_y / K_o$) and $x_{y,j}$ the intersection between the elastic branch and the displacement axis (initially $x_{y,j} = 0$).

It is a classical problem of dynamic non-linear analysis in the time domain, which can be solved by means of the usual numerical techniques. However, as the accelerograms are composed of linear branches, it is possible to solve motion equations in closed form, both in the elastic phase and the plastic phase, concentrating on the instants in which a change of phase occurs.

The integration of the motion equations can be performed by means of an algorithm proposed by Nigam and Jennings [26] for the elastic SDOF system and extended by Nau [27] to the elastic–perfectly plastic and elastic–strain hardening SDOF systems and, finally, by Cosenza [9] to the elastic–perfectly plastic SDOF system with geometric non-linearity.

5.3.2 Numerical analyses under simulated ground motion

Numerical analyses have been carried out by using a specifically developed computer program [10, 14]. In the first stage of research, ten artificial accelerograms were generated from the elastic design response spectrum, which corresponds to the one given in EC8 with the numerical parameters of the new Italian code GNDT [28]. Thirty different values, appropriately selected, of the design level ($\eta = F_y / MA$, where A is the peak ground acceleration) have been considered and the ductility demand ($\mu = x_{max}/x_y$) was computed by varying the period of vibration from 0.1 to 2.0 s. The analyses have been repeated for values of the stability coefficient γ equal to 0.0, 0.025, 0.05, 0.075, 0.1, 0.15 and 0.20.

The outcome of the analyses [14] is the reduction factor φ_Δ given as the ratio between the inelastic response spectrum in the absence of geometrical non-linearity ($\gamma = 0$), for a given value of the ductility demand, and the one computed including the P–Δ effect ($\gamma \neq 0$). This approach, therefore, allows evaluation of the q-factor $q(\gamma)$, including the P–Δ effect, from the one $q(\gamma = 0)$ computed neglecting geometrical non-linearity, as stated by the relationship

$$q(\gamma) = \frac{q(\gamma = 0)}{\varphi_\Delta} \qquad (5.14)$$

The reduction spectra, computed for a given value of the available ductility and averaged for the ten considered accelerograms, are plotted in Fig. 5.2 for some values of the stability coefficient γ. These spectra are characterized by the presence of a period value corresponding to a peak response, probably related to the shape of the elastic response spectrum adopted in earthquake generation, and by the existence of a period range ($T \geq 0.5$ s), where the reduction factor φ_Δ can be practically considered independent of T.

Therefore, in this range which is of particular importance for steel structures, the following relationship can be adopted:

$$\varphi_\Delta = \varphi_\Delta \, (\mu, \gamma) \qquad (5.15)$$

According to Cosenza, Faella and Piluso [14] the investigation on the quantitative assessment of equation (5.15) led to the formulation

$$\varphi_\Delta = \frac{1 + \psi_1(\mu - 1)^{\psi_2}\gamma}{1 - \gamma} \qquad (5.16)$$

where the coefficients ψ_1 and ψ_2 can be obtained from the analysis of results by numerical regression. The mathematical structure of equation (5.16) has been chosen so that φ_Δ tends to infinity in the case $\gamma = 1$, which represents the case of static instability. Moreover, equation (5.16) provides $\varphi_\Delta = 1$ for $\gamma = 0$, which corresponds to the absence of geometric non-linearity, and $\varphi_\Delta = 1/(1 - \gamma)$ for $\mu = 1$. This last result corresponds to the reduction of the yield level, due to the P–Δ effect, from F_y to $F_y/\varphi_\Delta = F_y(1 - \gamma)$ (Fig. 5.1).

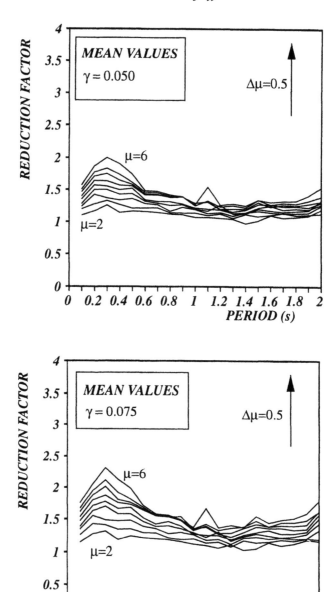

Fig. 5.2   Reduction factor spectra

It must be understood that the range of applicability of equation (5.16) is defined by the condition

$$\mu\gamma < 1 \tag{5.17}$$

which assures the absence of dynamic instability [9].

A regression analysis [14] has shown that, for $T \geq 0.5$ s, the mean values of φ_Δ can be obtained from equation (5.16) by setting ψ_1 to 0.62 and ψ_2 to 1.45. The corresponding curves are represented in Fig. 5.3, where the condition leading to dynamic instability is also shown. With reference to the characteristic values of φ_Δ, the regression analysis has provided values of $\psi_1 = 3.79$ and $\psi_2 = 0.75$; the corresponding representation is also given in Fig. 5.3.

Regarding the maximum q-factor reduction corresponding to the critical value of the period ($T \approx 0.3$ s), the mean values of φ_Δ are provided by equation (5.16) by using $\psi_1 = 2.64$ and $\psi_2 = 1.03$, while the characteristic ones are obtained for $\psi_1 = 5.70$ and $\psi_2 = 0.96$.

The reduction factor requires some examination in the low-period range. It has to be pointed out that the relationship between γ and T can be expressed as

$$\gamma = cost\ T^2 \tag{5.18}$$

This relationship can be obtained taking into account that both T and γ are dependent upon the mass and the stiffness of the system.

It is easy to show for an SDOF system that the constant in equation (5.18) is given by g/h, but in the general case of MDOF systems it is difficult to evaluate, being dependent upon both the elastic features of the structure and the collapse mechanism. In any event, equation (5.18) states that 'high' values of γ for 'low' values of T are not significant but, due to the absence of a precise evaluation of the relation between γ and T, the exclusion of non-significant values of T for given values of γ can only be qualitative.

5.3.3 Numerical analyses under historical earthquakes

In a second stage of the research [17], numerical analyses have been carried out by using, as seismic input, acceleration records of historical earthquakes (Elcentro NS 19/05/1940, Olympia N80E 13/04/1949, Ferndale N45E 11/09/1938, Vernon S82E 10/03/1933, Sturno NS 23/11/1980, Calitri EW 23/11/1980, Tolmezzo NS 06/05/1976, San

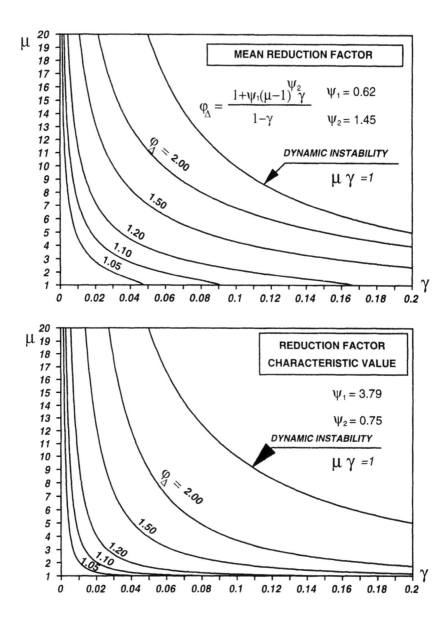

Fig. 5.3   Representation of the formulation for evaluating the reduction factor φ_Δ

Rocco EW 15/09/1976). These records have been all scaled to the same peak ground acceleration of 0.35g.

For each acceleration record, the seismic inelastic response of the mechanical model has been analysed by varying the period of vibration and the stability coefficient in the ranges previously specified. The required ductility has been evaluated for design levels η ranging from 0.02 to 2.50 with a very small step $\Delta \eta = 0.02$. Damage spectra have been derived both for $\gamma = 0$ and $\gamma \neq 0$, so that the corresponding spectra of the reduction factor φ_Δ have been obtained by considering the ratios between the spectra for $\gamma = 0$ and those for $\gamma \neq 0$.

The results of this analysis agree with the previous observations, derived from simulated accelerograms, that the reduction factor φ_Δ is substantially independent of the period, which gives rise only to random scatters. This result is shown in Fig. 5.4 with reference to the case $\gamma = 0.2$. Similar representations are obtained by varying the value of γ.

Therefore, the influence of T has been considered only from the statistical point of view and a relationship of the type (5.15) has still been assumed. Although a relationship of the type (5.16) was able to provide a good fit with the numerical simulation data, it was recognized that a better fit can be obtained by means of a formulation having an additional parameter ψ_3

$$\varphi_\Delta = \psi_3 + \frac{(1 - \psi_3)\,[1 + \psi_1(\mu - 1)^{\psi_2}\gamma]}{1 - \gamma} \tag{5.19}$$

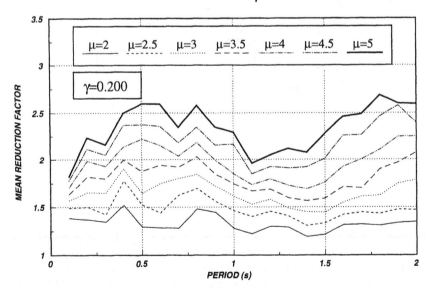

Fig. 5.4   Exemplifying reduction factor spectrum

The parameters ψ_1, ψ_2 and ψ_3 have been derived by a linear–logarithmic mixed regression. With reference to the mean values of the reduction factor φ_Δ, the values $\psi_1 = 1.4276$, $\psi_2 = 1.0415$ and $\psi_3 = 0.2790$ have to be assumed.

The characteristic values of φ_Δ, which take into account the scatters due to the random influence both of the period and of the acceleration record are provided by equation (5.19) with the parameters $\psi_1 = 4.3891$, $\psi_2 = 0.8282$ and $\psi_3 = 0.3082$. The degree of accuracy of the formulation (5.19) is shown in Fig. 5.5 where, with reference to the mean values of the reduction factor φ_Δ, the comparison between the values predicted through equation (5.19) and the ones obtained from the numerical simulations is presented. The damage criterion is based on the kinematic ductility.

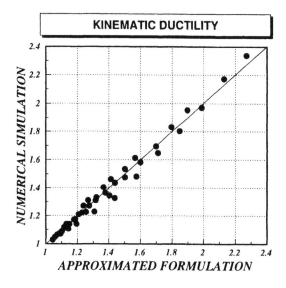

Fig. 5.5   Degree of reliability of formulation, equation (5.19)

5.3.4 Comparison between formulations

In the previous sections, the seismic response of the elastic–perfectly plastic SDOF system with geometrical non-linearity has been examined by using, as seismic input, both acceleration records generated in order to match a specified linear elastic design response spectrum (LEDRS) and acceleration records of historical earthquakes. In both cases, the analyses have led to the formulation of the reduction factor φ_Δ as a function of the stability coefficient γ and of the available ductility μ. In order to evaluate the possible influence of the features of the seismic input, the comparison between the results obtained for the SDOF system subjected to generated accelerograms and those obtained using historical earthquakes is given in Figs. 5.6 and 5.7 for four values of the available ductility, $\mu = 2$ to 5. In fact, the relation proposed by Cosenza *et al.* [14] has been derived using generated accelerograms, while that proposed by Faella *et al.* [17] has been obtained by analysing historical earthquakes. The comparison shows that the mean value of the reduction factor φ_Δ is scarsely affected by the features of the seismic input.

The mean values of the reduction factor φ_Δ evaluated according to the proposals of Cosenza *et al.* [14] and of Faella *et al.* [17] are compared in Figs. 5.6 and 5.7 with the formulations proposed by various authors [7, 8, 10] and with the Mexican Seismic Code [22].

The results are in good agreement with those provided in reference [8], where a study of the dependence of the q-factor upon μ, γ and T is reported. These curves cover the period range between 0.35 s (upper curve) and 2.0 s (lower curve), excluding, therefore, the lowest periods where some peak amplifications can arise. The small width of this band representing the period influence seems to confirm that φ_Δ can be considered independent of T with acceptable approximation.

From the analysis of Figs. 5.8 and 5.9, a greater scatter between the relation proposed by Cosenza *et al.* [14] and the one proposed by Faella *et al.* [17] is observed with reference to the characteristic values of φ_Δ, but this scatter is probably related to the number of accelerograms considered in the analysis. In the same figures, the results obtained by Bernal [7] are also shown. As this author has considered a limited number of historical earthquakes, the corresponding curve falls within the range between the mean and characteristic values.

Furthermore, with reference to Bernal's formulation [7], this has been obtained from the analysis of a limited number of historical earthquake records, so that the result can be affected by greater random scatters. Moreover, the analysis was carried out without introducing any limitation regarding the period range and, as a consequence, including the effects of critical values of the period, leading to peak amplifications of the ductility demand and, therefore, to a larger reduction of the q-factor. These factors justify the more conservative result obtained by Bernal, with respect to that of Cosenza *et al.* [14] (Figs. 5.6 and 5.7).

Finally, the formulation given in the Mexican Seismic Code, which is substantially equivalent to the simplified method suggested by Eurocode 8, as previously pointed out, seems to be excessively on the safe side (Figs. 5.6 and 5.7).

Overall stability effects

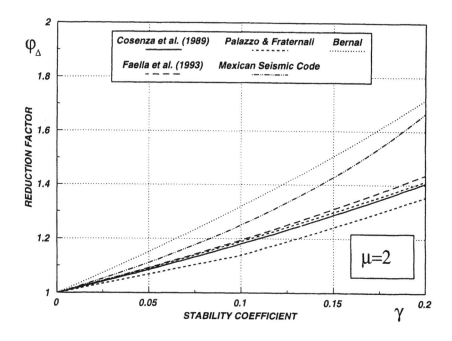

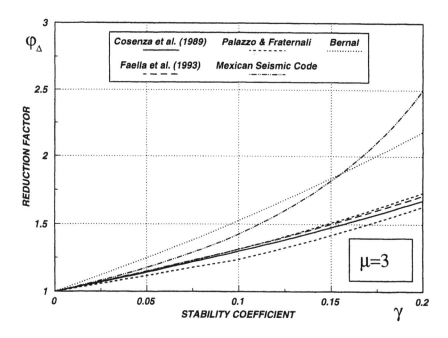

Fig. 5.6 Comparison between generated and historical earthquakes

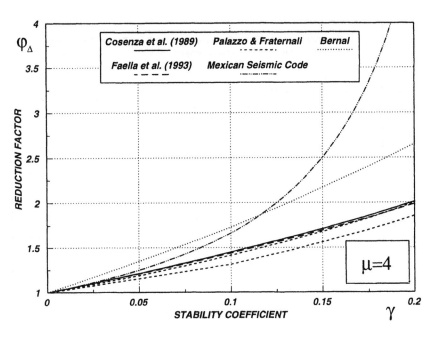

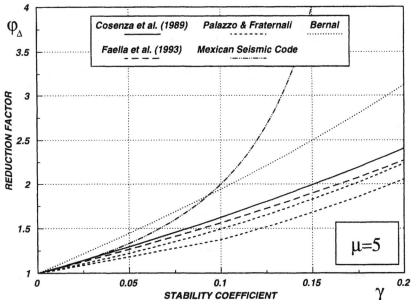

Fig. 5.7   Comparison between generated and historical earthquakes

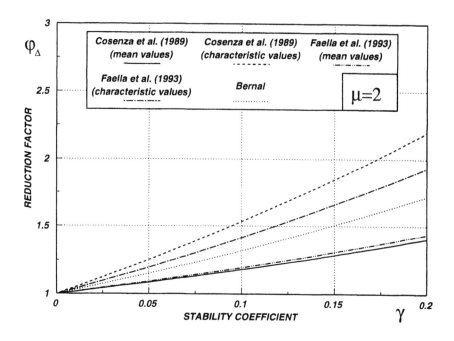

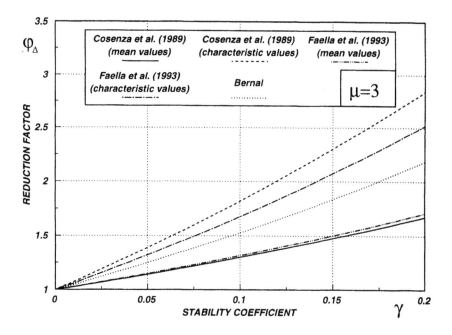

Fig. 5.8 Comparison among different formulations

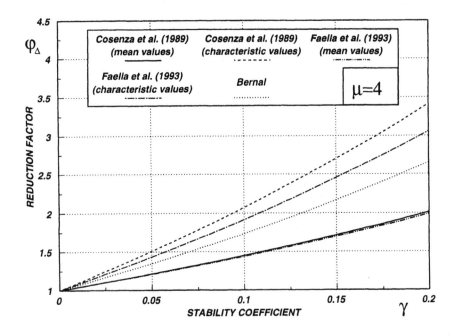

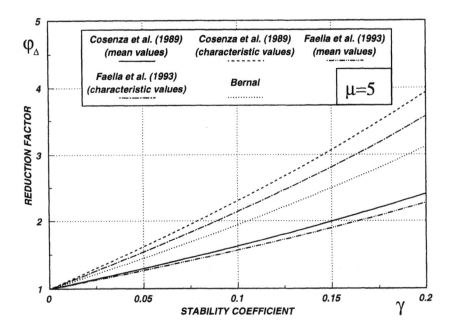

Fig. 5.9   Comparison among different formulations

5.4   INFLUENCE OF DAMAGE CHARACTERIZATION

As described in Chapter 1, the maximum displacement criterion leading to damage characterization through the required kinematic ductility, or simply the ductility μ, takes into account neither the number of plastic excursions nor the ductility required by each of them. For this reason, other damage parameters have been proposed by many authors [29–34]. Nowadays, the technical literature shows that parameters other than the kinematic ductility are used to characterize the structural damage. In the case of steel structures, these parameters are:

- the cyclic ductility μ_c
- the hysteretic ductility in one direction $\mu_h^{(1)}$
- the total hysteretic ductility μ_h
- the Park and Ang parameter $\mu_{P.A.}$
- the low-cycle fatigue μ_F.

The above parameters have already been briefly described in Chapter 1.

With reference to these parameters, the influence of the P–Δ effect on the seismic response of the SDOF system has been investigated [17]. The seismic input is represented by the historical earthquakes already considered in Section 5.3. Also, the same ranges of variation, of both the period of vibration and of the stability coefficient have been analysed.

Using the procedure previously described, the spectra of the reduction factor φ_Δ for a given value of the structural damage have been evaluated for the different damage criteria. The purpose of the analysis is to provide, for any damage criterion, a reduction factor φ_Δ of the q-factor, able to account for geometrical non-linearity. Therefore, the value of the q-factor in the presence of second-order effects can be expressed as

$$q\left(\gamma,\mu^*,T\right)=\frac{q\left(\gamma=0,\mu^*,T\right)}{\varphi_\Delta\left(\gamma,\mu^*,T\right)} \qquad (5.20)$$

where μ^* is the value of the parameter adopted for damage characterization (i.e. μ^* represents one of the parameters in the above list: μ, μ_c, $\mu_h^{(1)}$, $\mu_{P.A.}$ or μ_F).

The analyses have confirmed, by agreement with the results obtained with reference to the kinematic ductility [14, 17], that the reduction factor φ_Δ is substantially independent of the period T. This can be seen from the analysis of the mean spectra, some of which are represented in Figs. 5.10 to 5.13. Therefore, the reduction factor φ_Δ can be evaluated by means of a relationship of the type

$$\varphi_\Delta=\varphi_\Delta\left(\gamma,\mu^*\right) \qquad (5.21)$$

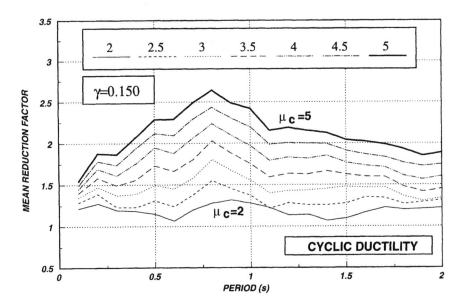

Fig. 5.10   Exemplifying reduction factor spectra

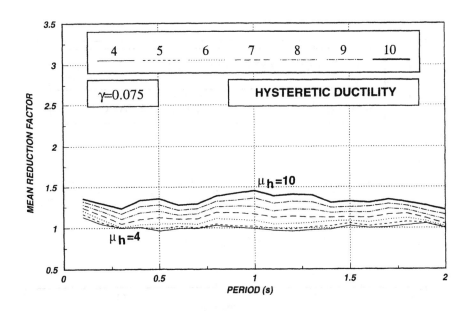

Fig. 5.11   Exemplifying reduction factor spectra

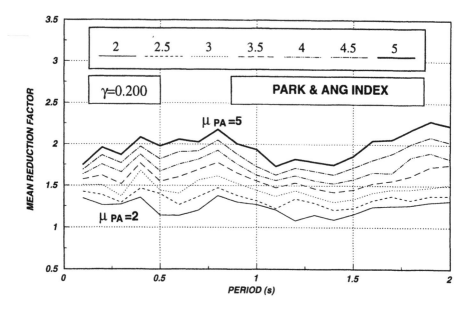

Fig. 5.12 Exemplifying reduction factor spectra

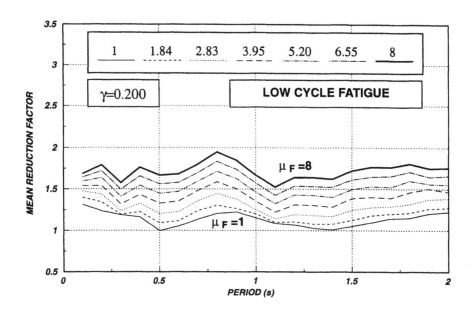

Fig. 5.13 Exemplifying reduction factor spectra

The reduction factor φ_Δ has been expressed [17] by means of the relationship

$$\varphi_\Delta = \psi_3 + \frac{(1-\psi_3)\,[1+\psi_1\,(\mu^*-k)^{\psi_2}\,\gamma]}{1-\gamma} \qquad (5.22)$$

which represents a generalization of equation (5.19).

With reference to the mean values of the reduction factor φ_Δ, the values of the parameters ψ_1, ψ_2, ψ_3 and k are given in Table 5.1 for the different damage characterization criteria.

As already mentioned, the parameters ψ_1, ψ_2 and ψ_3 have been computed by means of a linear–logarithmic mixed regression. With the same procedure, the characteristic values of φ_Δ, which account for the random influence both of the period of vibration and of the seismic input, can be determined through the parameters given in Table 5.2.

With reference to the examined damage criteria, Fig. 5.5 and Figs. 5.14–

Table 5.1 Values of parameters for evaluating, through equation (5.22), mean values for the reduction factor φ_Δ

Damage parameter	k	ψ_1	ψ_2	ψ_3
Kinematic ductility	1	1.4276	1.0415	0.2790
Cyclic ductility	0	0.2887	1.8273	0.0935
Hysteretic ductility in one direction	1	0.3045	1.3791	-0.4500
Total hysteretic ductility	1	0.1171	1.3788	-0.4687
Park and Ang parameter	1	0.9944	0.9542	0.2912
Low-cycle fatigue parameter	0	0.4010	0.5858	0.2101

Table 5.2 Values of parameters for evaluating, through equation (5.22), characteristic values for the reduction factor φ_Δ

Damage parameter	k	ψ_1	ψ_2	ψ_3
Kinematic ductility	1	4.3891	0.8282	0.3082
Cyclic ductility	0	1.4604	1.4751	0.2440
Hysteretic ductility in one direction	1	1.4155	1.2410	0.0065
Total hysteretic ductility	1	0.5990	1.2407	0.0074
Park and Ang parameter	1	3.2132	0.6659	0.2938
Low-cycle fatigue parameter	0	1.8164	0.3967	0.1962

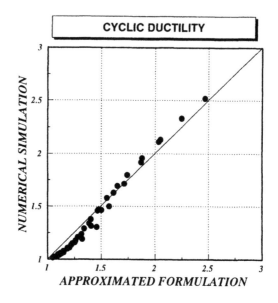

Fig. 5.14   Reliability of formulation, equation (5.22)

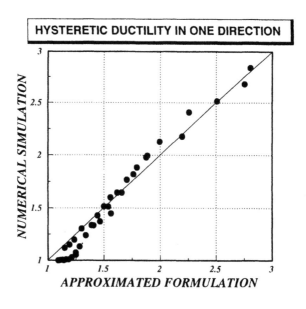

Fig. 5.15   Reliability of formulation, equation (5.22)

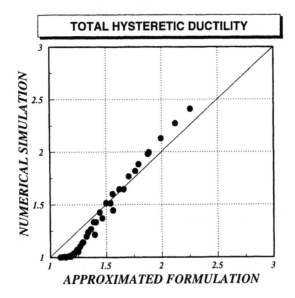

Fig. 5.16   Reliability of formulation, equation (5.22)

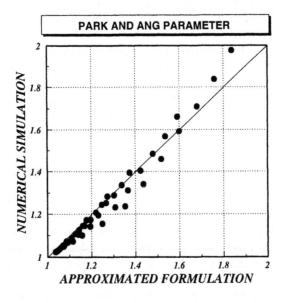

Fig. 5.17   Reliability of formulation, equation (5.22)

228 *Overall stability effects*

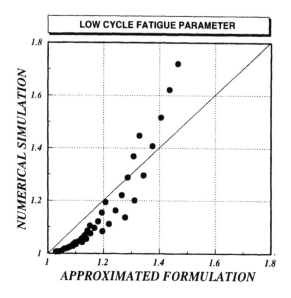

Fig. 5.18   Reliability of formulation, equation (5.22)

5.18 show the comparison between the mean values of the reduction factor φ_Δ predicted by the above formulations and those obtained from numerical simulations. In the case of the kinematic ductility (Fig. 5.5), of the cyclic ductility (Fig. 5.14) and of the Park and Ang parameter (Fig. 5.17), the reliability of the formulation (5.22) is excellent. The degree of reliability is slightly reduced in the case of the energy criteria (Figs. 5.15 and 5.16), but the formulation still provides very good results. Finally, in the case of the low-cycle fatigue criterion (Fig. 5.18), the scatters are more significant.

Finally, with reference to mean values of the reduction factor φ_Δ from 1.05 to 4, the results of the application of equation (5.22) are shown in Figs. 5.19–5.24, with respect to the defined damage parameters above.

5.5   OVERALL STABILITY EFFECTS IN MDOF SYSTEMS

The possibility of extending the results obtained for the SDOF system to multistorey frames has been investigated [18, 19]. Dynamic inelastic analyses, which have been carried out by means of the computer program DRAIN–2D [35], have been performed in order to analyse the seismic response of the frames shown in Fig. 5.25 taking into account the P–Δ effect. Three simulated accelerograms have been used. Each of them has been amplified until structural collapse has been reached, so that the ultimate value of the peak ground acceleration has been evaluated.

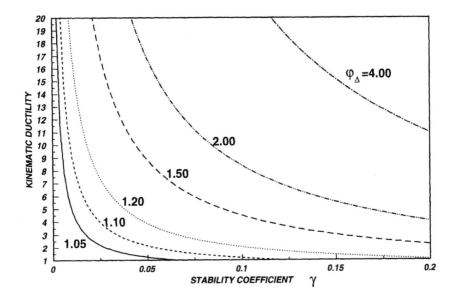

Fig. 5.19   Representation of formulation, equation (5.22)

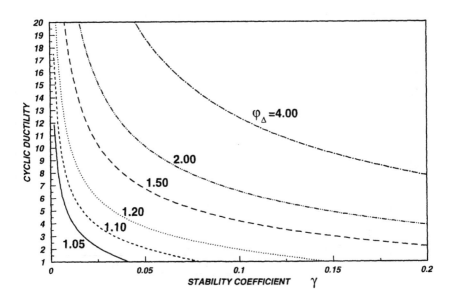

Fig. 5.20   Representation of formulation, equation (5.22)

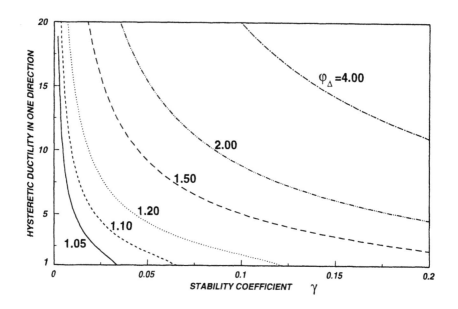

Fig. 5.21   Representation of formulation, equation (5.22)

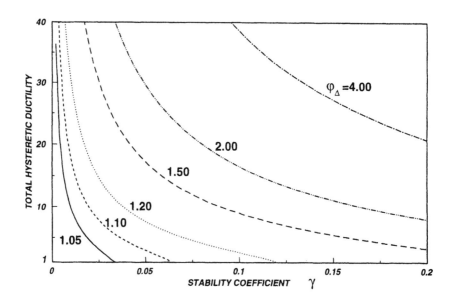

Fig. 5.22   Representation of formulation, equation (5.22)

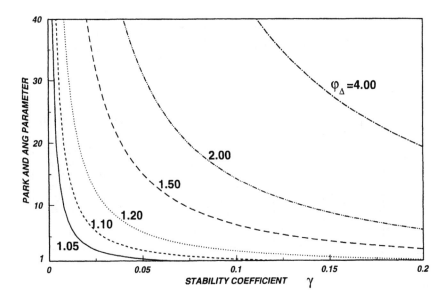

Fig. 5.23   Representation of formulation, equation (5.22)

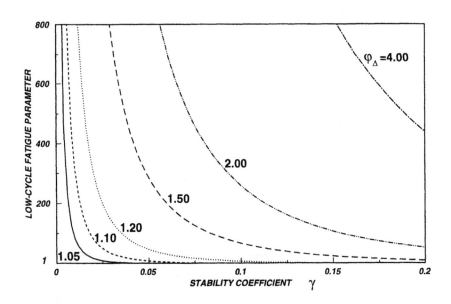

Fig. 5.24   Representation of formulation, equation (5.22)

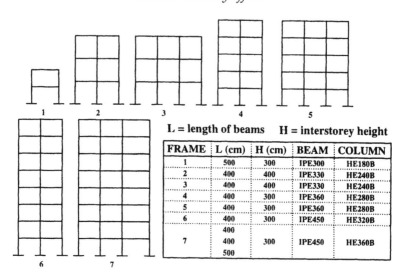

Fig. 5.25 Analysed frames

Successively, with the same amplified earthquake accelerogram, the structures have again been analysed by neglecting geometric non-linearity. As a consequence, a new value for the multiplier of the peak ground acceleration leading to collapse (which is greater than the former one) has been evaluated for each frame.

The comparison with the previous values, obtained taking into account geometrical non-linearity, provides the numerical simulation value φ_c of the reduction factor φ_Δ, due to the effect of vertical loads. These

Table 5.3 Comparison between the values φ_c obtained by means of numerical simulations and the values of the reduction factor φ_Δ of the q-factor (φ_m mean values and φ_k characteristic values) provided by the approximated formulation.

Frame	γ	μ	$\varphi_{mü...}$		
1	0.0853	3.185	1.273	1.728	1.550
2	0.0534	3.233	1.168	1.447	1.190
3	0.0560	3.236	1.177	1.470	1.200
4	0.0117	3.7303	1.043	1.106	1.085
5	0.0223	3.331	1.071	1.186	1.091
6	0.0372	2.581	1.085	1.245	1.110
7	0.0321	2.771	1.083	1.226	1.080

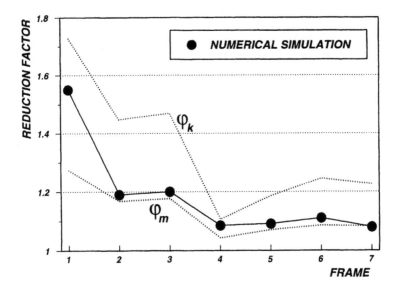

Fig. 5.26   Comparison between the values φ_c obtained by means of numerical simulations and the values of the reduction factor φ_Δ of the q-factor (φ_m mean values and φ_k characteristic values) provided by the approximated formulation.

values φ_c are given in Table 5.3 where they are compared with the mean values φ_m and the characteristic values φ_k from equation (5.16). In the same table the values of the available global ductility μ of the equivalent SDOF system and the slope γ of the softening branch of the α–δ curve, obtained by static inelastic analyses for the given frames, are also shown.

A comparison between the actual values of the reduction factor φ_c with those provided by equation (5.16) is given also in Fig. 5.26. A very good agreement with the forecasts based on the SDOF system (equation 5.16) can be observed for all the examined frames (MDOF systems). In fact, the computed reduction factor values fall within the interval between φ_k and φ_m and are mainly close to the lower bound.

5.6   INFLUENCE OF GEOMETRIC NON-LINEARITY

From the design point of view, it is important to determine in which cases the geometrical non-linearity is negligible. We can use the definition of a limiting value of the reduction factor. We define φ_{lim} as the mean value of the reduction factor, which separates cases in which P–Δ effects can be neglected ($\varphi_\Delta < \varphi_{lim}$) from cases in which geometrical non-linearity has to be included into the analysis ($\varphi_\Delta > \varphi_{lim}$).

By adopting the kinematic ductility as damage parameter, equation (5.16) provides maximum values of the stability coefficient γ, for any given value of the available ductility μ and with respect to a given limiting value $\varphi_\Delta = \varphi_{\lim}$. These maximum values of γ are given in Table 5.4. for a limiting level of the q-factor reduction equal to 1.05, 1.10 and 1.20, respectively.

Table 5.4   Maximum values of the stability coefficient γ for different values of the available ductility μ with respect to a given limit value $\varphi_{\lim}$

μ	$\varphi_{\lim} = 1.05$	$\varphi_{\lim} = 1.10$	$\varphi_{\lim} = 1.20$
2	0.030	0.058	0.110
4	0.012	0.024	0.047
6	0.007	0.013	0.026

Therefore, it can be seen that, even if a 20% q-factor reduction is considered negligible then, by assuming as an example the value 6 for the available ductility, the maximum value which can be reached by the stability coefficient γ is 0.026 (Table 5.4) and, therefore, according to equation (5.8) a value about of 40 is required for the critical elastic multiplier of vertical loads.

In other words the effects of geometrical non-linearity are, in general, not negligible. An exception is made for cases in which the ratio between the value of the design vertical loads and the one corresponding to their critical value is very small.

With reference to this problem it is useful to point out that UBC91 provides, as already mentioned, the value $\alpha_{cr} = 10$ as the upper limit which defines cases in which P–Δ effects have to be investigated. The corresponding value of γ is 0.1 which provides, for $\mu = 6$, a mean reduction factor of about 1.80 (Fig. 5.3). It seems that a 45% reduction (1/1.80) of the q-factor, due to overall stability effects, is far from negligible! And this result becomes even more impressive if we consider the characteristic values, leading to $\varphi_\Delta = 2.52$, which corresponds to a reduction of 60% (Fig. 5.3).

A more rational provision is given by Eurocode 8 and ECCS Recommendations, where control of the P–Δ effects is màde on the basis of the estimate of the inelastic displacements. In the case of moment-resisting frames, taking into account that $q = 5\alpha_u/\alpha_y \approx 5 \times 1.2 = 6$, second-order effects have to be taken into account when α_{cr} is approximately less than 60, which corresponds to γ approximately greater than 0.017. This last value, chosen by assuming approximately $q = \mu = 6$ (according to the ductility factor theory) corresponds to a reduction in the q-factor of about 10% (1/1.13, Table 5.4). Therefore, it can be concluded that the provision given by Eurocode 8 and ECCS Recommendations allows a more correct prediction of cases for which P–Δ effects can be neglected, while the UBC91 provisions are too unsafe.

5.7 REFERENCES

[1] *A. De Luca, C. Faella, V. Piluso:* 'Stability of Sway Frames: Different Approaches Around the World', ICSAS91, International Conference on Steel and Aluminium Structures, Singapore, 22–24 May, 1991.

[2] *R. Husid:* 'The Effect of Gravity on the Collapse of Yielding Structures with Earthquake Excitation', IV WCEE, Santiago, Chile, 1969.

[3] *C.K. Sun, G.V. Berg, R.D. Hanson:* 'Gravity Effect on Single Degree Inelastic System', Journal of Mechanical Division, ASCE, February, 1973.

[4] *F. Braga, A. Parducci:* 'Plastic Deformations Required to R/C Structures During Strong Earthquakes – Influence of the Mechanical Decay by Taking into Account P-Δ Effect', VI European Conference on Earthquake Engineering, Dubrovnik, Yugoslavia, September 18–22, 1978.

[5] *S. Mahin, V. Bertero:* 'An evaluation of inelastic seismic design spectra', Journal of the Structural Division, ASCE, September, 1981.

[6] *G. Al Sulaimami, J.M. Roesset:* 'Design Spectra for Degrading Systems', ASCE, Journal of Structural Engineering, Vol. 115, 1985.

[7] *D. Bernal:* 'Amplification Factors for the Inelastic Dynamic P-Δ Effects in Earthquake Analysis', Earthquake Engineering and Structural Dynamics, Vol. 15, 1987.

[8] *B. Palazzo, F. Fraternali:* 'L' Influenza dell' Effetto P-Δ sulla Risposta Sismica di Sistemi a Comportamento Elastoplastico: Proposta di una Diversa Formulazione del Coefficiente di Struttura', Giornate Italiane della Costruzione in Acciaio, CTA, Trieste, Ottobre, 1987.

[9] *E. Cosenza:* 'Duttilità Globale delle Strutture Sismoresistenti in Acciaio', Ph.D. Thesis, Napoli, Aprile, 1987.

[10] *E. Cosenza, A. De. Luca, C. Faella, V. Piluso:* 'A Rational Formulation for the q-factor in Steel Structures', paper 8-4-11, Vol. V, IX WCEE, Tokyo-Kyoto, August 2–9, 1988.

[11] *I. Mitani, M. Makino:* 'Post Local Buckling Behaviour and Plastic Rotation Capacity of Steel Beam-Columns', 7th. WCEE, Istanbul, 1980.

[12] *H. Akiyama:* 'Earthquake-Resistant Limit State Design for Buildings', University of Tokyo Press, 1985.

[13] *F.M. Mazzolani, V. Piluso:* 'Evaluation of the Rotation Capacity of Steel Beams and Beam-Columns', 1st COST C1 Workshop, Strasbourg, 28–30 October, 1992

[14] *E. Cosenza, C. Faella, V. Piluso:* 'Effetto del Degrado Geometrico sul Coefficiente di Struttura', IV Convegno Nazionale, L'Ingegneria Sismica in Italia, Milano, Ottobre, 1989

[15] *R. Landolfo, F.M. Mazzolani, M. Pernetti:* 'L'influenza dei criteri di dimensionamento sul comportamento sismico dei telai in acciaio', IV Convegno Nazionale, L'Ingegneria Sismica in Italia, Milano, Ottobre, 1989

[16] *R. Landolfo, F.M. Mazzolani:* 'The consequence of the design criteria on the seismic behaviour of steel frames', 9th European Conference on Earthquake Engineering, Moscow, 11–16 September 1990.

[17] *C. Faella, O. Mazzarella, V. Piluso:* 'L'influenza della non-linearità geometrica sul danneggiamento strutturale', VI Convegno Nazionale, L'Ingegneria Sismica in Italia, Perugia, 13–15 Ottobre 1993.

[18] *C.A. Guerra, F.M. Mazzolani, V. Piluso:* 'Overall Stability Effects in Steel Structures Under Seismic Loads', 9th European Conference on Earthquake Engineering, Moscow, 11–16 September 1990.

[19] *F.M. Mazzolani, V. Piluso:* 'P-Δ Effect in Seismic Resistant Steel Structures', SSRC Annual Technical Session & Meeting, Milwaukee, April, 1993.

[20] *UBC91:* 'Uniform Building Code', International Conference of Building Officials, Edition 1991.

[21] *Commission of the European Communities:* 'Eurocode 3: Design of Steel Structures', 1992.

[22] *IAEE:* 'Earthquake Resistant Regulations: A World List', compiled by the International Association for Earthquake Engineering, 1984.

[23] *Commission of the European Communities:* 'Eurocode 8: European Code for Seismic Regions', Draft of October 1993.

[24] *ECCS (European Convention for Constructional Steelwork):* 'European Recommendations for Steel Structures in Seismic Zones', 1988.

[25] *M.R. Horne:* 'An approximate method for calculating the elastic critical loads of multistorey plane frames', Structural Engineer, N.53, 242, 1975.

[26] *N. Nigam, P. Jennings:* 'Digital Calculation of Response Spectra from Strong Motion Earthquake Records', EERL Report, California Institute of Technology, Pasadena, June, 1968.

[27] *J.M. Nau:* 'Computation of Inelastic Response Spectra', Journal of Engineering Mechanics, ASCE, January, 1981.

[28] *CNR-GNDT:* 'Norme Tecniche per le Costruzioni in Zone Sismiche', Gruppo Nazionale per la Difesa dai Terremoti, Dicembre, 1984.

[29] *Y.J. Park, A.H.S. Ang:* 'Mechanic Seismic Damage Model for Reinforced Concrete', Journal of Structural Engineering, ASCE, April 1985.

[30] *H. Krawinkler, M. Zohrei:* 'Cumulative Damage in Steel Structures Subjected to Earthquake Ground Motion', Computer & Structures, Vol. 16, N.1–4, 1983.

[31] *J.E. Stephens, J.T.P. Yao:* 'Damage Assessment Using Response Meas-

urements', Journal of Structural Engineering, ASCE, Vol. 113, April, 1987.

[32] *E. Cosenza, G. Manfredi, R. Ramasco:* 'An Evaluation of the Use of Damage Functionals in Earthquake Engineering Design', 9th European Conference on Earthquake Engineering, Moscow, September, 1990.

[33] *E. Cosenza, G. Manfredi, R. Ramasco:* 'The Use of Damage Functionals in Earthquake Engineering: A Comparison between Different Methods', Earthquake Engineering and Structural Dynamics, Vol. 22, 1993.

[34] *P. Fajfar:* 'Equivalent Ductility Factors, taking into account Low-Cycle Fatigue', Earthquake Engineering and Structural Dynamics, Vol. 21, pp. 837–848, 1992.

[35] *G.H. Powell:* 'Drain–2D User's Guide', Earthquake Engineering Research Center, University of California, Berkeley, September 1973.

6

Behaviour of connections

6.1 CONNECTION ROLE IN SEISMIC-RESISTANT FRAMES

Steel structures are more and more extensively used in regions of high seismic risk, because of their excellent performance in terms of strength and ductility. It is mainly the mechanical behaviour of materials and structural elements which determines the fulfilment of the design requirements set by seismic codes. Dissipative structures are designed by allowing the yielding of some zones of its members, the so-called dissipative zones. During a catastrophic earthquake these zones must dissipate the earthquake input energy by means of hysteretic ductile behaviour in the plastic range. The formation of appropriate dissipative mechanisms is related to the structural typology. Moment-resisting frames have a large number of dissipative zones, located near the beam-to-column connections, which dissipate energy by means of cyclic bending behaviour. Therefore, the ductility of seismic-resistant steel frames is also deeply influenced by the behaviour of their connections.

Two different approaches can be taken in the design of beam-to-column connections of seismic-resistant steel frames. The first is based on the location of the dissipative zones at the beam ends, so that the earthquake input energy is dissipated through the cyclic plastic bending of the beam ends. The second approach relies on the energy dissipation through the cyclic plastic bending of the connections, i.e. dissipative zones are located in the connections rather than in the beams. It is clear that, in the first case, connections should possess sufficient strength so that plastic hinges can be formed at the beam ends providing ductility to the frame. In the second case, the key parameters of the connection behaviour are their ductility and energy dissipation capacity under cyclic loading.

Up to now, in seismic codes there is a general agreement in adoption of the first design approach. As an example, Eurocode 8 [1] recommends that connections in dissipative zones shall have sufficient overstrength to allow for yielding of connected parts, taking into account the maximum value of their yield strength. As a consequence, the connections of the dissipative parts by means of butt welds or full penetration welds is suggested, because they are deemed to satisfy the overstrength criterion. As an alternative, the use of fillet weld connections or bolted connections is allowed, provided that their yield strength is at least 1.20 times that of the connected part.

Even if participation by the connections in the seismic energy dissipation, according to the second approach, is not forbidden, it is strongly limited in everyday practice, because experimental control of the effectiveness of such connections under cyclic loading is required.

In the first design approach, the connection parameters affecting the global behaviour of the frame are the rotational stiffness and the ultimate bending resistance. In the opposite approach, two additional parameters have to be considered: the connection rotation capacity and its energy dissipation capacity.

This chapter covers the classification of beam-to-column connections and then the analysis and modelling of their behaviour under monotonic and cyclic loading conditions.

6.2   FRAME CLASSIFICATION

According to the structural system providing the strength against lateral displacements, frames are classified in Eurocode 3 [2] as braced and unbraced frames. The term 'braced frames' is used when a very stiff system of bracing elements is provided in order to withstand the total amount of horizontal forces. In the opposite case, the term unbraced frame is adopted.

A further classification is provided on the basis of frame sensitivity to second-order effects in the elastic range. The term 'non-sway frame' is used when the in-plane lateral stiffness is sufficient to make it acceptable to neglect the geometrical second-order effects; this requirement is satisfied, according to Eurocode 3, when the critical elastic multiplier of vertical loads is greater than 10. When the previous condition is not fulfilled, whether for braced or for unbraced frames, the term 'sway frame' is used.

A steel frame belonging to the above categories, whether considered as a plane or spatial structural system, essentially consists of linear members joined together by connections.

Some definitions are necessary, because of the indiscriminate everyday use of the terms 'joints' and 'connections'. The most rigorous definition [3–5] (Fig. 6.1), for connection is the physical component which mechanically fastens the beam to the column, and it is concentrated at the location where the fastening action occurs, while the joint is the connection plus

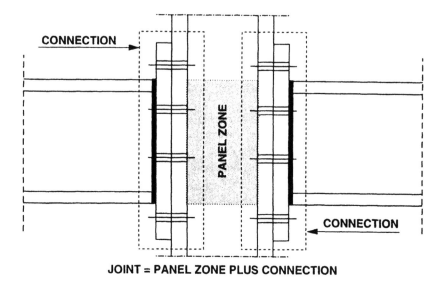

JOINT = PANEL ZONE PLUS CONNECTION

Fig. 6.1 Distinction between joint and connection

the corresponding zone of interaction between the connected members, namely the panel zone of the column web.

Thus the deformation of the 'joint' normally contains the deformation of the 'connection', which significantly contributes to the inelastic response of the structural system.

Prior to the introduction of the concept of semirigidity, two extreme assumptions were made when designing steel frames. The first assumption was that all ends of the members converging at the joint were subjected to the same rotation and the same displacements, due to the rigid and monolithic behaviour of the joint. In the opposite assumption, joints were assumed incapable of transmitting moments and therefore permitted free rotation. The first type are referred to as continuous frames, the second as nominally pinned frames. Modern design codes introduced the concept that real joints behave in an intermediate way, between these two extreme cases of rigid or simple framing.

The design of a structure has to be based on the actual load versus deformation characteristics of the·joints. The structural system is consequently classified as semicontinuous (semirigid or partially restrained) construction and its response to the external loads is in general influenced by the structural properties of both members and connections, which can be expressed by strength, stiffness and deformation capacity. Assuming that the bending behaviour is predominant, the above con-

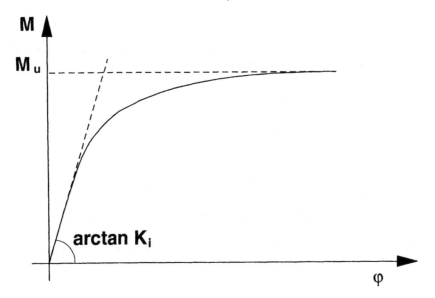

Fig. 6.2 Moment versus rotation curve

nection parameters are given by the moment versus rotation relationship (Fig. 6.2).

In order to include semirigid joint effects in frame design it is necessary to represent the M–φ curve of the connection in a convenient analytical form. For seismic design purposes the M–φ relationship is cyclical and must include all degradation effects which arise as the number of cycles increases.

Interesting applications of semirigid joints are considered nowadays, because of their convenience from the point of view of constructional aspects (simplicity means economy). This new concept can be introduced more generally within the framework of mechanical imperfections of the 'industrial frame', thus leading to a more realistic interpretation of the overall load-carrying capacity of the structure. Following a recent proposal [6], the industrial frame is a structure affected by global imperfections and it is formed by 'imperfect bars' (the 'industrial bar' concept was developed during the 1960s) joined together by 'imperfect joints' (the semirigid joint concept developed during the 1980s). At present the use of semirigid joints is not codified in recommendations dealing with seismic-resistant structures, probably due to a lack of knowledge and experience which therefore prevents the development of a consolidated calculation method.

6.3   CLASSIFICATION OF JOINTS

As we are dealing with moment-resisting frames, the emphasis is on the beam-to-column connections. There are several technological systems that provide this connection (Figs. 6.3 and 6.4). From the values of the non-dimensional stiffness ($\overline{K} = KL/EI_b$, where L and I_b are, respectively, the length and the inertia moment of the connected beam) (Fig. 6.5) [7], non-linear behaviour ranging from the quasi-perfectly rigid (fully welded, extended end plates) to the flexible (double web angle) is possible. The intermediate positions correspond to a range of semirigidity; some common types are top and bottom flange splices, tee stubs, flush end plate, flange and web angles, and flange angles only.

The EC3 provisions [2] introduce in a code, for the first time, a correlation between the type of structure and the method of global analysis, based on the type of connections. For semicontinuous framing the following distinctions have to be drawn, according to the joint properties involved in the global structural analysis.

a. Elastic analysis has to be based on reliably predicted linear moment–rotation relationships for the used connections.
b. Rigid–plastic analysis has to be based on the design moment resistances of connections which have been demonstrated to have sufficient rotational capacity.
c. Elastic–plastic analysis has to be based on the non-linear moment–rotation relationship for the used connections.

In the case of elastic design, the classification by rigidity leads to the three main categories of nominally pinned, rigid and semirigid connections.

Nominally pinned connections are assumed to transfer the shear and eventually the normal force from the beam to the column. Moreover, they must be able to rotate without developing significant moments, which might adversely affect the resistance of the columns.

Rigid connections transmit all end reactions, and their deformation is sufficiently small that their influence on the moment distribution in the structure or on its overall deformation may be neglected.

Semirigid connections are designed to provide a predictable degree of interaction between members, based on the design moment–rotation characteristics of the joints.

The first two categories are the traditional ones. The nominally pinned connections are widely used when lateral stiffness of the structure is guaranteed by appropriate bracing systems. In the case of moment-resisting frames the rigid connections often lead to relatively expensive constructional details. The intermediate third category has been introduced to fill the gap between pinned and rigid connections and is now accepted in the updated codes (i.e. EC3).

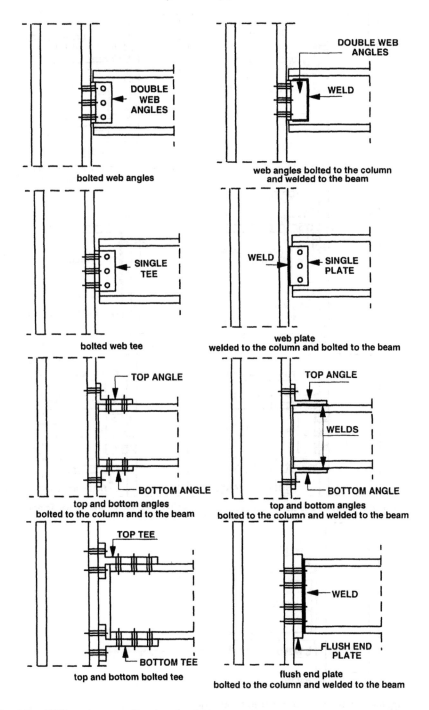

Fig. 6.3 Different connection typologies

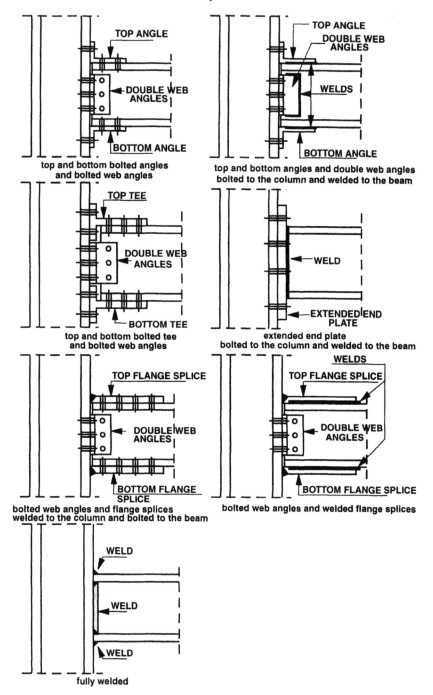

Fig. 6.4   Different connection typologies

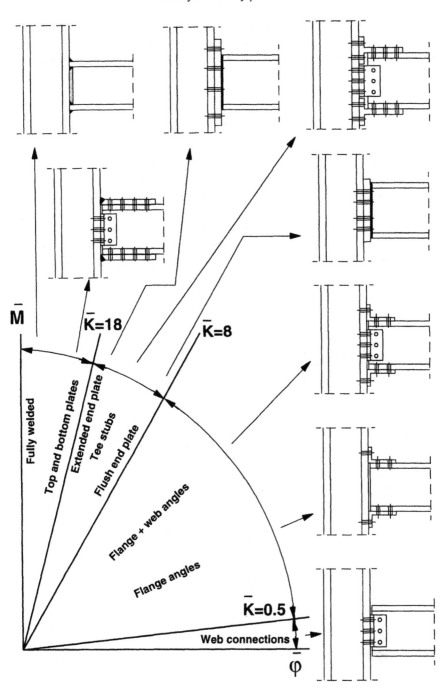

Fig. 6.5 Typical values of the non-dimensional stiffness of different connections

In the case of elastic design, the semirigid connection is taken into account in the calculation model by a rotational spring which is characterized by the elastic constant K [8]. The typical values of K are usually expressed through the ratio $\bar{K}$ between the rotational stiffness K of the connection and the flexural stiffness $E\,I_b/L$ of the connected beam (Fig. 6.5) [9].

In the application of plastic design, according to EC3, the connections can be classified (Fig. 6.6) as full strength and partial strength.

Full-strength connections have a design resistance at least equal to that of the connected member (case A). A plastic hinge will be formed in the adjacent member, but not in the connection. If the rotation capacity is limited (case B), an extra reserve of strength should be required to account for possible overstrength effects in the member. On the contrary, the design resistance of partial strength connections is less than that of the connected member (case C). A plastic hinge will be formed in the connection, so sufficient rotation capacity is required (case D). This means that case C is unsuitable, because its rotation capacity could be exceeded under design loads.

Remembering the present state of codification in the field of seismic-resistant structures, full-strength connections are usually preferred and an overstrength of 1.2 is suggested for fillet welds and bolted connections both in ECCS Recommendations and in Eurocode 8 [1, 10]. The above

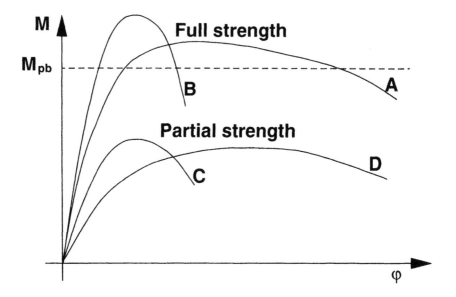

Fig. 6.6   Definition of full-strength and partial-strength connections

classification is based only on a qualitative point of view, which is universally accepted. However, for practical applications, a quantitative classification is necessary.

Different classification systems have been proposed in the technical literature. The main difficulty in the development of a classification system is to establish criteria able to make it suitable for serviceability as well as ultimate limit states design. The principal interests are the connection stiffness and the connection strength respectively, for the serviceability limit state and the ultimate limit state. In addition, there are other properties, such as rotation capacity and energy dissipation capacity, which play a fundamental role in seismic-resistant frames.

Since the connection response is non-linear, it would be logical to have regions in the M–φ diagram defined through non-linear boundary curves. However, for practical purposes, the use of linearized boundary curves is preferred. This simplification is justified because the initial linear branch is able to reflect the conditions related to the serviceability limit state, while the horizontal plateau is related to the ultimate limit state conditions. Furthermore, a bilinear approximation is used in many computer programs for analysing semirigid frames [11].

Beam-to-column connections can be classified on the basis of their rotational stiffness K and their flexural resistance, i.e. the peak value M_u of the design moment versus rotation curve.

According to Eurocode 3 [2], the boundary curves of the classification diagram are expressed through the non-dimensional parameters

$$\overline{K} = \frac{K_i\, L}{E\, I_b} \tag{6.1}$$

$$\overline{m} = \frac{M_u}{M_{pb}} \tag{6.2}$$

$$\overline{\varphi} = \varphi\, \frac{E\, I_b}{M_{pb}\, L} \tag{6.3}$$

where K_i is the initial rotational stiffness of the connection, and M_{pb}, I_b and L are, respectively, the plastic moment, the moment of inertia and the length of the connected beam.

With reference to the rotational stiffness, the beam-to-column connections can be classified as

- nominally pinned, for $\overline{K} \leq 0.5$
- semirigid, for $0.5 < \overline{K} < \overline{K}^*$
- rigid, for $\overline{K} \geq \overline{K}^*$.

The value of $\overline{K}^*$ has been computed so that it assures for $\overline{K} \geq \overline{K}^*$ a reduction, with respect to the ideal infinitely rigid frame, of the critical elastic multiplier of the vertical loads not greater than 5%, for any given value of the ratio between the flexural stiffness of the beams and the one of the columns. Therefore, $\bar{K}^*$ depends on the type of frame, braced or unbraced, assuming a value of 8 in the first case and of 25 in the second case. The latter is the situation for moment-resisting frames.

With reference to the flexural resistance, the beam-to-column connections can be classified as

- nominally pinned, for $\bar{m} < 0.25$
- partial strength, for $0.25 \leq \bar{m} < 1$
- full strength, for $\bar{m} \geq 1$

In addition, for full-strength connections characterized by $\bar{m} \geq 1.2$, control of the rotation capacity is not necessary.

The boundary curve between rigid and semirigid connections, according to the Eurocode 3 classification diagram, is a trilinear curve. The first branch is given by

$$\bar{m} = \bar{K}^* \bar{\varphi} \ \text{ for } \ \bar{\varphi} \leq \frac{2}{3\bar{K}^*} \tag{6.4}$$

with $\bar{K}^* = 8$ in the case of braced frames and $\bar{K}^* = 25$ in the case of unbraced frames. The second branch is given by

$$\bar{m} = \frac{25\,\bar{\varphi} + 4}{7} \ \text{ for } \ \frac{2}{3\,\bar{K}^*} \leq \bar{\varphi} \leq 0.12 \tag{6.5}$$

for unbraced frames, and by

$$\bar{m} = \frac{20\,\bar{\varphi} + 3}{7} \ \text{ for } \ \frac{2}{3\,\bar{K}^*} \leq \bar{\varphi} \leq 0.20 \tag{6.6}$$

for braced frames.

Finally, the third branch is given by $\bar{m} = 1$ for $\bar{\varphi} > 0.12$ and $\bar{\varphi} > 0.20$, respectively, for unbraced and braced frames. The corresponding classification diagrams are given in Fig. 6.7 for both unbraced and braced frames. The three regions define the rigid, semirigid and nominally pinned (or flexible) connections. Although the above terms are usually adopted, it would be preferable to use more precise terms, such as rigid-full-strength connections and semirigid-partial-strength connections for the first two regions respectively.

However, the case of connections whose moment–rotation curve lies into two different classification regions is not uncommon. Some examples

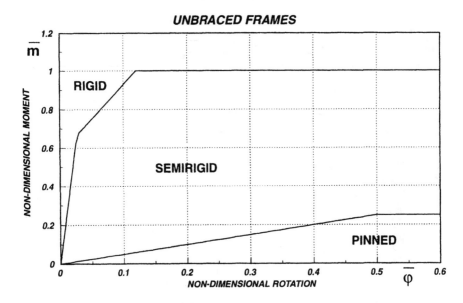

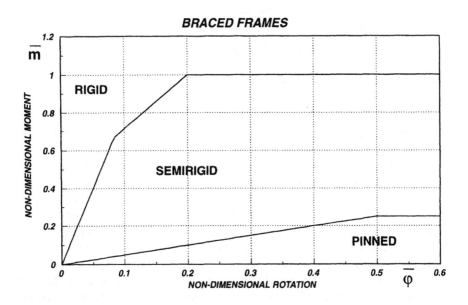

Fig. 6.7   Connection classification according to Eurocode 3

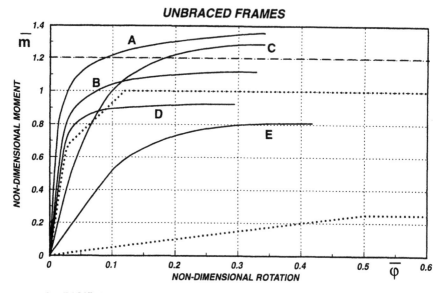

Fig. 6.8   Examples of connection classification

of different types of connection M–φ curves are given in Fig. 6.8, with the corresponding classifications.

The classification adopted by Eurocode 3 is, at least, independent of the geometrical properties of the column. Different methods have been proposed in order to make the classification independent of beam length [12, 13, 14]. Bjorhovde *et al.* [12] have introduced the concept of equivalent reference length of the beam, L_e, which represents the length of the beam whose flexural stiffness EI_b/L_e is equal to the rotational stiffness of the connection. It is evident that the stiffer the connection, the shorter the equivalent reference length of the beam will be, as it is confirmed by the results obtained by the same authors from the analysis of experimental data (Table 6.1, where *d* represents the depth of the beam). The authors propose a classification method based on a reference length of the beam equal to *5d*, representing a mean value. Therefore, in order to classify a beam-to-column connection, the moment–rotation curve has to be represented in the $\overline{m}$–$\overline{\varphi}^{\,*}$ plane, where

Table 6.1 Equivalent reference length and moment capacity of typical connections

Extended end plate	Flush end plate	Top/seat angles and web angles	Header plate	Double web angles
$1d < L_e < 2d$	$2d < L_e < 5d$	$4d < L_e < 7d$	$L_e \approx 10d$	$L_e \approx 15d$
$\overline{m} \approx 0.9$	$\overline{m} \approx 0.6$	$0.45 < \overline{m} < 0.6$	$\overline{m} \approx 0.2$	$\overline{m} \approx 0.15$

$$\overline{\varphi}^* = \varphi \, \frac{E \, I_b}{5 \, M_{pb} \, d} \qquad (6.7)$$

The classification diagram proposed by these authors for both unbraced and braced frames is given in Fig. 6.9. The limiting values of the equivalent length of the beam are assumed to be $2d$ and $10d$, while the limiting values of $\overline{m}$ are 0.7 and 0.2. On the basis of their test data, the authors also give the following relationship between the rotation capacity and the non-dimensional ultimate moment [12]

$$\overline{\varphi}_u^* = \frac{5.4 - 3 \, \overline{m}}{2} \qquad (6.8)$$

where $\overline{\varphi}_u^*$ is the ultimate value of the non-dimensional rotation (6.7).

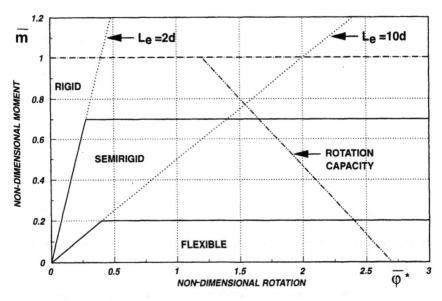

Fig. 6.9 Classification proposal according to Bjorhovde, Brozzetti and Colson [12]

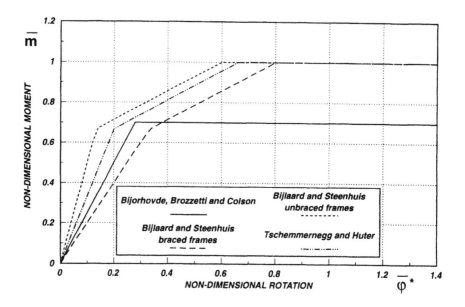

Fig. 6.10   Comparison among different classification proposals

Classification diagrams, which do not require a knowledge of the beam length, have also been proposed by Bijlaard and Steenhuis [13] and by Tschemmernegg and Huter [14]. The first authors [13] propose the use of the Eurocode 3 classification by assuming a constant ratio between beam length and beam depth ($L/d = 25$ for unbraced and $L/d = 20$ for braced frames). However, in order to have a unique classification both for braced and unbraced frames, the second authors [14] propose values for the L/d variable with the non-dimensional ultimate moment $\overline{m}$ (from $L/d = 12$ for $\overline{m} = 0$ to $L/d = 16.5$ for $\overline{m} = 1$, in the case of braced frames, and from $L/d = 37.5$ for $\overline{m} = 0$ to $L/d = 27.5$ for $\overline{m} = 1$, in the case of unbraced frames).

With reference to the boundary curve which separates the rigid and semirigid regions, a comparison between proposals independent of the beam length is given in Fig. 6.10.

Sometimes, and usually in the case of M–φ curves based on raw test results, a change of the initial slope can arise (Fig. 6.11) due to settling of the connection during the first part of the loading phase. The initial rotation φ_0 does not represent a real connection deformation because, in real structures, it would have been removed during the construction, but it plays a role in the successive loading cycles. The proper stiffness value is also represented in the same figure.

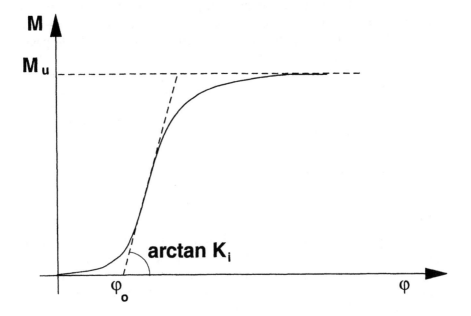

Fig. 6.11   Moment–rotation curve with a change of initial slope

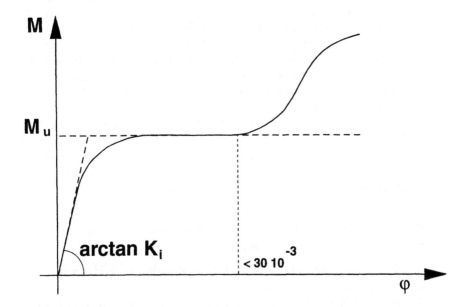

Fig. 6.12   Moment–rotation curve with intermediate plateau

In the case of connections made with pretensioned bolts, as required for slip-resistant joints, the M–φ curve is characterized by an intermediate plateau (Fig. 6.12) corresponding to the occurrence of the bolt slippage. At the end of the plateau, connection stiffness is partially regained. This type of behaviour can also be found in some connections, such as double web angle connections, where large deformations can give rise to the collision between the beam and the column face. In these cases the ultimate moment of the connection should be cautelatively assumed equal to the one corresponding to the plateau [12].

6.4 CYCLIC BEHAVIOUR

6.4.1 Type of cycles

From the point of view of cyclic behaviour, the joint can be stable or unstable; it can be considered stable if it exhibits the same behaviour as the monotonic test, even if the number of cycles increases. On the other hand the behaviour can be considered unstable when the stiffness decreases with the number of cycles [15].

Under cyclic loading, joints can be characterized by three types of behaviour (Fig. 6.13).

a. Joints of the first category exhibit a stable behaviour, characterized by hysteresis loops having the same area inside the curve which remains constant with increasing number of cycles (Fig. 6.13a).
b. Joints of the second category exhibit an unstable behaviour due to permanent deformations in holes and bolts, thus reducing the stiffening effect of the tightening force (Fig. 6.13b). The slope of the hysteresis curves characterizing the stiffness of the ith cycle is continuously decreasing.
c. Joints of the third category exhibit an unstable behaviour characterized by bolt slippage. This phenomenon significantly modifies the shape of the curve by reducing the dissipated energy for the same values of deformations. The increasing deterioration is due to the permanent deformations of holes and shanks (Fig. 6.13c).

Cases **b** and **c** lead to collapse due to deterioration of stiffness.

For seismic considerations a chart of prerequisites [16] can be attempted, and in particular the following.

- A seismic moment connection must be detailed to withstand forces which act in either direction.
- The problem of low-cycle fatigue, which is associated with cyclic loading in the range of large plastic deformations, becomes a factor to be considered in seismic moment connections.

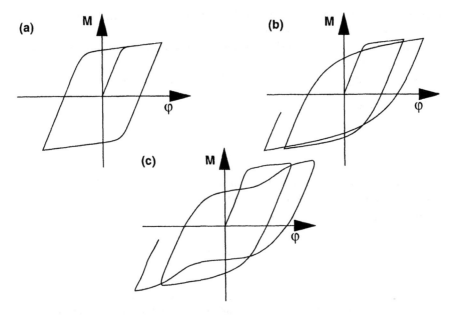

Fig. 6.13 Typical hysteresis loops of structural joints

- Since seismic loadings are random in nature, a probabilistic rather than a deterministic approach of analysis is necessary to assess the behaviour of a seismic moment connection.
- In order to avoid a local collapse mechanism, the design of a framed structure must be done in such a way as to confine plastic hinges to be developed in the beams rather than in the columns.

Considering the shape of hysteresis loops, the ideal cyclic behaviour of the connection must guarantee a sufficient level of strength without stiffness deterioration for a sufficient number of cycles.

The evaluation of an acceptable degree of degradation of the design parameters requires a quantitative analysis of their effects on the overall behaviour of the structure.

6.4.2 Behaviour parameters

The choice of behaviour parameters is fundamental to the interpretation of the overall behaviour of this structural component, as well as for the comparison of the test results between the connected members. They have to characterize the degree of ductility, the amount of energy absorption, and the deterioration of stiffness and strength during the loading process.

The behaviour parameters can be defined in two different ways,

dimensional or non-dimensional. The dimensional parameters provide
information mainly for use in the comparison of the same parameters
coming from the connected members, and also to evaluate the non-dimensional ones.

The following non-dimensional behaviour parameters can be defined.

a. Kinematic ductility parameter

$$\mu_i = \frac{v_i}{v_y} \tag{6.9}$$

the ratio between the maximum value of the displacement at the end of
each cycle and the conventional elastic one.

b. Energetic efficiency parameter

$$E_i = \frac{A_i}{F_y(v_i - v_y)} \tag{6.10}$$

the ratio between the energy dissipated by the structural element in one
half-cycle and the energy which could be dissipated in the same cycle in
ideal elastic–perfectly plastic conditions (where A_i is the area of the ith
semi-cycle).

c. Strength degradation parameter

$$D_{S_i} = \frac{F_i}{F_y} \tag{6.11}$$

the ratio between the peak level of the external action and the
conventional elastic limit.

d. Stiffness degradation parameter

$$D_{k_i} = \frac{D_{S_i}}{\mu_i} = \frac{F_i v_y}{v_i F_y} \tag{6.12}$$

which corresponds to the ratio between the secant stiffness for a given
cycle and the conventional elastic stiffness.

The above parameters can also be used to define the ultimate limit
states which conventionally correspond to the end of the test. In fact, the
structural components can be submitted during the test to conditions
which are not of interest from the physical point of view. In particular,
an ultimate limit state is reached when a significant degradation of
strength, stiffness or energy dissipation capacity arises.

In this way the ultimate states can be identified, as follows.

• The deterioration due to geometrical or mechanical effect produces an intolerable reduction of the peak load carrying capacity during the cyclic test, i.e. a limiting value of D_{S_i} is attained.

- The excess of deformation produces a reduction of the stiffness which becomes unacceptable when it leads to unstable effects on the overall structure or introduces too large a residual deformation, i.e. a limiting value of D_{k_i} is reached.
- The structural component is unable to fulfil its functions when the energetic efficiency parameter is reduced below a given amount, i.e. a limiting value of E_i occurs.

Therefore, the definition of the ultimate limit state conditions corresponds to the definition of the limit values of D_{S_i}, D_{k_i} and E_i, which identify the end of the test. The interpretation of these parameters and of their limiting values requires an understanding of their influence on the overall behaviour of the structure.

Since this is a complex task, a first step towards solution can be limited to an analysis of typical joint behaviour, by comparison of test results.

6.4.3 ECCS Recommendations

As far as the evaluation of behavioural parameters is concerned, many tests have recently been carried out with the aim of reducing the existing knowledge gap.

It was observed that experiments on joints under cyclic loads are carried out at several research organisations often under different testing procedures. This prevents comparison of the various results coming from different laboratories. The need to standardize such experimental procedures was clear and the ECCS Committee TC 13 decided to work out recommended testing procedures for assessing the behaviour of structural steel elements or connections under cyclic loads [17].

In this provision the testing procedure is detailed together with the parameters to be measured in order to provide the so-called behaviour functions.

The non-dimensional parameters are obtained through limit elastic quantities (Fig. 6.14a) which are used to compare the generic true cycle (Fig. 6.14b) with the ideal elasto-plastic cycle having the same amplitude (Fig. 6.14c). According to ECCS Recommendations, the experimental cyclic tests should be carried out by considering successive groups of cycles with an increasing displacement amplitude. Each group of cycles is composed of three cycles having the same amplitude. By denoting with i the index of the generic cycle of each group of three cycles to be performed with the same selected displacement amplitude, the ECCS test standards suggest the following parameters.

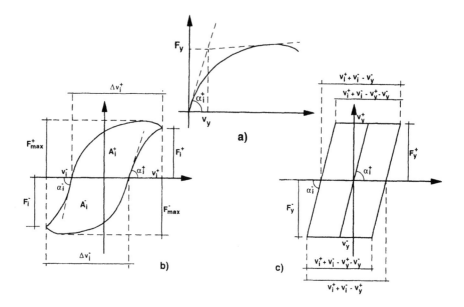

Fig. 6.14 Basic data in the ECCS Recommendations for cyclic tests

a. Partial ductility

$$\mu_i^+ = \frac{v_i^+}{v_y^+} \text{ and } \mu_i^- = \frac{v_i^-}{v_y^+} \tag{6.13}$$

This parameter represents the ratio between the absolute value of the maximum positive (or negative) displacement in the ith cycle and the absolute value of the yield displacement in the corresponding side of the cycle. So, the higher this ratio, the greater is the structure's capacity to withstand large deformations out of the elastic range.

b. Full ductility (or cyclic ductility)

$$\mu_{c_i}^+ = \frac{\Delta v_i^+}{v_y^+} \text{ and } \mu_{c_i}^- = \frac{\Delta v_i^-}{v_y^-} \tag{6.14}$$

This parameter represents the ratio between the absolute value of the maximum displacement amplitude in the positive force range (or negative) in the ith cycle and the corresponding yield displacement. For this parameter the same considerations as for partial ductility apply.

c. Full ductility ratio

$$\psi_i^+ = \frac{\Delta v_i^+}{v_i^+ + v_i^- - v_y^-} \text{ and } \psi_i^- = \frac{\Delta v_i^-}{v_i^+ + v_i^- - v_y^+} \tag{6.15}$$

This parameter represents the ratio between the absolute value of the maximum displacement amplitude in the positive force range (or negative) in the ith cycle and the corresponding displacement in a perfect elasto-plastic behaviour. So, the higher this ratio, the greater is the deterioration of the structure (for instance due to loss of stiffness, slip etc.).

d. Resistance ratio (or strength degradation ratio)

$$\varepsilon_i^+ = \frac{F_i^+}{F_y^+} \text{ and } \varepsilon_i^- = \frac{F_i^-}{F_y^-} \tag{6.16}$$

This parameter represents the ratio between the force corresponding to the maximum positive or negative displacement in the ith cycle and the yield force in perfect elasto-plastic behaviour.

e. Stiffness ratio (or stiffness degradation ratio)

$$\zeta_i^+ = \frac{\tan\alpha_i^+}{\tan\alpha_y^+} \text{ and } \zeta_i^- = \frac{\tan\alpha_i^-}{\tan\alpha_y^-} \tag{6.17}$$

This parameter is the ratio between the stiffness of the tested structure in the ith cycle and the initial stiffness. A small value for this ratio (<1) indicates a large loss of stiffness of the structure. This can be caused by global and local buckling phenomena, by the Bauschinger effect exhibited by steel subjected to inelastic load reversals or by the residual curvature during previous cycles.

f. Absorbed energy ratio

$$\eta_i^+ = \frac{A_i^+}{F_y^+(v_i^+ + v_i^- - v_y^+ - v_y^-)} \text{ and } \eta_i^- = \frac{A_i^-}{F_y^-(v_i^+ + v_i^- - v_y^+ - v_y^-)} \tag{6.18}$$

This parameter represents the ratio between the energy absorbed by the structure in a real half-cycle and the energy absorbed in the half-cycle corresponding to perfect elasto-plastic behaviour with the same displacement amplitude.

All these parameters are defined as the ratio between the value found in the cyclic testing procedure and that corresponding to a reference test, in which it is assumed to exhibit perfect elasto-plastic behaviour. The optimum condition is therefore reached when the behaviour of the structure closely follows the ideal perfect elasto-plastic behaviour, i.e. when the values of these parameters are near to unity. A small value for these parameters (<1) can be assumed as the end of the test, because this indicates a substantial loss of resistance, stiffness or energy dissipation.

It has to be recognized that, although the parameters that are to be recorded during the test are universally accepted, some doubts have been raised over the loading program. In fact, it has been shown that the collapse under cyclic loading can be interpreted as a low-cycle fatigue phenomenon. For this relatively recent trend, failure is expressed through a fatigue curve which relates to the plastic amplitude of the semi-cycles or reversals (Chapter 3)

$$N_{rf} = \frac{1}{B} \left(\frac{\Delta v_p}{v_y} \right)^{-b} \tag{6.19}$$

where N_{rf} is the number of reversals to failure in a constant amplitude (Δv_p) cyclic test.

As a consequence, many constant amplitude tests should be carried out in order to obtain, by regression analysis, the parameters B and b of the fatigue curve (6.19).

However, it is easy to recognize that, if the assumption of validity of Miner's rule on linear damage cumulation is made, the ECCS loading program allows also determination of the parameters of the fatigue curve. In both cases at least two experimental tests on the same structural detail are necessary, but a greater number is preferred to account for random scatters.

6.5   PREDICTION OF MONOTONIC BEHAVIOUR

6.5.1 General

As discussed previously, the required level of information about semirigid and/or partial strength behaviour of connections is dependent on the type of global structural analysis to be performed.

The most accurate knowledge of connection behaviour is obtained through experimental tests, but this technique is too expensive for everyday practice and is usually reserved for research purposes only.

Considerable experimental testing has been conducted throughout the world, leading to the development of computerized data banks [18, 19]. However, their use is mainly devoted to the validation of models, aimed at the prediction of joint behaviour from its geometrical and mechanical properties, rather than to daily design practice. In fact, the designer has only a low probability of finding in the data bank the specific joint he is studying, due to the great variety of connection configurations (Figs.

6.3 and 6.4) and of panel zone stiffening details (Fig. 6.15). This situation justifies the use of suitable models for predicting joint behaviour.

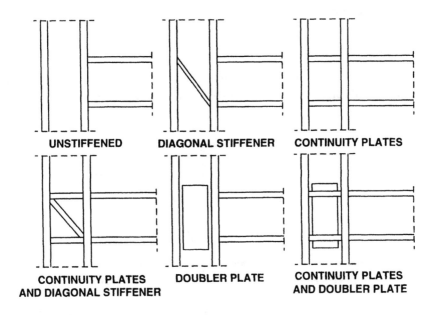

Fig. 6.15 Different stiffening details

Such models can be classified into four main categories:

- mathematical models or curve fitting
- analytical models
- mechanical models
- finite elements.

It should be clear that the degree of accuracy obtained through a predictive model has to be judged not merely on the basis of the refinement level attained in fitting the true moment–rotation curve, but rather on the influence that this level of approximation has on the overall frame performance. Moreover, it has to be taken into account that the need for very accurate representations of joint behaviour decreases as the extension of the structure increases. This also means that the use of sophisticated numerical procedures, such as the finite element method, is justified only in research activity in order to perform parametric analyses that are able to demonstrate the role of the key parameters of the connections, and to improve predictions obtained by simpler methods.

These models relate only to connection behaviour under monotonic loading, but it is clear that all connections, including those in non-seismic areas, can be subjected to unloading and reloading. For this reason, and particularly in the case of seismic loading, simulation of the cyclic behaviour of the connection is recognized to be a fundamental matter, and is discussed in Section 6.7.

6.5.2 Mathematical models

Figure 6.16 provides a compendium of past attempts to represent in a mathematical form the M–φ relationship of the connection.

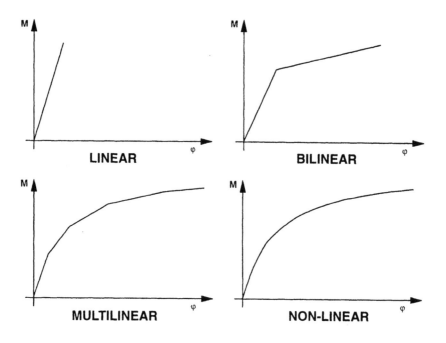

Fig. 6.16   Different mathematical models for the moment–rotation curve

The most simple model, but also the least accurate, is the linear model which overestimates the connection stiffness at finite rotations. A significant improvement is obtained through the bilinear model. Despite not being able to account for the continuous changes of stiffness in the knee region of the connection M–φ curve, this model is used in many computer programs for analysing semirigid frames [11]. In order to overcome this limitation, trilinear [20] and multilinear [21] models have been proposed.

A very high degree of accuracy can be obtained by non-linear models. Some of the important ones are now summarized.

The first non-linear model to be proposed is the odd-power polynomial model, expressing the M–φ curve through the relationship [22]

$$\varphi = C_1 \, (kM) + C_2 \, (kM)^3 + C_3 \, (kM)^5 \qquad (6.20)$$

where k is a parameter which depends on the main geometrical properties of the connection, and C_1, C_2 and C_3 are curve-fitting constants.

The main drawback of this formulation is that, in some cases, the slope of the M–φ curve may become negative for some values of M [23]. This is physically unacceptable and, in addition, can cause numerical difficulties in the analysis of semirigid frames using the tangent stiffness formulation. In order to overcome these difficulties, a modified procedure has been developed [24]. The parameter k of equation (6.20) is expressed as

$$k = P_1^{\alpha_1} \, P_2^{\alpha_2} \, \dots \, P_n^{\alpha_n} \qquad (6.21)$$

where P_i is the generic geometric connection parameter affecting the M–φ curve, while the coefficients α_i are obtained through a curve-fitting process.

A different approach to avoid the problem of negative stiffness arising in the polynomial representation is given by the cubic B-spline model [25]. This method is based on the subdivision of the experimental M–φ curve into a number of subsets.

In order to fit each subset of data accounting for continuities of first and second derivatives of the knot points between two successive subsets, a cubic B-spline curve is used leading to a formulation of the type [4]

$$\varphi = \varphi_o + \sum_{j=0}^{m} b_j \, (< M - M_j >)^3 \qquad (6.22)$$

in which m is the number of knot points between two elementary parts of the M–φ curve and

$$< M - M_j > \, = M - M_j \qquad \text{for } M > M_j$$

$$< M - M_j > \, = 0 \quad \text{for } M < M_j \qquad (6.23)$$

where M_j is the upper bound moment of the jth part of the curve, while φ_o represents the initial rotation (usually $\varphi_o = 0$) and the coefficients b_j are obtained by least-squares curve fitting. This curve-fitting process requires a large quantity of data.

A simpler approach is the use of a power expression of the Goldberg and Richard type [26, 27]

$$M = \frac{(K_i - K_p)\ \varphi}{\left[1 + \left(\varphi/\varphi_y\right)^n\right]^{1/n}} + K_p\ \varphi \text{ with } \varphi > 0 \text{ and } M > 0 \qquad (6.24)$$

where K_p is the strain-hardening connection stiffness and φ_y the reference plastic rotation, equal to M_u/K_i.

In its elastic–perfectly plastic form, equation (6.24) gives

$$M = \frac{K_i\ \varphi}{\left[1 + \left(\varphi/\varphi_y\right)^n\right]^{1/n}} \text{ with } \varphi > 0 \text{ and } M > 0 \qquad (6.25)$$

or, equivalently,

$$\varphi = \frac{M}{K_i\left[1 - \left(M/M_u\right)^n\right]^{1/n}} \text{ with } \varphi > 0 \text{ and } M > 0 \qquad (6.26)$$

Since the model has only four parameters (three in its elastic–perfectly plastic form), it is not as accurate as the cubic B-spline model, but the input data requirement is drastically reduced.

As an alternative, an equation of the Ramberg–Osgood type [28] can be used. According to Ang and Morris [29], a reasonably good representation of a non-linear M–φ curve is obtained through the relationship

$$\frac{\varphi}{\varphi_{k_o}} = \frac{kM}{(kM)_o}\left[1 + \left|\frac{kM}{(kM)_o}\right|^{n-1}\right] \qquad (6.27)$$

where k is a standardization constant dependent on the connection, while φ_{k_o} and $(kM)_o$ are defined in Fig. 6.17. The exponent n controls the shape of the curve.

A model good at representing non-linear connection behaviour is the exponential model of Lui and Chen [30]. It has the form

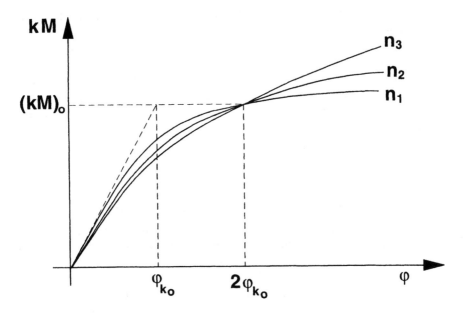

Fig. 6.17 Ang and Morris mathematical model

$$M = \sum_{j=1}^{m} C_j \left[1 - \exp\left(\frac{-|\varphi|}{2 j a} \right) \right] + M_o + K_p \ |\varphi| \qquad (6.28)$$

where M_o is the initial moment, K_p is the strain-hardening connection stiffness and a and C_j are modelling parameters. This is a multiparameter model requiring $(m + 3)$ parameters, where m is the number of curve-fitting constants (C_j); usually, for a sufficient degree of accuracy, $m = 4$ to 6.

Finally, a mathematical model of the exponential type has also been proposed by Yee and Melchers [31]. Their proposal, like the power model, requires only parameters directly related to the mechanical behaviour of the connection, reserving the calibration against test data for only one empirical coefficient c

$$M = M_p \left[1 - \exp\left(\frac{-\varphi (K_i - K_p + c\,\varphi)}{M_p} \right) \right] + K_p \ \varphi \qquad (6.29)$$

Here, M_p is a parameter related to the joint's plastic resistance.

The advantage of mathematical models is the possibility of representing with extreme accuracy any shape of the M–phi curve. On the down side, they are often limited to the joint configurations used for the calibrating formulae.

In addition, they are not able to show the contribution of each component to the overall joint response. Therefore, their application outside the range of the calibration data has to be avoided, expecially for cases in which the collapse mechanism is strongly affected by the geometrical and/or mechanical properties of the connection.

6.5.3 Analytical models

In order to predict the moment–rotation curve of different connection typologies directly from the geometrical and mechanical properties, several authors have applied the basic concepts of elastic structural analysis and limit design to simplified models of various types of beam-to-column connections. This approach starts with the observation of test behaviour to identify the main deformation sources and the collapse mechanism of the connections. Therefore, on the basis of experimental evidence, a simplified connection model is assumed for predicting the initial connection stiffness by an elastic analysis. In addition, the observed plastic mechanism is modelled to predict the ultimate moment through the balance between internal and external work. The reliability of the results obtained through the assumed models is controlled by verifying the agreement with test data. Finally, the description of the M–φ relationship is completed by a curve fitting using the predicted initial stiffness and ultimate moment capacity and a shape parameter.

Among the researchers which have devoted much effort towards predicting the response of the connection directly from its geometrical and mechanical properties, the pride of place goes to Chen *et al.* [32–34], for their extensive research on the semirigid behaviour of connections with angles; only their final formulae for predicting the initial stiffness and the ultimate moment of connections with angles are given herein. For top and seat angles with double web angle connections (Fig. 6.18), the following relationships are given [32, 34].

The connection initial stiffness is given by

$$K_i = \frac{3\,E\,I_t\,d_1^2}{g_1\,(g_1^2 + 0.78\,t_f^2)} + \frac{6\,E\,I_a\,d_3^2}{g_3\,(g_3^2 + 0.78\,t_a^2)} \tag{6.30}$$

where I_t and I_a are the inertia moments of the leg adjacent to the column face of the top angle and of the web angle respectively

$$I_t = \frac{L_t\,t_t^3}{12} \text{ and } I_a = \frac{L_p\,t_a^3}{12} \tag{6.31}$$

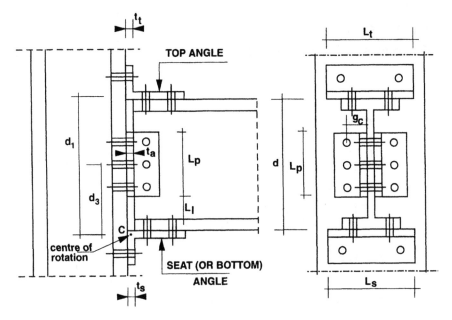

Fig. 6.18   Top and seat angle connection with double web angles

where t_t and t_a are the thickness of the top angle and of the web angle.

In addition, g_1 and g_3 represent the distance between the nut edge and the middle line of the angle leg adjacent to the beam, where g_1 refers to the top angle and g_3 to the web angle (Fig. 6.19). Finally, L_t and L_p are the length of the top angle and of the web angle, respectively. Moreover, with reference to top and seat angles, d_1 is the distance between the middle lines of the legs adjacent to the beam flanges; d_3 is the distance between the centre of the web angles and the middle line of the seat angle leg adjacent to the beam flange.

The ultimate bending moment is given by

$$M_u = \sigma_y \frac{L_s\, t_s^2}{4} + \frac{V_{pt}\,(g_1 - k_t)}{2} + V_{pt}\, d_2 + 2\, V_{pa}\, d_4 \qquad (6.32)$$

where L_s and t_s are the length and the thickness of the seat angle and k_t the distance between the heel of the top angle and the toe of the fillet (Fig. 6.19). Moreover, in equation (6.32) the following notation has been used

$$d_2 = d + \frac{t_s}{2} + k_t \qquad (6.33)$$

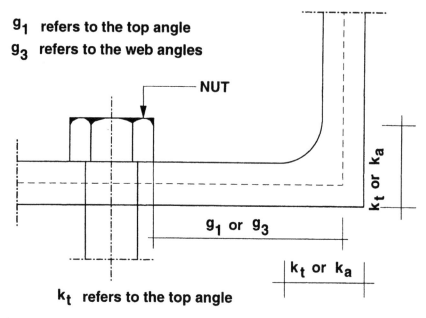

g_1 refers to the top angle

g_3 refers to the web angles

NUT

g_1 or g_3

k_t or k_a

k_t or k_a

k_t refers to the top angle

Fig. 6.19 Geometrical angle parameters

$$d_4 = \frac{2\,V_{pu} + \dfrac{\sigma_y\, t_a}{2}}{3\left(V_{pu} + \dfrac{\sigma_y\, t_a}{2}\right)}\, L_p + L_l + \frac{t_s}{2} \tag{6.34}$$

where L_p and t_a are the length and the thickness of the web angles, L_l is the distance between the lower edge of the web angles and the compression flange of the beam, while V_{pu} is obtained from the relationship

$$\left(\frac{2\,V_{pu}}{\sigma_y\, t_a}\right)^4 + \frac{g_c - k_a}{t_a}\left(\frac{2\,V_{pu}}{\sigma_y\, t_a}\right) = 1 \tag{6.35}$$

where g_c is the distance from the web of the beam and the centre of the bolt holes in the web angle leg adjacent to the column face, while k_a has the same meaning as k_t, but refers to the web angles.

Finally, in equation (6.32), V_{pa} and V_{pt} can be calculated from

$$V_{pa} = \frac{V_{pu} + \dfrac{\sigma_y\, t_a}{2}}{2}\, L_p \tag{6.36}$$

and

$$\left(\frac{2\,V_{pt}}{\sigma_y\,L_t\,t_t}\right)^4 + \frac{g_1 - k_t}{t_t}\left(\frac{2\,V_{pt}}{\sigma_y\,L_t\,t_t}\right) = 1 \tag{6.37}$$

For top and seat angle connections [34], as in equation (6.30) the first term of the second member represents the contribution of the top and seat angles only, the initial stiffness of the connection is given by

$$K_i = \frac{3\,E\,I_t\,d_1^2}{g_1\,(g_1^2 + 0.78\,t_t^2)} \tag{6.38}$$

The ultimate moment is

$$M_u = \sigma_y\,\frac{L_s\,t_s^2}{4} + \frac{V_{pt}\,(g_1 - k_t)}{2} + V_{pt}\,d_2 \tag{6.39}$$

In fact, in equation (6.32) the last term of the second member represents the contribution of the web angles. The value of V_{pt} is still provided by equation (6.37).

For single web angle connections [33, 34], the initial stiffness is given by

$$K_i = \frac{G\,t_a^3}{3}\,\frac{4.2967\,\cosh\dfrac{4.2967\,g_1}{L_p}}{\dfrac{4.2967\,g_1}{L_p}\,\cosh\dfrac{4.2967\,g_1}{L_p} - \sinh\dfrac{4.2967\,g_1}{L_p}} \tag{6.40}$$

while the ultimate moment is given by

$$M_u = \frac{2\,V_{pu} + \dfrac{\sigma_y\,t_a}{2}}{6}\,L_p^2 \tag{6.41}$$

where V_{pu} is still provided by equation (6.35).

Finally, in the case of double web angle connections the initial stiffness is twice that provided by equation (6.40) and the ultimate bending moment is twice that provided by equation (6.41).

Once the connection initial stiffness K_i and the ultimate moment capacity M_u have been evaluated, the complete moment–rotation curve can be predicted by using the power model (6.26), where the shape parameter has to be determined to obtain a best fit with the experimental data.

According to reference [35], the shape parameter n can be computed as follows.

- For single web angle connections

$$n = 0.520 \ \log_{10} \frac{M_u}{K_i} + 2.291 \geq 0.70 \qquad (6.42)$$

- For double web angle connections

$$n = 1.322 \ \log_{10} \frac{M_u}{K_i} + 3.952 \geq 0.60 \qquad (6.43)$$

- For top and seat angle connections

$$n = 2.003 \ \log_{10} \frac{M_u}{K_i} + 6.070 \geq 0.30 \qquad (6.44)$$

- For top and seat angle connections with double web angles

$$n = 5.483 \log_{10} \frac{M_u}{K_i} + 14.745 \geq 0.80 \qquad (6.45)$$

It has to be established that the above relationships for beam-to-column connections with angles do not include the deformation sources arising from the key elements of the column, as for instance the flexural deformation of the column flange. In fact, in the above studies, it is assumed that the connection lies on a rigid support. In addition, in many cases the deformation of the compression zone of the column web and the shear deformation of the column web can contribute significantly to the overall response of the joint. In other words, it has to be taken into account that above relations refer to the M–φ curve of the connection rather than to the joint as a whole.

A complete study, also including the column deformations, is that of Yee and Melchers [31] dealing with extended end plate connections. They recognize five contributions to the overall deformation of the joint:

- flexural deformation of the end plate
- flexural deformation of the column flange
- bolt extension
- shear deformation of the panel zone
- deformation due to the compression zone of the column web.

The initial stiffness is provided by

$$K_i = \frac{(d_b - t_{bf})^2}{\Delta_{ep} + \Delta_{cf} + \Delta_b + \Delta_s + \Delta_{wc}} \qquad (6.46)$$

where d_b is the overall depth of the beam and t_{bf} the thickness of the beam flanges, while the sum at the denominator represents the different contributions, as listed above.

In the case of stiffened connections with snug tightened bolts, the authors give

$$\Delta'_{ep} = \frac{Z_{ep}}{E} \left[\frac{1}{8} - \frac{q_s}{2} \left(\frac{3}{4} \, \alpha_{ep} - \alpha_{ep}^3 \right) \right] \qquad (6.47)$$

$$\Delta'_{cf} = \frac{Z_{cf}}{E} \left[\frac{1}{8} - \frac{q_s}{2} \left(\frac{3}{4} \, \alpha_{cf} - \alpha_{cf}^3 \right) \right] \qquad (6.48)$$

where

$$\alpha_{ep} = \alpha_{cf} = \frac{a}{2 \, (a + b)} \qquad (6.49)$$

with a and b defined as in Fig. 6.20, and

$$Z_{ep} = \frac{16 \, (a + b)^3}{b_{ep} \, t_{ep}^3} \qquad (6.50)$$

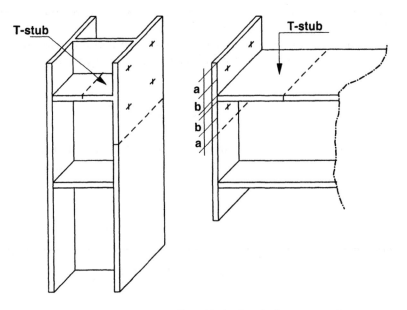

Fig. 6.20   T-stub orientation for stiffened bolted end plate connections

$$Z_{cf} = \frac{16\,(a\,+\,b)^3}{b_{cf}\,t_{cf}^3} \tag{6.51}$$

where b_{ep} and b_{cf} are the width of the end plate and of the column flange respectively, while t_{ep} and t_{cf} are the thicknesses of the same elements.

In addition, the parameter q_s is given by

$$q_s = \frac{Z_{ep}\left(\frac{3}{2}\,\alpha_{ep} - 2\,\alpha_{ep}^3\right) + Z_{cf}\left(6\,\alpha_{ep}^2 - 8\,\alpha_{ep}^3\right)}{\left[Z_{ep}\left(\frac{3}{2}\,\alpha_{cf} - 2\,\alpha_{cf}^3\right) + Z_{cf}\left(6\,\alpha_{cf}^2 - 8\,\alpha_{cf}^3\right)\right] + \dfrac{k_1 + 2\,k_4}{2\,A_s}} \tag{6.52}$$

where A_s is the bolt shaft area and

$$k_1 = L_s + 1.43\,L_t + 0.71\,L_n \tag{6.53}$$

$$k_4 = 0.10\,L_n + 0.20\,L_w \tag{6.54}$$

where L_s is the shaft length, L_t is the threaded length, L_n is the nut height and L_w is the thickness of the washer.

As equations (6.47) and (6.48) represent, respectively, the contributions of the flexural deformation of the column flange and of the end plate evaluated, taking into account the compatibility condition requiring that at a bolt line the sum of the end plate deflection and the column flange deflection has to be equal to the bolt elongation, in equation (6.46) it has to be assumed that

$$\Delta_{ep} + \Delta_{cf} + \Delta_b = \Delta'_{ep} + \Delta'_{cf} \tag{6.55}$$

Equations (6.47) and (6.48) are also valid in the case of stiffened connections with pretensioned bolts, provided that the parameter q_s is substituted by q_t

$$q_t = \frac{Z_{ep}\left(\frac{3}{2}\,\alpha_{ep} - 2\,\alpha_{ep}^3\right) + Z_{cf}\left(6\,\alpha_{ep}^2 - 8\,\alpha_{ep}^3\right)}{\left[Z_{ep}\left(\frac{3}{2}\,\alpha_{cf} - 2\,\alpha_{cf}^3\right) + Z_{cf}\left(6\,\alpha_{cf}^2 - 8\,\alpha_{cf}^3\right)\right] + \dfrac{k_2\,k_3}{2\,A_s\,(k_2 + k_3)}} \tag{6.56}$$

where

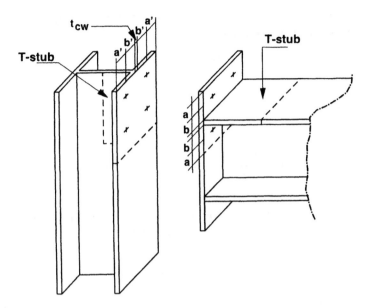

Fig. 6.21 T-stub orientation for unstiffened bolted end plate connections

$$k_2 = L_s + 1.43\,L_t + 0.91\,L_n + 0.40\,L_w \tag{6.57}$$

$$k_3 = \frac{t_{ep} + t_{cf}}{5} \tag{6.58}$$

For unstiffened connections (Fig. 6.21), the only difference with respect to the stiffened case is basically the orientation of the column flange T-stub. Therefore, equations (6.47) and (6.48) are still valid (with q_s for snug tightened bolts and q_t for pretensioned bolts) provided that the following substitutions are made

$$\alpha_{cf} = \frac{a'}{b_{cf} - t_{cw}} \tag{6.59}$$

and

$$Z_{cf} = \frac{\left(b_{cf} - t_{cw}\right)^3}{\left[(a + b) + \dfrac{t_{bf}}{2}\right] t_{cf}^{\,3}} \tag{6.60}$$

where t_{cw} and t_{bf} are the thicknesses of the column web and of the beam flange, respectively.

With respect to the contribution from the shear deformation of the panel zone, this can be computed as

$$\Delta_s = \frac{2\,(d_b - t_{bf})\,(1 + v)}{E\,d_c\,t_{cw}}$$
(6.61)

for unstiffened panels, and

$$\Delta_s = \frac{1}{E \left[\dfrac{A_{st}\,(d_c - t_{cf})^2}{L_b^3} + \dfrac{d_c\,t_{cw}}{2\,(1 + v)\,(d_b - t_{bf})} \right]}$$
(6.62)

when a diagonal stiffener is provided, where A_{st} is the area of the stiffener, d_c the overall depth of the column, and

$$L_b = \sqrt{(d_b - t_{bf})^2 + (d_c - t_{cf})^2}$$
(6.63)

Finally, with reference to the contribution due to the deformation of the compressed zone of the column web, this can be neglected when a continuity plate (stiffener) is located at the beam compression flange level. In the opposite case, it can be assumed equal to

$$\Delta_{wc} = \frac{1 - v^2}{E\,t_{cw}}$$
(6.64)

The strain-hardening can be significant in unstiffened connections where web shear yielding occurs. Similar behaviour is exhibited when web buckling of the column compression zone arises. For these unstiffened connections the rotational stiffness K_p of the 'hardening branch' of the M–φ curve can be computed from

$$K_p = \frac{(d_b - t_{bf})^2}{\Delta_p}$$
(6.65)

where

$$\Delta_p = \Delta'_{ep} + \Delta'_{cf} + \Delta_{wc} + \Delta_{sp}$$
(6.66)

when shear yielding of the column web occurs, and

$$\Delta_p = \Delta'_{ep} + \Delta'_{cf} + \Delta_{wcb} + \Delta_{sp}$$
(6.67)

when web buckling of the column compression zone arises.

In equation (6.66) Δ_{sp} represents the contribution of the column web after shear yielding and is provided by [31]

$$\Delta_{sp} = \frac{50\,(d_b - t_{bf})^3}{3\,E\,I_c} + \frac{50\,(d_b - t_{bf})(1 + \nu)}{E\,d_c\,t_{cw}} \tag{6.68}$$

where I_c is the inertia moment of the column.

In equation (6.67), Δ_{wcb} represents the contribution of the column compression zone after web buckling. It is given by

$$\Delta_{wcb} = \frac{2.45}{E\,t_{cw}} \tag{6.69}$$

Equation (6.66) takes into account that at the loading stage corresponding to shear yielding the column flange and the end plate have not reached the yield state. Equation (6.67) takes into account that, when column web buckling occurs in the compression zone, the reduction of the effective shear-resisting area of the web leads to the contemporary occurrence of shear yielding.

With respect to the ultimate moment of the connection, this is governed by the weaker connection element. The following failure modes have to be taken into account:

- tension bolt failure
- formation of a plastic mechanism in the end plate
- formation of a plastic mechanism in the tension zone of the column flange
- shear yielding of the column web
- column web buckling
- column web crippling.

The resistance of the connection elements can be computed by methods already available in the technical literature [31, 36].

Furthermore, the authors propose the use of the predicted values of K_i, K_p and M_u for estimating the complete M–φ curve by means of the exponential model (6.29) and by adopting, on the basis of comparisons with available experimental data, the following values of the shape parameter c:

- $c = 0$ for stiffened connections with snug tightened bolts
- $c = 3.5$ for stiffened connections with pretensioned bolts
- $c = 1.5$ for unstiffened connections.

With the same philosophy, Johnson and Law [37] developed a method for predicting the initial stiffness and plastic moment capacity of flush end plate connections.

All these studies have shown, through comparison between predicted curves and experimental test results, that the connection stiffness can be obtained, neglecting the interaction effects, by superimposing the flexibilities of the joint components. This approach has been recently included in Eurocode 3 (Annex J) [36].

In particular, Annex J of Eurocode 3 [36] presents a general method for predicting the mechanical properties of beam-to-column joints for double T-shaped beams and columns. The beam is assumed to be connected to the column flange. The model deals with welded connections and bolted end-plate connections including extended end plates and flush end plates. The column panel zone can be both unstiffened and stiffened with continuity plates and/or diagonal stiffeners. In addition, a simple method is also provided for estimating the rotational capacity of connections, whose collapse is governed by a brittle failure mode. Some parts of the method can be also applied to the relevant parts of other connection types.

As a general conclusion, the analytical methods are seen to be the most interesting and promising, because they allow an approximate evaluation of the key parameters of the M–φ curve without resort to testing, even though they still require a curve fitting when the full moment–rotation curve is desired. However, up to now, the suggested procedures should be applied within the range for which the validity of the procedure has been confirmed by comparison with experimental test results.

6.5.4 Mechanical models

Mechanical models are based on the simulation of the joint/connection by using a set of rigid and flexible components. The non-linearity of the reponse is obtained by means of inelastic constitutive laws adopted for the deformable elements. These constitutive laws can be provided by test data or analytical models.

Mechanical models have been developed by several researchers, with reference to the connection as well as to the whole joint, with the aim of directly showing the response of the joint/connection.

Figure 6.22 shows the mechanical model adopted by Wales and Rossow [38] for simulating the behaviour of double web angle connections under bending moment and axial load. This model has been extended by Chmielowiec and Richard to the case of top and bottom angle with double web angles [39] (Fig. 6.23).

The response of unstiffened welded joints is affected both by deformations due to the transmission of the load from the beam flanges to the column and by panel zone deformation. Such behaviour can be simu-

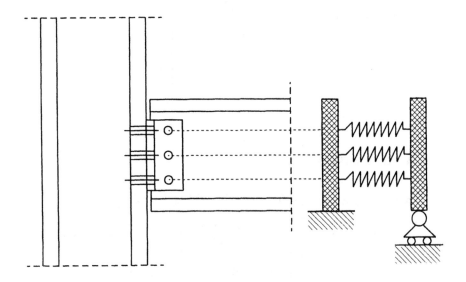

Fig. 6.22   Mechanical model for web angle connections

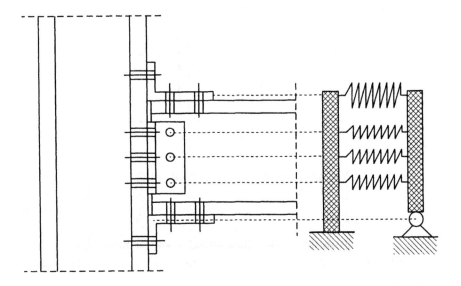

Fig. 6.23   Mechanical model for top and seat angle connections with web angles

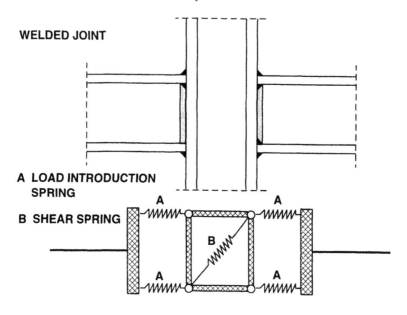

Fig. 6.24   Mechanical model of a welded joint

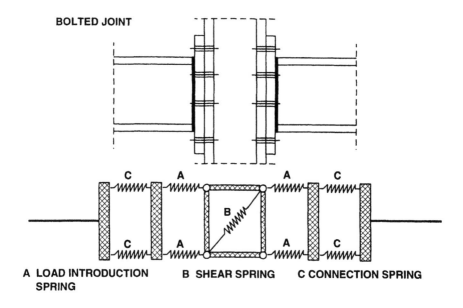

Fig. 6.25   Mechanical model of a bolted joint

lated by the mechanical model, developed by Tschemmernegg *et al.*, represented by Fig. 6.24 [40–42].

The model is characterized by two non-linear springs, namely the load introduction spring and the shear spring. The first one accounts for the deformation due to the load transmitted by the beam flanges, the second one simulates the shear deformation of the panel zone.

Several tests have been carried out to define the characteristics of the springs, in order to use the model to provide the response of all combinations of beams and columns made by European rolled sections. The model has been extended to the case of end plate bolted connections [42]. In this case new sources of deformation are taken into account by means of additional springs, namely connection springs (Fig. 6.25).

Mechanical models are suitable for modelling joint/connection response, provided that a knowledge of the constitutive law of springs is available. These constitutive laws can be obtained through experimental tests or by means of analytical models. In this second case the mechanical model represents only a tool for superimposing the effects of the key deformation sources, so that the global approach is nearer to the analytical one.

6.5.5 Finite element analysis

The finite element technique seems, in principle, to be the most suitable tool to investigate the response of a joint. Nevertheless, it has to be recognized that, in spite of the continuous progress, some of the requirements needed for an accurate simulation are still today unsolved.

In fact, moment–rotation curves represent the result of a very complex interaction among the elementary parts constituting the connection. In particular, the analysis of steel joints requires modelling of the following features [4]:

- geometrical and material non-linearities of the elementary parts of the connection;
- bolt pre-load and its response under a general stress distribution;
- interaction between bolts and plate components: i.e. shank and hole, head or nut contact;
- compressive interface stresses and friction resistance;
- slip due to bolt-to-hole clearance;
- variability of contact zones;
- welds;
- imperfections (residual stresses and so on).

As already pointed out, up to now, some of the basic mechanisms of these interactions are yet to be fully understood. The first requirement has now been completely fulfilled. In fact, the analysis of isolated plates has reached a high degree of accuracy thanks to the ability to account for the spread of plasticity, the strain-hardening, the instability effects and the representation of large strain and/or displacements. On the contrary, all other requirements need a level of refinement not yet attained. As a consequence, on one hand, the finite element technique already represents a sufficiently accurate tool for the modelling of welded beam-to-column joints [43] and, on the other hand, it is a very sophisticated approach whose potential for modelling bolted connections [44] is largely unexplored.

6.6 TEST RESULTS UNDER CYCLIC LOADS

Some years ago researchers from Napoli University and Milan Polytechnic carried out a broad programme dealing with the analysis of the behaviour of steel beam-to-column joints subjected to alternate loading conditions [45, 46]. This research was within the field of activity of the Committee TC13 'Seismic Design' of the European Convention for Constructional Steelwork (ECCS) which, for many years, has investigated the application of steel structures in seismic regions.

This research was mainly devoted to the experimental analysis of the behaviour of the main joint typologies in common use. For this purpose 14 specimens were designed in order to cover a wide range of possible solutions for connecting I beams to I columns, with different degrees of rigidity varying from fully welded joints to bolted angle connections. Experiments were performed using the testing equipment at the Structural Engineering Department of Milan Polytechnic [47].

This equipment (Fig. 6.26) allows the application of a quasi-static horizontal cyclic displacement at the top of the vertical member (beam), and a constant axial load in the horizontal member (column).

The applied horizontal load is measured by a dynamometer, which forms an integral part of the equipment. It consists of a round bar connected through a cylindrical hinge to the jack and through a spherical hinge to the specimen. A strain gauge bridge is set in the middle of the bar. The displacement component is measured by a special device which utilizes an inductive transducer. Both signals (load and displacement), sent by the strain gauge bridge and the transducer, are passed through digital amplifiers to an *x*–*y* recorder, allowing real-time control of the test.

The applied axial load is measured by a 1000 kN load cell set between the counter piece and the specimen. Readings are recorded through a digital amplifier.

The experimental programme was designed to analyse the cyclic behaviour of beam-to-column joints, using common types of rigid and

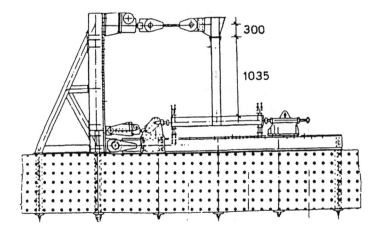

300

1035

Fig. 6.26 Testing equipment

semirigid connections. Joint details were designed to obtain the same flexural strength of the connected beam.

The 14 specimens can be grouped into four main categories (Figs. 6.27–6.33), which cover the practical range of behavioural variability of load versus displacement relationships. For each category a 'base joint', identified by the number 1, was assumed as the least rigid. Other joints of the same category were obtained by introducing different local stiffening elements to the base joint.

Type A. The connection is obtained by three plate splices, which are welded to the column and bolted to the flanges and to the web of the beam. Variations from the basic type A1 are obtained by introducing a diagonal stiffener in the panel zone (A2), by adding inner reinforcing cover plates for the beam flanges (A3), or by means of both these variations (A4).

Type B. Angles are bolted both to the column and to the beam. Variations from the basic type B1 consist in stiffening the column web at the end of the leg of top and bottom angles (B2). In case B3 both angles connnecting the beam flanges are reinforced with transversal triangular plates. B4 has both the variations of B2 and B3.

Type C. End plate joints with rigid column stub. Variations from the basic type C1 are derived by introducing full-depth stiffeners (C2, C4)

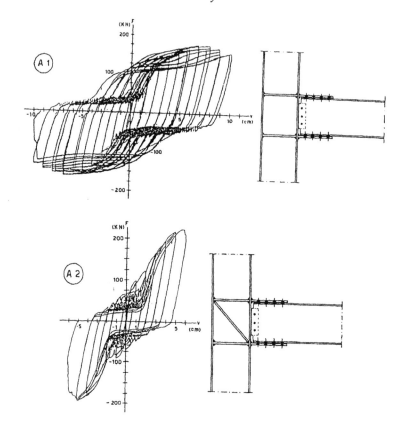

Fig. 6.27   Specimens A1 and A2

or partial depth stiffeners (C3) in the beam web or also increasing the thickness of the end plate (C3, C4).

Type D. Fully welded joints of current type (D1) and with doubler plates on the column web (D2).

In all cases both beams and columns are double-T IPE 300 shaped, and made of Fe360 steel.

Cyclic tests have been performed following the experimental procedure suggested by ECCS Recommendations [17]. An axial load of about 0.30 times the squash load was applied to the column.

Figures 6.27 to 6.33 also show the hysteresis loops obtained for the 14 specimens in the load versus displacement representations. The comparison among the different results emphasizes the role played by the main parameters which characterize the behaviour of the joint under alternate loading conditions [48].

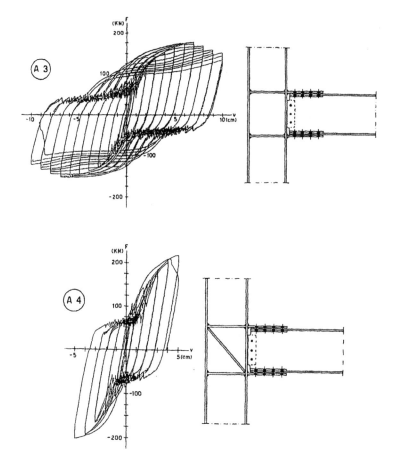

Fig. 6.28 Specimens A3 and A4

From their test performance, it can be seen that all the joints exhibited a largely sufficient ductility. The energy absorption capacity is very different from one type to another, as can be observed from their hysteresis loops (Figs. 6.27–6.33). As expected, type D gives the largest hysteretic area and its behaviour is very close to that of a rigid joint. Type C shows a continuous degradation of the stiffness, due mainly to the axial deformation of bolts and to the flexural deformation of the end plate. Type A is characterized by a progressive reduction of the hysteretic area, because of slippage in the plate joint due to bolt-hole clearance. Type B behaves in the worst way from all points of view because, in addition to the previous sources of degradation, permanent deformation of the angles also arises. Starting from the base types A1, B1, C1 and D1, ad-

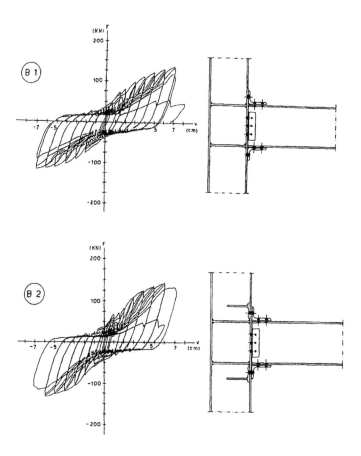

Fig. 6.29   Specimens B1 and B2

ditional elements can be located to stiffen some parts, as in common practice. For instance, diagonal stiffeners in the panel zone of the column web, triangular plates stiffening the connecting angles, vertical stiffeners in the web of the beam, and so on.

From the experimental evidence, some general considerations can be drawn about the use of additional stiffeners in the main connection types.

- If stiffeners are added to those parts of the connection which are most responsible for its flexibility, the amount of energy absorption is decreased but the load-carrying capacity is increased;
- If the added elements do not substantially modify the evolution of the deformation mechanism but, on the contrary, increase the local strength

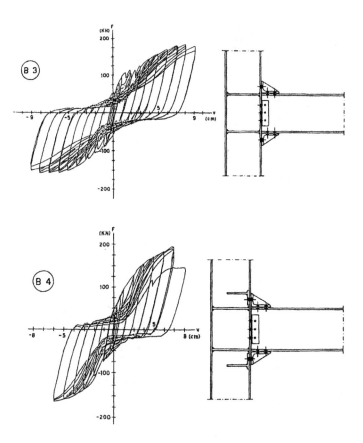

Fig. 6.30 Specimens B3 and B4

of the connection components (as for example an increase of thickness), then there will be an increase of energy absorption and strength, provided that the original type of collapse is ductile.

These general considerations are shown in Fig. 6.34 and Fig. 6.35, where for each joint both the non-dimensional strength and the non-dimensional dissipated energy are provided. The non-dimensional strength has been defined as the ratio between the maximum force F_{max}, applied at the end of the cantilever specimen during the test and the force F_p leading to the yielding of the beam ($F_p = M_{pb}/L$, where M_{pb} is the plastic moment of the beam computed on the basis of the actual value of the yield stress of the

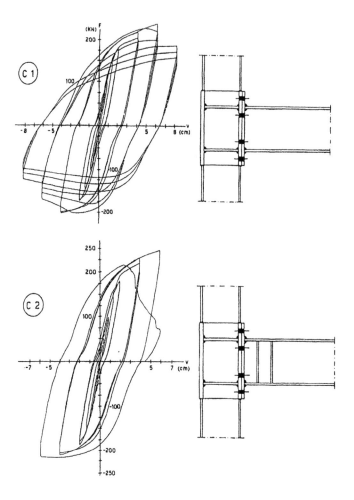

Fig. 6.31   Specimens C1 and C2

beam flanges and L the length of the cantilever specimen). The non-dimensional dissipated energy has been defined as

$$\overline{E} = \frac{E}{F_p\,v_p} \tag{6.70}$$

where v_p is the end displacement corresponding to F_p and evaluated for the ideally rigid conditions ($V_p = F_p L^3 / 3EI_b$).

Within the A category, comparison between specimens A2 and A1

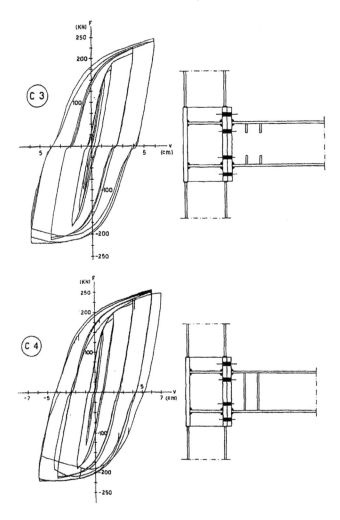

Fig. 6.32 Specimens C3 and C4

and between A4 and A3 points out that the introduction of a diagonal stiffener in the panel zone has led to the increase of the joint strength, but its energy dissipation capability is considerably reduced. In addition, the comparison between specimen A3 and A1, whose panel zone is unstiffened, shows that the use of additional flange splices provides a slight increase of strength and energy dissipation. Finally, in the case of joints having a stiffened panel zone (A4 and A2), the additional flange splices

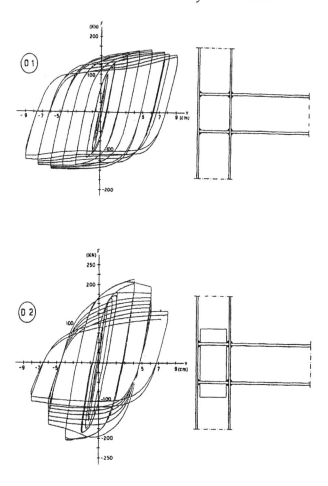

Fig. 6.33   Specimens D1 and D2

provide a very slight increase of strength and a very slight reduction of energy dissipation.

With reference to category B, comparison between B2 and B1 and between B4 and B3 shows that the stiffening elements of the column web provide an increase of strength and a reduction of the energy dissipation capacity. By comparing B3 to B1 and B4 to B2, it can be observed that the introduction of a triangular plate stiffening the top and seat angles provides an increase of strength, but a reduction in the energy dissipation capacity of the joint.

In the case of extended end plate connections, comparison between C2 and C1 shows that the use of additional ribs to stiffen the beam web

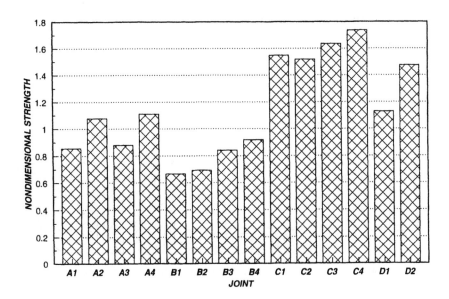

Fig. 6.34 Non-dimensional strength of tested specimen

produces a significant decrease of the energy dissipation. Comparison between C3 and the previous type C specimen shows that an increase of end plate thickness provides an increase of both strength and energy dissipation (compare C3 to C2), but the energy increase does not compensate for the loss due to the use of additional partial depth stiffeners in the beam web. A further stiffening of the beam web (specimen C4) gives rise to an increase of both strength and energy dissipation, but the energy dissipation capacity is still less than the one of the base joint C1.

Finally, in the case of welded joints, from the comparison between D2 and D1 it can be shown that the use of doubler plates stiffening the panel zone provides an increase in strength and a reduction in the energy dissipation capacity.

The cyclic behaviour of beam-to-column connections designed to attain a flexural strength greater than that of the connected beam has also been investigated [49]. Both beam-to-column flange and beam-to-column web connections have been tested.

With reference to beam-to-column flange connections, attention has been focused on three types of connections: fully welded connections, bolted extended end plate connections and welded flange-bolted web connections.

The last type (Fig. 6.36) represents the most widely used moment-

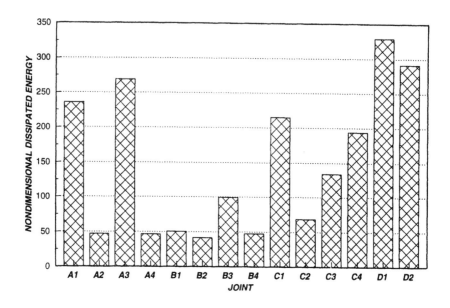

Fig. 6.35   Non-dimensional dissipated energy of tested specimen

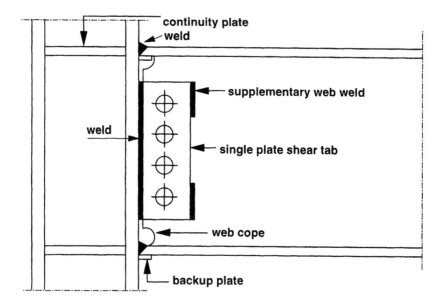

Fig. 6.36   The most common connection typology for seismic-resistant MR frames in the USA

connection detail in current USA practice for seismic-resistant frames. The beam web is field-bolted to a single plate shear tab which is shop-welded to the column. The beam flanges are welded to the column using complete penetration welds. Web copes are required to accommodate the backup plate at the top flange and to permit making the bevel weld at the bottom flange.

The experimental results [49] have confirmed the high ductility of fully welded beam-to-column flange connections and the good performances of bolted extended end plate connections. On the contrary, welded flange-bolted web connections may not give a satisfactory performance when used for beams in which the web accounts for a substantial portion of the beam flexural strength. For this reason, the use of supplementary welds between the beam web and the shear tab for beam sections with $Z_f / Z \leq 0.70$ has been suggested (where Z is the total plastic section modulus of the beam and Z_f is the plastic section modulus of the beam flanges only).

The supplementary web welds have to be able to develop at least 20% of the flexural strength of the beam web. However, a successive experimental program [50, 51] has not shown a clear influence of the Z_f / Z ratio and/or the web connection detail on the performance of welded flange-bolted web connections. Rather, the variability of the connection performance seems to be ascribed to the beam flange welds. Notwithstanding, specimens with the supplementary web welds performed somewhat better than the analogous ones without web welds. Consequently, the authors [50, 51] suggest keeping the web weld recommendation in the code requirements.

In the case of beam-to-column web connections, a similar detail is often used. Continuity plates are located at the beam flange level by means of welds along the column web and the column flanges. A shear tab is welded to the column web and to the continuity plates. The beam web is field-bolted to the shear tab and, successively, the beam flanges are connected to the continuity plates by means of complete penetration single-bevel welds. The cyclic behaviour of this type of connection has been experimentally investigated [49]. Well detailed beam-to-column web connections are able to provide sufficient ductility for their use in seismic-resistant frames, in particular, this can be achieved with the aid of welded reinforcing ribs connecting the continuity plates and the beam flanges.

The above conclusions on the performance of different beam-to-column connection typologies are based on their plastic rotation capacity which has to be compared with a bench mark of about ±0.02 radians. This value represents a reasonable estimate of the plastic rotation demand in steel moment-resisting frames subjected to severe earthquakes [49].

The cyclic behaviour of partial strength connections of the types shown in Fig. 6.37 has been experimentally investigated [52, 53]. Attention has been focused on the connection only. In fact, the specimen consists of a long beam stub attached through the connection to be tested to a rigid counterbeam, so that the testing conditions are able to represent the case of beam-to-column joints with negligible panel zone deformability. The test results show that this type of connection can provide a satisfactory ductility, but a significant reduction of the energy dissipation capacity is exhibited (Fig. 6.38) in comparison with the full strength rigid welded joint results shown in Fig. 6.35.

As the experimental results have shown that a very satisfactory ductility and energy dissipation capacity is obtained through the plastic hinge formation in the beam, some researchers [54] have proposed a reduction of the beam flange width near the beam-to-column connections in order to force the formation of the plastic hinge in the beam rather than in the connection and/or in the panel zone. An experimental program including this type of beam-to-column connection detail has been developed [55, 56].

The complete experimental programme was devoted to both bare steel joints and to composite joints, with and without slab. In particular, for bare steel joints, the cyclic behaviour of bolted web-welded flange connections, end-plate connections and fully welded connections has been investigated.

The experimental results have confirmed that, in fully welded connections, the reduction of the beam flange causes a little loss of strength, but allows the formation of the plastic hinge in the beam leading to a very ductile behaviour.

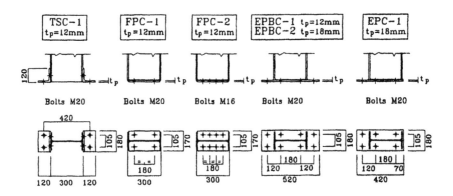

Fig. 6.37   Partial strength connections tested in references [52 and 53]

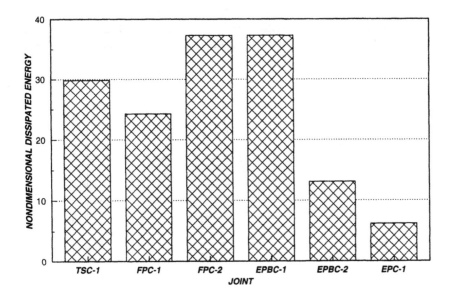

Fig. 6.38 Energy dissipation capacity of partial strength connections

6.7 PREDICTION OF CYCLIC BEHAVIOUR

The problem of simulating cyclic behaviour of the connection typologies which are more widely used in seismic-resistant steel structures is of primary importance, expecially for low-cycle fatigue effects. In fact, from the seismic point of view, particular attention has to be paid to those phenomena which cause progressive deterioration of the strength, stiffness and energy dissipation capacity of the joint.

Experimental investigations on the cyclic behaviour of beam-to-column joints have identified the main causes of deterioration for different connection typologies. The next step is the formulation of models able to simulate the joint moment–rotation relationship.

As for monotonic behaviour (Section 6.5), prediction of the cyclic behaviour of the connection should be possible using mathematical models, analytical models, mechanical models and finite element analyses. However, research effort has been devoted mainly only to the development of mathematical models [57–62] and, in some cases, to the extension of mechanical models [63–66].

The proposed mathematical models are characterized by varying degrees of accuracy (Fig. 6.39). Many of these models are multilinear (Fig. 6.39a, b, d–f). Some of them assume a continuous and stable behaviour which globally simulates the connection (Fig. 6.39c). Others simplify even more complex behaviour, with possible discontinuity due to the bolt-hole clearance (Fig. 6.39b) and non-symmetry of the relationship in the tensile and compressive region (Fig. 6.39f).

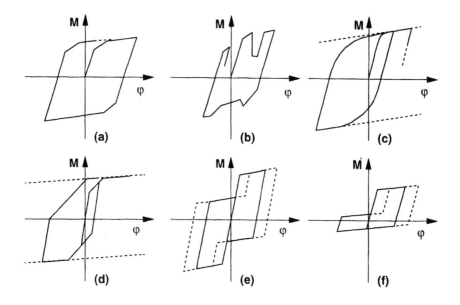

Fig. 6.39   Mathematical models for the cyclic behaviour of beam-to-column connections

From the critical analysis of existing models, a more refined proposal has been developed [61, 62].

The proposed analytical model (Fig. 6.40) allows representation of the load history for a generic joint by means of a moment–rotation relationship which can be divided into four branches:

a. load increasing from O to M^+
b. load decreasing from M^+ to O
c. load increasing in the negative direction from O to M^-
d. load decreasing from M^- to O

The M–φ relationships for each branch are as follows.

Branch a
This branch is divided into three different parts in which the following hold:

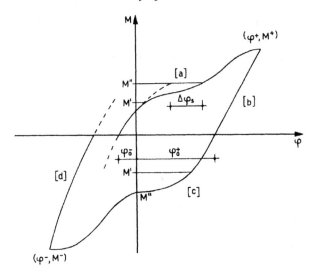

Fig. 6.40 Mathematical model of De Martino, Faella and Mazzolani [61]

for $M \leq M'$

$$\varphi = \varphi_0^- + \frac{M}{K(\Omega)} + C_1 \left(\frac{M}{M_0(\Omega)} \right)^{C_2} \tag{6.71}$$

for $M' \leq M \leq M''$

$$\varphi = \varphi_0^- + \frac{M}{K(\Omega)} + C_1 \left(\frac{M}{M_0(\Omega)} \right)^{C_2} + \frac{\Delta \varphi_s(\Omega)}{2}$$

$$+ \left(\frac{\Delta\varphi_s(\Omega)}{2} - C_3 \right) \rho \ |\rho|^{s-1} + C_3 \ \rho \tag{6.72}$$

for $M \geq M''$

$$\varphi = \varphi_0^- + \Delta\varphi_s(\Omega) + \frac{M}{K(\Omega)} + C_1 \left(\frac{M}{M_0(\Omega)} \right)^{C_2} \tag{6.73}$$

where:

- φ_o^- is the residual rotation of the previous cycle at the beginning of the branch (in the first cycle $\varphi_o^- = 0$);
- $K(\Omega)$ is the initial stiffness of the cycle which depends on the energy Ω dissipated in previous cycles;
- M_o is the conventional elastic limit;
- C_1 and C_2 are the two parameters defining the Ramberg–Osgood type component of the loading curve;
- C_3 is a parameter defining the shape of the slip phase whose amplitude is $\Delta\varphi_s(\Omega)$;
- ρ is a parameter governing the slip phase, being

$$\rho = \frac{2M - M' - M''}{M'' - M'} \tag{6.74}$$

The following relationship holds for $K(\Omega)$

$$K(\Omega) = K\left[1 - \frac{\Delta K}{K}\left(\frac{\Omega}{\Omega_{max}}\right)^m\right] \tag{6.75}$$

where Ω_{max} is the overall energy dissipated by the joint up to collapse, ΔK is the decrease of joint stiffness measured in the cycle before the occurrence of collapse, and m is the exponent which best expresses the decrease of stiffness as far as the dissipated energy increases.

Since experimental results have often shown a variation of the slip phase, not only $\Delta\varphi_s$, but also M' and M'' have to be generally considered dependent on the dissipated energy Ω.

Branch b

The unloading curve from M^+ to O is given by

$$\varphi = \varphi^+ + \frac{(M - M^+)}{K(M,\Omega)} \tag{6.76}$$

where

$$K(M,\Omega) = K(\Omega)\left[1 - \frac{\Delta'K}{K(\Omega)}\frac{M^+ - M}{M^+} + \frac{\Delta''K}{K(\Omega)}\right] \tag{6.77}$$

The residual rotation after unloading can be derived by the previous equation by setting $M = 0$

$$\varphi_o^+ = \varphi^+ - \frac{M^+}{K(0,\Omega)} \tag{6.78}$$

Depending on the values of $\Delta'K$ and $\Delta''K$, equations (6.76) and (6.77) can represent different types of unloading behaviour. The case $\Delta'K = \Delta''K = 0$ corresponds to the linear unloading. In the case $\Delta'K < \Delta''K$ the residual rotation is greater than that corresponding to the linear unloading, while for $\Delta'K > \Delta''K$ the residual rotation is less than that corresponding to the linear unloading. In fact, from equation (6.77) for $M = 0$

$$K(0,\Omega) = K(\Omega) - \Delta'K + \Delta''K \qquad (6.79)$$

Branch c

The three parts into which this branch can be divided are governed by the following equations:

for $|M| \leq M'$

$$\varphi = \varphi_o^+ + \frac{M}{K(\Omega)} - C_1 \left| \frac{M}{M_0(\Omega)} \right|^{C_2} \qquad (6.80)$$

for $M' \leq |M| \leq M''$

$$\varphi = \varphi_o^+ + \frac{M}{K(\Omega)} - C_1 \left| \frac{M}{M_0(\Omega)} \right|^{C_2} - \frac{\Delta\varphi_s(\Omega)}{2}$$

$$- \left(\frac{\Delta\varphi_s(\Omega)}{2} - C_3 \right) \rho \, |\rho|^{s-1} - C_3 \, \rho \qquad (6.81)$$

for $|M| ⸴ M''$

$$\varphi = \varphi_o^+ - \Delta\varphi_s(\Omega) + \frac{M}{K(\Omega)} - C_1 \left(\frac{M}{M_0(\Omega)} \right)^{C_2} \qquad (6.82)$$

where φ_o^+ is the residual rotation of the previous cycle at the beginning of the branch [c] (in the first cycle $\varphi_o^+ = 0$), ρ is the parameter governing the slip phase which, in this case, is given by

$$\rho = \frac{|\, 2M \,| - M' - M''}{M'' - M'} \qquad (6.83)$$

Branch d

The unloading curve from M^- to O is given by

$$\varphi = \varphi^- + \frac{(M - M^-)}{K(M,\Omega)} \qquad (6.84)$$

where

$$K(M,\Omega) = K(\Omega) \left[1 - \frac{\Delta'K}{K(\Omega)} \frac{M^- - M}{M^-} + \frac{\Delta''K}{K(\Omega)} \right] \qquad (6.85)$$

The residual rotation after unloading can be derived from the previous equation by setting $M = 0$

$$\varphi_0^- = \varphi^- - \frac{M^-}{K(0,\Omega)} \qquad (6.86)$$

The comparison between the results of tests and the numerical model can be carried out on the basis of the main behavioural parameters. An

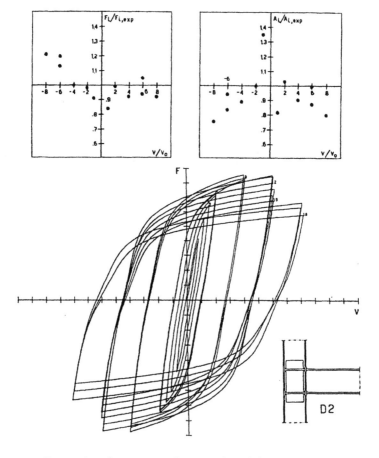

Fig. 6.41   Comparison between mathematical model and experimental results

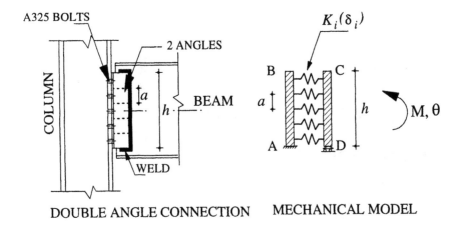

Fig. 6.42   Mechanical model for the cyclic moment–rotation response of double
angle connections [66]

example is given in Fig. 6.41 which refers to the joint type D2 (Section
6.6).

The simulated hysteresis curve is plotted in the lower part of the figure. The upper diagrams give the ratios between simulated and
experimental values for the strength F and the area A of a half-cycle at
each value of the imposed displacement v. The corresponding scatters
are usually less than 15%. This approach has been applied [67] to all
specimens tested in [46], leading to the conclusion that this refined
mathematical model is always in good agreement with the experimental
results and, therefore, it can be satisfactorily used in the simulation of
joint behaviour in semirigid framed structures.

As already seen for the prediction of monotonic behaviour, mathematical models are, in principle, able to closely fit any shape of M–φ
curve, but they suffer from the disadvantage that they cannot be extended outside the range of calibration data.

By using the approach introduced by Wales and Rossow [38], other
researchers have modelled the cyclic moment–rotation response for double angle connections [63, 64, 66].

The joint is idealized as two rigid bars connected by an assembly of
non-linear springs representing the response of the double angles. Double angles are divided in slices, defined by the middle lines between two

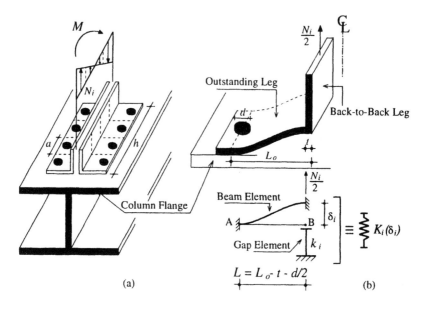

Fig. 6.43 Modelling angle behaviour

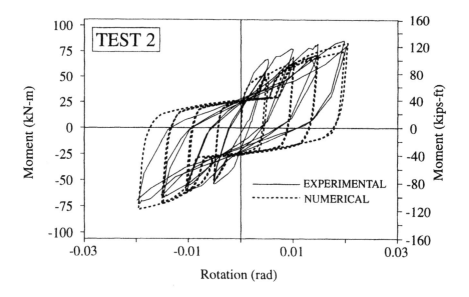

Fig. 6.44 Comparison between mechanical model results and experimental test results

adjacents bolts connecting the angle to the column (Fig. 6.42). The main problem in predicting cyclic moment–rotation behaviour is represented by the anisotropy of the angle slice behaviour. In fact, the gap between the outstanding leg and the column flange alternately opens and closes giving rise to significant boundary non-linearity. This behavioural feature has been accounted for by introducing gap elements which are activated when the gap between the outstanding leg and the column flange closes (Fig. 6.43).

Test results [68, 69] have shown that the angle behaviour is primarily dependent on the flexural response of the outstanding leg. Therefore, the modelling effort has been particularly devoted to the description of this flexural response by means of an inelastic beam element with distributed plasticity [70, 71]. The kinematic hardening model has been used for the cyclic plasticity of the material [66].

As an example, the comparison between experimental results and the simulation model is given in Fig. 6.44 for a specimen tested elsewhere [68].

A similar approach [65] has also been followed which includes the panel zone deformation through the shear spring model suggested by Tschemmernegg and Humer [40, 41].

6.8 REFERENCES

[1] *Commission of the European Communities:* 'Eurocode 8: Structures in Seismic Regions', Draft of October 1993.

[2] *Commission of the European Communities:* 'Eurocode 3: Design of Steel Structures', 1990.

[3] *F.S.K. Bijlaard, D.A. Nethercot, J.W.B. Stark, F. Tschemmernegg, P. Zoetemeijer:* 'Structural Properties of Semi-rigid Joints in Steel Frames', IABSE Survey, Periodica 2/1989.

[4] *D.A. Nethercot, R. Zandonini:* 'Methods of Prediction of Joint Behaviour: Beam-to-column connections', in 'Structural Connections: Stability and Strength', edited by R. Narayanan, Elsevier Applied Science Publishers, 1990, Chap. 2, pp. 22–62.

[5] *P.A. Kirby, R. Zandonini, J.B. Davison:* 'On the Determination of Moment Rotation Characteristics of Beam-to-column Joints', RILEM Workshop: Needs in Testing Metals, Naples, May 1990.

[6] *E. Cosenza, A. De Luca, C. Faella, F.M. Mazzolani:* 'Imperfections Sensitivity of Industrial Steel Frames', in 'Steel Frames Advances', edited by R. Narayanan, Elsevier Applied Science Publishers, 1987.

[7] *E. Cosenza, A. De Luca, C. Faella:* 'Inelastic Buckling of Semirigid Sway Frames', in 'Structural Connection: Stability and Strength', edited by R. Narayanan, Elsevier Applied Science Publishers, 1989, Chap. 9.

[8] *M.H. Ackroid, M.H. Gerstle:* 'Behaviour of Type-2 Steel Frames', ASCE Journal of the Structural Division, 108, ST7, 1982.

[9] *A. Astaneh, M. Nader:* 'Proposed Code Provisions for Seismic Design of Steel Semi-Rigid Frames', 1st COST C1 Workshop, Strasbourg, 28–30 October, 1992.

[10] *ECCS (European Convention for Constructional Steelwork):* 'European Recommendations for Steel Structures in Seismic Zones', Technical Working Group 1.3 (now TC13): 'Seismic Design', Doc.N.54, 1988.

[11] *J.W.B. Stark, F.S.K. Bijlaard:* 'Structural Properties of Connections in Steel Frames', in 'Connections in Steel Structures: Behaviour, Strength and Design', Elsevier Applied Science Publishers, London, pp. 186–194, 1988.

[12] *R. Bjorhovde, J. Brozzetti, A. Colson:* 'Classification System for Beam-to-Column Connections', Journal of Structural Engineering, ASCE, Vol. 116, N.11, November, 1990.

[13] *F.S.K. Bijlaard, C.M. Steenhuis:* 'Prediction of the influence of connection behaviour on the strength, deformations and stability of frames, by classification of connections', 2nd International Workshop, Pittsburgh, April, 1991.

[14] *F. Tschemmernegg, M. Huter:* 'Classification of beam-to-column joints', COST C1 Working Group Meeting, Liege, 14–16 June, 1993.

[15] *E. Cosenza, A. De Luca, A. De Martino, C. Faella, F.M. Mazzolani:* 'Inelastic Behaviour of Framed Structures under Cyclic Loads', Euromech Colloquium 174, Palermo, October 1983.

[16] *W.F. Chen, E.M. Lui:* 'Beam-to-column Moment-resisting Connections', in 'Steel Framed Structures: Stability and Strength' edited by R. Narayanan, Elsevier Applied Science Publishers, 1989, Chap. 6, pp 115–134.

[17] *ECCS-CECM-EKS (European Convention for Constructional Steelwork):* 'Recommended Testing Procedure for Assessing the Behaviour of Structural Steel Elements under Cyclic Loads', Doc. ECCS TWS 1.3 N.45/86, 1986.

[18] *K. Weinand:* 'SERICON: databank on joints in building frames', 1st COST C1 Workshop, Strasbourg, October 28–30, 1992.

[19] *N. Kishi, W.F. Chen:* 'Data Base for Beam-to-Column Connections', I and II, Purdue University, West Lafayette, Indiana, 1986.

[20] *P.D. Moncarz, K.H. Gerstle:* 'Steel Frames with Nonlinear Connections', Journal of Structural Division, ASCE, Vol. 107, ST8, pp. 1427–41, 1981.

[21] *C. Poggi, R. Zandonini:* 'Behaviour and Strength of Steel Frames with Semi-Rigid Connections', in 'Connection Flexibility and Steel Frame Behaviour, ed. W.F. Chen, American Society of Civil Engineers, 1985.

[22] *M.J. Frye, G.A. Morris:* 'Analysis of flexibly connected steel frames', Canadian Journal of Civil Engineering, N.2, pp. 280–91, 1975.

[23] *J.B. Radziminski, A. Azizinamini:* 'Prediction of Moment-Rotation Behaviour of Semi-Rigid Beam-to-Column Connections', in 'Connection in Steel Structures: Behaviour Strength and Design', Elsevier Applied Science Publishers, London, 1988.

[24] *A. Azizinamini, J.H. Bradburn, J.B. Radziminski:* 'Static and Cyclic Behaviour of Steel Beam-Column Connections', Structural Research Studies, Dept. of Civil Engineering, University of South Carolina, Columbia, March 1985.

[25] *S.W. Jones, P.A. Kirby, D.A. Nethercot:* 'Modelling of Semirigid Connection Behaviour and its Influence on Steel Column Behaviour', in 'Joints in Structural Steelwork', ed. J.H. Howlett *et al.*, Pentech Press, London, pp. 573–587, 1981.

[26] *J.E. Goldberg, R.M. Richard:* 'Analysis of Nonlinear Structures', Journal of the Structural Division, ASCE, Vol. 89, ST, August, 1963.

[27] *B.J. Abbott, R.M. Richard:* 'Versatile elastic–plastic stress-strain formula', Journal of Engineering Mechanics Division, ASCE, Vol. 101, EM4, pp. 511–15, 1975.

[28] *W. Ramberg, W.R. Osgood:* 'Description of Stress-Strain Curves by 3 Parameters', Technical Report 902, National Advisory Committee for Aeronautics, 1943.

[29] *K.M. Ang, G.A. Morris:* 'Analysis of 3–Dimensional Frames with Flexible Beam-Column Connections', Canadian Journal of Civil Engineering, Vol. 11, pp. 245–54, 1984.

[30] *E.M. Lui, W.F. Chen:* 'Analysis and Behaviour of Flexibly Jointed Frames', Engineering Structures, Vol. 8, pp. 107–15, 1986.

[31] *K.L. Yee, R.E. Melchers:* 'Moment-Rotation Curves for Bolted Connections', Journal of Structural Engineering, ASCE, Vol. 112, pp. 615–35, 1986.

[32] *W.F. Chen, N. Kishi, K.G. Matsuoka, S.G. Nomachi:* 'Moment-Rotation Relation of Top and Seat Angle with Double Web Angle Connections', in 'Connections in Steel Structures: Behaviour, Strength and Design', Elsevier Applied Science, London, 1988.

[33] *W.F. Chen, N. Kishi, K.G. Matsuoka, S.G. Nomachi:* 'Moment-Rotation Relation of Single/Double Web Angle Connections' in 'Connections in Steel Structures: Behaviour, Strength and Design', Elsevier Applied Science, London, 1988.

[34] *N. Kishi, W.F. Chen:* 'Moment-Rotation Relations of Semirigid Connections with Angles', Journal of Structural Engineering, ASCE, Vol. 116, N.7, July, 1987.

[35] *J.Y. Liew, D.W. White, W.F. Chen:* 'Limit States Design of Semi-Rigid Frames Using Advanced Analysis: Part 1 – Connection Modeling and

Classification', Journal of Constructional Steel Research, Vol. 26, pp. 1–27, 1993.

[36] *CEN/TC250/SC3–PT9:* 'Eurocode 3, Part 1: Joints in Building Frames (Annex J)', 3rd Draft, CEN/TC250/SC3 meeting, Dublin, September 1993.

[37] *R.P. Johnson, C.L.C. Law:* 'Semi-rigid Joints for Composite Frames', In 'Joints in Structural Steelwork', ed. J.H. Howlett *et al.* Pentech Press, London, pp. 3.3–3.19, 1981.

[38] *M.W. Wales, E.C. Rossow:* 'Coupled Moment-Axial Force Behaviour in Bolted Joints', Journal of Structural Engineering, ASCE, Vol. 109, pp. 1250–66, 1983.

[39] *M. Chmielowiec, R.M. Richard:* 'Moment Rotation Curves for Partially Restrained Steel Connections', Report to AISC, University of Arizona, p.127, 1987.

[40] *F. Tschemmernegg:* 'On the Nonlinear Behaviour of Joints in Steel Frames', in 'Connection in Steel Structures: Behaviour, Strength and Design', ed. R. Bjorhovde *et al.,* Elsevier Applied Science Publishers, London, pp. 158–65, 1988.

[41] *C. Humer, F. Tschemmernegg:* 'A non-linear joint model for the design of structural steel frames', Costruzioni Metalliche, N.1, 1988.

[42] *F. Tschemmernegg, C. Humer:* 'The design of structural steel frames under consideration of the non-linear behaviour of joints', Journal of Constructional Steel Research, Vol. 11, pp. 73–103, 1988.

[43] *K.V. Patel, W.F. Chen:* 'Nonlinear Analysis of Steel Moment Connections', ASCE, Journal of Structural Engineering, Vol. 110(8), pp. 1861–75, 1984.

[44] *S.L. Lipson, M.I. Hague:* 'Elasto-plastic Analysis of Single-angle Bolted-Welded Connections Using the Finite Element Method', Computers and Structures, 9(6), 533–45, 1978.

[45] *G. Ballio, L. Calado, A. De Martino, C. Faella, F.M. Mazzolani:* 'Steel Beam-to-column Joints under Cyclic Loads: Experimental and Numerical Approach', 8th ECEE, Lisbon, September 1986.

[46] *G. Ballio, L. Calado, A. De Martino, C. Faella, F.M. Mazzolani:* 'Cyclic Behaviour of Steel Beam-to-column Joints: Experimental Research', Costruzioni Metalliche, N.2, 1987.

[47] *G. Ballio, R. Zandonini:* 'An experimental equipment to test steel structural members and subassemblages subject to cyclic loads', Ingegneria Sismica, N.2, 1985.

[48] *A. De Martino:* 'Hysteretic Behaviour of Beam-to-Column Joints: Comparison and Interpretation of Experimental Results', 9th WCEE, Tokyo-Kyoto, August 1988.

[49] *K. Tsai, E.P. Popov:* 'Steel Beam-Column Joints in Seismic Moment Resisting Frames', Report N. UCB/EERC–88/19, Earthquake Engin-

eering Research Center, University of California, Berkeley, November, 1988.

[50] *M.D. Engelhardt, A.S. Husain:* 'Cyclic Tests on Large Scale Steel Moment Connections', Tenth World Conference on Earthquake Engineering, Madrid, Balkema Rotterdam, pp. 2885–2890, 1992.

[51] *M.D. Engelhardt, A.S. Husain:* 'Cyclic-Loading Performance of Welded Flange-Bolted Web Connections', Journal of Structural Engineering, ASCE, Vol. 119, pp. 3537–3550, No.12, December 1993.

[52] *C. Bernuzzi, R. Zandonini, P. Zanon:* 'Semi-Rigid Steel Connections under Cyclic Loads', Proceedings of the 1st World Conference on Constructional Steel Design, Acapulco, Mexico, 6–9 December, 1992.

[53] *C. Bernuzzi:* 'Cyclic Response of Semi-Rigid Steel Joints', Proceedings of the 1st COST C1 Workshop, pp. 194–209, Strasbourg, 28–30 October 1992.

[54] *G. Ballio, A. Plumier, B. Thunus:* 'The influence of concrete on the cyclic behaviour of composite joints', IABSE Symposium, Brussels, 1990.

[55] *G. Ballio, Y. Chen:* 'An Experimental Research on Beam to Column Joints: Exterior Connections', XIV Congresso C.T.A., Viareggio, 24–27 Ottobre 1993.

[56] *G. Ballio, Y. Chen:* 'An Experimental Research on Beam to Column Joints: Interior Connections', XIV Congresso C.T.A., Viareggio, 24–27 Ottobre 1993.

[57] *P.D. Moncarz, K.H. Gerstle:* 'Steel Frames with Nonlinear Connections', Journal of Structural Division, ASCE, Vol. 107, ST8, pp. 1427–1441, 1981.

[58] *K. Takanashi, H. Tanaka, H. Taniguchi:* 'Influence of slipping at high strength bolt connections on dynamic behaviour of frames', Institute of Industrial Science, Tokyo.

[59] *B. Kato:* 'Beam-to-Column Connection Research in Japan', Journal of Structural Division, ASCE, Vol. 108, ST2, 343–359, 1982.

[60] *U. Andreaus, G. Ceradini, P. D'Asdia:* 'Un modello per la rappresentazione di strutture dotate di vincoli interni non perfettamente bilaterali', AIMETA, VI Congresso Nazionale, Genova, pp. 302–311, 1982.

[61] *A. De Martino, C. Faella, F.M. Mazzolani:* 'Simulation of Beam-to-column Joint Behaviour under Cyclic Loads', Costruzioni Metalliche, N.6, 1984.

[62] *F.M. Mazzolani:* 'Mathematical Model for Semi-rigid Joints under Cyclic Loads', in 'Connections in Steel Structures: Behaviour, Strength and Design', ed. R. Bjorhovde *et al.*, Elsevier Applied Science Publishers, London, pp. 112–120, 1988.

[63] *M. De Stefano, A. De Luca:* 'Mechanical Models for Semirigid Connections', Proceedings of the 1st World Conference on Constructional Steel Design, Acapulco, Mexico, 6–9 December, 1992.

[64] *M. De Stefano, A. De Luca:* 'A Mechanical Model for Simulating the

Cyclic Flexural Response of Double Angle Connections', Proceedings of the 1st COST C1 Workshop, pp. 382–393, Strasbourg, 28–30 October 1992.

[65] *P.J. Madas, A.S. Elnashai:* 'A component-based model for beam-to-column connections', Tenth World Conference on Earthquake Engineering, Madrid, Balkema Rotterdam, pp. 4495–4499, 1992.

[66] *M. De Stefano, A. De Luca, A. Astaneh:* 'Modeling of Cyclic Moment-Rotation Response of Double-Angle Connections', Journal of Structural Engineering, ASCE, Vol. 120, N.1, January, 1994.

[67] *G. Ballio, A. De Martino, F.M. Mazzolani:* 'The Semi-rigid Behaviour of Beam-to-column Bolted Joints', International Colloquium on Bolted and Special Structural Joints, USSR, Moscow, 15–20 May, 1989

[68] *A. Astaneh, M. Nader, L. Malik:* 'Cyclic Behaviour of Double Angle Connections', Journal of Structural Engineering, ASCE, Vol. 115, N.5, 1989.

[69] *J.H. Shen, A. Astaneh:* 'Behavior and Hysteresis Model of Steel Semi-rigid Connections with Bolted Angles', Report UCB/EERC–93, Earthquake Engineering Research Center, University of California, Berkeley, 1993.

[70] *M. De Stefano, A. Astaneh:* 'Axial Force-Displacement Behavior of Steel Double Angles', Journal of Constructional Steel Research, Vol. 20, pp. 161–181, 1991.

[71] *M. De Stefano, A. Astaneh, A. De Luca, I. Ho:* 'Behaviour and Modeling of Double Angle Connections subjected to Axial Loads', Structural Stability Research Council, Annual Technical Session, Chicago, 1991.

7

Seismic behaviour of semirigid frames

7.1 FRAME VERSUS CONNECTION BEHAVIOUR

In the previous chapter the behaviour of beam-to-column connections has been analysed with reference both to monotonic loading conditions and to cyclic loading conditions. It has been shown that this behaviour is generally non-linear, with different degrees of deterioration as far as the number of cycles increases. Particular attention has also been paid to the connection classification. Two main parameters are involved in the connection classification, the flexural strength and the rotational stiffness.

With reference to the flexural strength, three connection classes have been identified:

- full strength connections
- partial strength connections
- nominally pinned connections.

In addition, with reference to the rotational stiffness, the following classification has been made:

- rigid or fully restrained connections
- semirigid or partially restrained connections
- nominally pinned connections.

Excluding the case of nominally pinned connections, which do not correspond to the case of moment-resisting frames, four fundamental cases can be recognized:

a. full-strength rigid connections
b. full-strength semirigid connections
c. partial-strength rigid connections
d. partial-strength semirigid connections

Such a distinction is necessary because the seismic behaviour of moment-resisting frames is considerably affected both by the strength and by the stiffness of the beam-to-column connections. However, the term

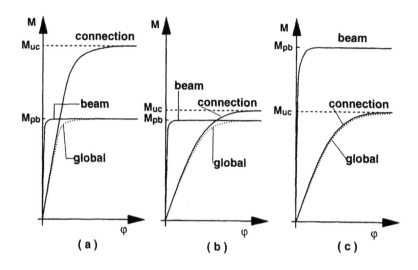

Fig. 7.1   Global behaviour due to the beam-to-connection resistance ratio

semirigid frame is often used indiscriminately both for frames with full-strength semirigid connections (case **b**) and for frames with partial-strength connections (cases **c** and **d**). On the contrary, the term rigid frame is often adopted to denote frames with full-strength rigid connections, but the distinction between cases **a** and **c** should be made.

It is evident that full-strength rigid connections (case **a**) represents the reference case, i.e. the case in which the beam-to-column connection exhibits the ideal behaviour.

In particular, from the point of view of the location of the dissipative zones, completely different behaviours are developed for full-strength connections and for partial-strength connections. In fact, in order to simulate the behaviour of the beam-to-column joints, both the connection and the cross-section of the connected member have to be taken into account.

Three different types of behaviour can be recognized depending upon the ultimate moment of the connection and on the plastic moment of the connected beam [1]: $M_{uc} > M_{pb}$ (Fig. 7.1a), $M_{uc} \approx M_{pb}$ (Fig. 7.1b), $M_{uc} < M_{pb}$ (Fig. 7.1c), where M_{uc} and M_{pb} are the ultimate moment of the connection and the plastic moment of the connected beam, respectively.

In the first case, the shape of the moment versus rotation global (beam plus connection) curve can be quite accurately represented with a bilinear elastic–perfectly plastic model. A possible plastic hinge will develop at the beam end, while the connection remains in the elastic range. Plastic rotations involve only the beam end.

In the third case, the accurate modelling of the shape of the moment versus rotation global curve requires a non-linear representation, because it is coincident with the connection curve. Moreover, the dissipative zone is located within the connection, while the beam remains in the elastic range. In this case the connection has to be able to experience large plastic rotations.

Finally, in the intermediate case, representing a transition condition, non-linear modelling is still required. Furthermore, the yielding occurs both in the connection elements and at the end of the beam.

In the European seismic code, Eurocode 8 [2], it is required that connections in dissipative zones have to possess sufficient overstrength to allow for yielding of the ends of connected members. It is deemed that the above design condition is satisfied in the case of welded connections with butt welds or full penetration welds. However, for fillet weld connections and for bolted connections the design resistance of the connection has to be at least 1.20 times the plastic resistance of the connected member. This means that the use of full-strength connections is recommended and dissipative zones have to be located at the member end rather than in the connections.

The use of partial-strength connections, i.e. the contribution of the connections in dissipating the earthquake input energy, is not forbidden. However, it is strongly limited because, in such a case, experimental control of the effectiveness of such connections in dissipating energy is required.

Codified rules for evaluating the behaviour factor to be used in design are not still available for either full-strength or partial-strength semirigid frames.

Although numerous experimental tests on the behaviour of beam-to-column connections under monotonic and cyclic loads have been carried out by many researchers, the seismic behaviour of semirigid frames has not been exhaustively investigated.

Recently, Astaneh and Nader [4, 5] investigated the behaviour of a one-storey one-bay steel structure with rigid, semirigid and flexible connections subjected to a variety of base excitations using the shaking table of the Earthquake Engineering Research Center of the University of California at Berkeley. The same authors have investigated, by numerical simulations, the seismic response of four-storey, seven-storey and ten-storey semirigid frames [6]. Moreover, the seismic inelastic behaviour of an instrumented six-storey semirigid steel building has been analysed [7].

On the basis of the above analyses, a first proposal for a recommended design procedure has been developed [5, 6]. In particular, as in semirigid frames with partial-strength connections large plastic deformations have to be experienced by the connecting elements, the connection design has to be done by forcing the yielding to occur at a desiderable location within the connection. It is suggested [6] that connections should be designed so that yielding occurs only in the plate elements, such as angles, splices and end-plates, while the fasteners, such as bolts and welds, have to remain in elastic range. In addition, design equations for calculating the base shear and the corresponding distribution over the height have been proposed, including equations for estimating the period of vibration and the design value of the behaviour factor [6]. Nevertheless, the above proposal cannot be considered exhaustive due to the limited number of structural schemes and ground motions that have been analysed. This is confirmed by the great number of parameters governing the seismic behaviour of moment-resisting frames. In addition, this number increases where the connection behaviour is concerned.

For these reasons, and with the aim of clarifying all parameters affecting the seismic inelastic behaviour of full/partial strength semirigid frames, a simplified model has been introduced for full-strength semirigid connections [8], and extended [9] to the case of partial-strength semirigid connections. This approach, which will be described in the following sections, also has the advantage of allowing the use of the large amounts of information concerning the seismic response of SDOF systems and, therefore, includes the effects of the random variability of the ground motion.

7.2 ANALYSED SUBSTRUCTURE

Study of the seismic response of steel frames, including the connection behaviour, can be carried out using the substructure representation in Fig. 7.2. It represents a subassemblage which has been extracted from an actual frame by assuming that beams are subjected to double curvature bending with zero moment in the midspan section and by considering each beam as belonging to two storeys. In this way their mechanical properties have been halved. It is believed that this model is representative of frames failing in global mode. The flexural stiffness of the columns of the original frame, from which the substructure has been derived, is equal to EI_c/h; the flexural stiffness of the beams is equal to EI_b/L. The moment versus rotation curve of the connection is modelled with a bilinear relation which is completely defined by means of only two parameters: the elastic rotational stiffness K_φ and the ultimate moment M_{uc}. It is evident that, with reference to the elastic range, the comparison between the semirigid substructure and the rigid one is governed by the following non-dimensional parameters

$$\zeta = \frac{E\,I_b\,/\,L}{E\,I_c\,/\,h} \tag{7.1}$$

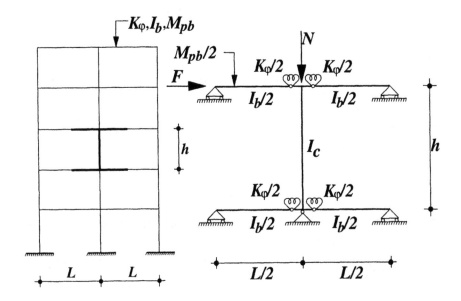

Fig. 7.2 The simplified model: substructure extracted from an actual frame

and

$$\overline{K} = \frac{K_\varphi L}{E I_b} \tag{7.2}$$

where ζ is the beam-to-column stiffness ratio and $\overline{K}$ is the non-dimensional rotational stiffness of the connection.

In addition, with reference to the inelastic behaviour, the non-dimensional ultimate flexural resistance of the connection

$$\overline{m} = \frac{M_{uc}}{M_{pb}} \tag{7.3}$$

provides the distinction between the two fundamental cases, i.e. the full-strength connections for $\overline{m} > 1$ and the partial strength connections for $\overline{m} < 1$.

The connection influence on the elastic behaviour of the model will be investigated in Section 7.3, as it is common to both full-strength and partial-strength connections. The analysis of the parameters influencing the inelastic behaviour of the model will be covered in Section 7.4.

7.3　CONNECTION INFLUENCE ON ELASTIC PARAMETERS

7.3.1 Period of vibration

In order to investigate the influence of the connection rotational stiffness on the period of vibration, it is sufficient to evaluate the influence of K on the lateral stiffness of the substructure. In fact, it is easy to recognize that the ratio between the period of vibration of the semirigid model T_k and that of the rigid model T_∞ is given by

$$\frac{T_k}{T_\infty} = \left(\frac{K_{l_\infty}}{K_{l_k}} \right)^{1/2} \tag{7.4}$$

where K_{l_k} and K_{l_∞} are the lateral stiffness of the semirigid model and the rigid model, respectively.

The study of the substructure subjected to a horizontal force F applied at the top level provides the relationship [9]

$$\begin{bmatrix} \dfrac{6EI_b}{L}\dfrac{K}{K+6} + \dfrac{6EI_c}{h} & -\dfrac{6EI_c}{h^2} \\[3ex] -\dfrac{6EI_c}{h^2} & \dfrac{6EI_c}{h^3} \end{bmatrix} \left\{ \begin{array}{c} \theta \\ \delta \end{array} \right\} = \left\{ \begin{array}{c} 0 \\ \dfrac{F}{2} \end{array} \right\} \tag{7.5}$$

where θ is the rotation of the nodes and δ the top sway displacement.

The lateral stiffness of the substructure is derived as F/δ, leading to the relationship

$$K_{l_k} = \frac{12EI_c}{h^3}\frac{K\zeta}{K+6+K\zeta} \tag{7.6}$$

which, for $K \to \infty$, gives

$$K_{l_\infty} = \frac{12EI_c}{h^3}\frac{\zeta}{1+\zeta} \tag{7.7}$$

Therefore, from equation (7.4) [9]

$$\frac{T_k}{T_\infty} = \left(\frac{K(1+\zeta)+6}{K(1+\zeta)} \right)^{1/2} \tag{7.8}$$

This relationship is represented in Fig. 7.3. It shows that the connection deformability produces an increase in the period of vibration. This is, generally, an advantageous effect, because it corresponds to the shifting

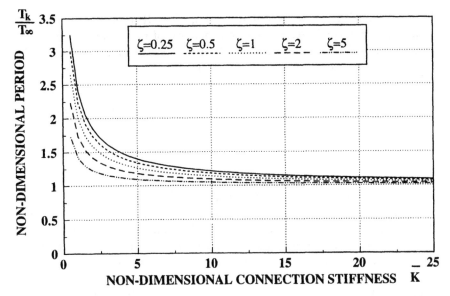

Fig. 7.3   Connection influence on the period of vibration

in a period range of the elastic design response spectrum where the spectral acceleration is reduced.

7.3.2 Stability coefficient

The frame sensitivity to second-order effects is expressed through the stability coefficient γ (Chapter 2). It represents the slope of the softening branch of the behavioural curve relating the multiplier of the horizontal forces α to the non-dimensional top sway displacement δ/δ_1 (where δ_1 is the top displacement under the horizontal forces corresponding to $\alpha = 1$). In the case of single degree of freedom systems, such as the substructure herein analysed, the stability coefficient can be expressed as (Chapter 2)

$$\gamma = \frac{N}{K_l h} \tag{7.9}$$

where N is the vertical load and K_l is the lateral stiffness of the substructure.

As a consequence, the ratio between the stability coefficient of the semirigid model γ_k and that of the rigid model γ_∞ is given by [9]

$$\frac{\gamma_k}{\gamma_\infty} = \frac{K_{l_\infty}}{K_{l_k}} = \frac{\overline{K}(1+\zeta)+6}{\overline{K}(1+\zeta)} \tag{7.10}$$

This relationship is represented in Fig. 7.4. It points out that the connection deformability leads to an increase of the frame sensitivity to second-order effects. This phenomenon is undesirable because it gives rise to an amplification of the ductility demand under severe ground motion.

7.4 CONNECTION INFLUENCE ON INELASTIC PARAMETERS

7.4.1 General

In the previous sections, it has been shown that the connection deformability has two effects on the elastic behaviour of the substructure, and therefore on the frame's elastic behaviour. In fact, on one hand, it determines an increase of the period of vibration and, on the other hand, it leads to an increase of frame sensitivity to second-order effects. From the seismic point of view, the first effect is desirable as it locates the structure in a more favourable zone of the elastic design response spectrum. However, the increase of the frame sensitivity to second-order effects is undesirable, due to the amplification of the negative effects of the geometric non-linearity under seismic loads. The role of these two important effects has been widely investigated [8].

As far as inelastic behaviour is concerned, the distinction between full-strength and partial-strength connections becomes determinant. In fact,

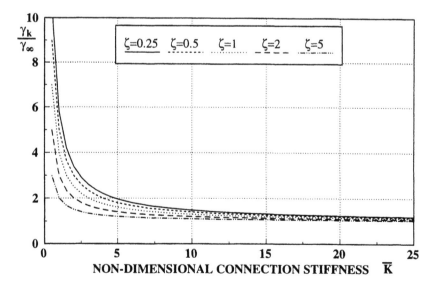

Fig. 7.4   Connection influence on the model sensitivity to second- order effects

it determines the location of the dissipative zones, which are concentrated at the beam ends in the first case and in the connecting elements in the second case. According to the type of connection, the ultimate strength and/or the available ductility are involved. In the case of a full-strength connection, a decrease of the available ductility arises, due to the increase of lateral deformability of the structure, while the ultimate resistance of the frame is slightly decreased due to second-order effects [10]. For partial-strength connections, however, a decrease of the lateral load resistance of the frame occurs, while the connection influence on the available ductility is strictly related to the ratio between the connection plastic rotation capacity and the beam plastic rotation capacity [10].

7.4.2 Frames with full-strength connections

Figure 7.5 shows the lateral load versus top displacement curve both for a substructure with full-strength rigid connections and for the same substructure with full-strength semirigid connections. First of all, it can be seen that the first-order yield resistance is the same in the two cases:

$$F_{y_k}^{(FS)} = F_{y_\infty}^{(FS)} \qquad (7.11)$$

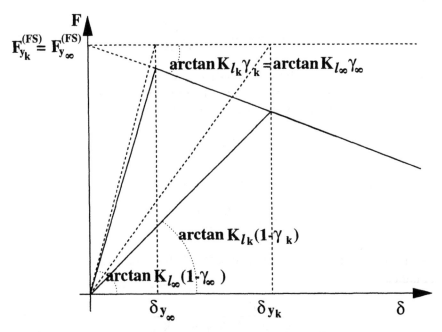

Fig. 7.5 Full strength connections: influence on the lateral load versus top displacement curve of the substructure

Due to second-order effects, the yield resistance decreases, assuming the following values:

$$F''^{(FS)}_{y_k} = F^{(FS)}_{y_k} (1 - \gamma_k) \tag{7.12}$$

$$F''^{(FS)}_{y_\infty} = F^{(FS)}_{y_\infty} (1 - \gamma_\infty) \tag{7.13}$$

Therefore, where $\gamma_k > \gamma_\infty$ the actual yield resistance of the substructure is decreased, due to the amplification of second-order effects produced by the connection deformability. The slope of the softening branch of the lateral load versus top displacement curve is unaffected, as also shown by equation (7.10).

The available ductility of the substructure can be expressed as

$$\mu_k^{(FS)} = 1 + \frac{\theta_{pb} h}{\delta_y} \tag{7.14}$$

where θ_{pb} is the plastic rotation that the beam is able to withstand.

The maximum plastic rotation that a member is able to develop is usually derived by means of a three-point bending test (Fig. 7.6). As a consequence, it can be expressed as

$$\theta_{pb} = R \; \theta_y = R \; \frac{M_{pb} \; L}{4 \, E \, I_b} \tag{7.15}$$

where R is the plastic rotation capacity of the beams.

In addition, equation (7.5) provides the following relationship between the member end rotation and the top displacement

$$\theta = \frac{\overline{K} + 6}{\overline{K} + 6 + \overline{K} \; \zeta} \frac{\delta}{h} \tag{7.16}$$

The bending moment at the end of the beam is given by

$$M = \frac{3 \, E \, I_b}{L} \frac{\overline{K}}{\overline{K} + 6} \theta \tag{7.17}$$

By imposing the yield condition, $M = M_{pb}/2$, the following relationship is obtained

$$\frac{h}{\delta_y} = \frac{6 \, E \, I_b}{M_{pb} L} \frac{\overline{K}}{\overline{K} + 6 + \overline{K} \; \zeta} \tag{7.18}$$

Therefore, by substituting equations (7.18) and (7.15) in equation

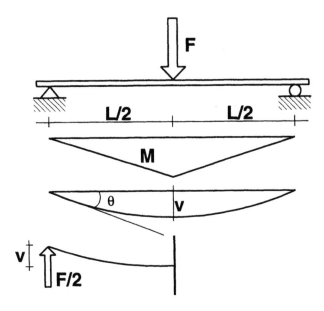

Fig. 7.6   Three-point bending test for evaluating beam rotation capacity

(7.14), the available ductility of the substructure with full-strength semirigid connections can be expressed as [10]

$$\mu_k^{(FS)} = 1 + \frac{3}{2} \overline{R} \frac{\overline{K}}{\overline{K} + 6 + \overline{K} \zeta} \qquad (7.19)$$

which, for $\overline{K} \rightarrow \infty$, gives

$$\mu_\infty^{(FS)} = 1 + \frac{3}{2} \frac{\overline{R}}{1 + \zeta} \qquad (7.20)$$

This relationship represents the ductility of the model with full-strength rigid joints.

The above equations show that connection deformability leads to a reduction of the available ductility. This phenomenon is illustrated in Fig. 7.7, for $\overline{R} = 2$ and $\overline{R} = 6$, through the $\mu_k^{(FS)}/\mu_\infty^{(FS)}$ versus $\overline{K}$ relationship. In particular, comparison between the cases for $\overline{R} = 2$ and $\overline{R} = 6$ shows that loss of ductility due to connection deformability increases as the beam rotation capacity increases, i.e. as the ductility $\mu_\infty^{(FS)}$ increases. Finally, considering equation (7.10), it is interesting to note that the

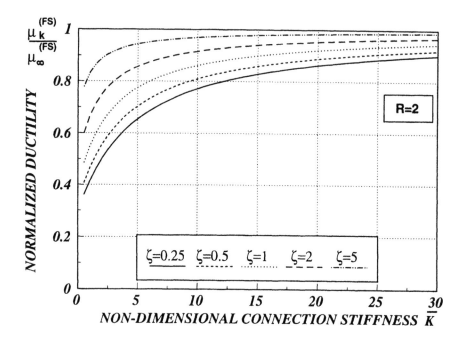

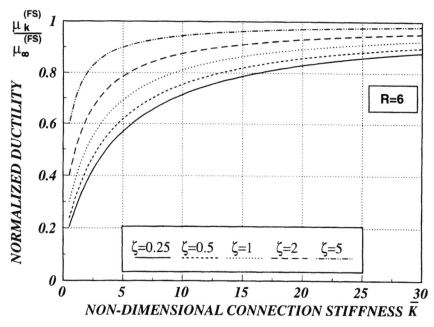

Fig. 7.7   Model with full-strength connections: influence of connection stiffness on available ductility

ductility of the model with full-strength semirigid connections can also be expressed as

$$\mu_k^{(FS)} = 1 + (\mu_\infty^{(FS)} - 1) \frac{K_{l_k}}{K_{l_\infty}} \qquad (7.21)$$

which explicitly shows the effect of the increase of the model lateral deformability, due to connection flexibility. In fact, for $K_{l_k}/K_{l_\infty} = 1$ it gives $\mu_k^{(FS)} = \mu_\infty^{(FS)}$, while connection deformability leads to the condition $K_{l_k}/K_{l_\infty} < 1$ and, therefore, to a ductility reduction (i.e. $\mu_k^{(FS)} < \mu_\infty^{(FS)}$).

7.4.3 Frames with partial-strength connections

In the cases of frames with partial-strength semirigid connections, a significant decrease of the yield resistance of the substructure arises, while the connection influence on the available ductility is strictly related not only to the connection deformability, but also to the ratio between the connection plastic rotation capacity and the beam plastic rotation capacity. This is due to the fact that, in this case, yielding is located in the connecting elements rather than at the beam ends.

Figure 7.8 shows the lateral load versus top displacement curve of the substructure with partial-strength semirigid connections and its comparison with the same model having full-strength rigid connections.

The reduction of the first-order yield resistance is given by

$$\frac{F_{y_k}^{(PS)}}{F_{y_\infty}^{(FS)}} = \overline{m} = \frac{M_{uc}}{M_{pb}} \qquad (7.22)$$

The actual yield resistance decreases further due to second-order effects. In fact, for the model with partial-strength semirigid connections, it is given by

$$F''^{(PS)}_{y_k} = F^{(PS)}_{y_k} (1 - \gamma_k) \qquad (7.23)$$

while, for the model with full-strength rigid connections, it is still given by equation (7.13).

The available ductility of the model with partial-strength connections is obtained taking into account that, in this case, yielding occurs in the connection. Therefore, it can be expressed as

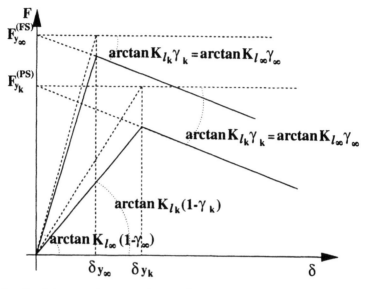

Fig. 7.8   Partial-strength connections: influence on the lateral load versus top displacement curve of the substructure model

$$\mu_k^{(PS)} = 1 + \frac{\varphi_{pc}\, h}{\delta_y} \tag{7.24}$$

where φ_{pc} is the maximum plastic rotation that the connection is able to withstand.

In addition, the yielding condition can be written as

$$\frac{h}{\delta_y} = \frac{6\,E\,I_b}{M_{yc}\,L}\,\frac{\overline{K}}{\overline{K}+6+\overline{K}\,\zeta} \tag{7.25}$$

where M_{yc} is the yielding resistance of the connection.

By substituting equations (7.25) and (7.2) into equation (7.24), and taking into account that $\varphi_{yc} = M_{yc} / K_\varphi$ is the yield rotation of the connection, the following relationship is obtained:

$$\mu_k^{(PS)} = 1 + \left(\frac{\varphi_{uc}}{\varphi_{yc}} - 1\right)\frac{6}{\overline{K}+6+\overline{K}\,\zeta} \tag{7.26}$$

where $\varphi_{uc} = \varphi_{pc} + \varphi_{yc}$ is the ultimate rotation of the connection.

According to Bjorhovde *et al.* [11], the ultimate rotation of the connections can be expressed as (Chapter 6)

$$\varphi_{uc} = \frac{5.4 - 3\,\overline{m}}{2}\,\frac{5\,M_{pb}\,d_b}{EI_b} \tag{7.27}$$

where d_b is the depth of the beam. Also the connection stiffness can be expressed as [11]

$$K_\varphi = \frac{EI_b}{L_e} = \frac{EI_b}{\eta\, d_b} \tag{7.28}$$

where the equivalent beam length of the connection $L_e = \eta\, d_b$, i.e. the length of a beam having a flexural stiffness equal to the rotational stiffness of the connection.

Therefore, the yield rotation of the connection can be expressed as

$$\varphi_{yc} = \frac{\overline{m}\, M_{pb}}{E\, I_b}\, \eta\, d_b \tag{7.29}$$

By substituting equations (7.27) and (7.29) into equation (7.26), the available ductility of the substructure with partial-strength semirigid connections can be expressed through the relationship

$$\mu_k^{(PS)} = 1 + \left(\frac{5}{\eta}\, \frac{5.4 - 3\,\overline{m}}{2\,\overline{m}} - 1 \right) \frac{6}{\overline{K} + 6 + \overline{K}\, \zeta} \tag{7.30}$$

Combination of equations (7.2) and (7.28) gives

$$\eta = \frac{L}{d_b\, \overline{K}} \tag{7.31}$$

Consequently, the ductility of the model can be expressed as [10]

$$\mu_k^{(PS)} = 1 + \left(\frac{5\, d_b\, \overline{K}}{L}\, \frac{5.4 - 3\,\overline{m}}{2\,\overline{m}} - 1 \right) \frac{6}{\overline{K} + 6 + \overline{K}\, \zeta} \tag{7.32}$$

This equation shows that the available ductility of the substructure model with partial-strength semirigid joints depends on the beam depth-to-span ratio d_b/L, the beam-to-column stiffness ratio ζ, the non-dimensional stiffness of the connection $\overline{K}$ and the corresponding non-dimensional ultimate moment $\overline{m}$.

7.4.4 Ductility of the substructure

The analysis of the ductility of the substructure model can be simplified by taking into account that the stiffness of the connection and its ultimate moment are related. In fact, from the analysis of experimental data collected in the SERICON data bank [12], the condition [10]

$$\overline{m} = 2.1054\, \eta^{-0.4376} \tag{7.33}$$

has been derived for extended end plate connections to a stiffened column, and

$$\overline{m} = 1.1038 \, \eta^{-0.5711} \tag{7.34}$$

for extended end plate connections to an unstiffened column.

In addition, from analysis of the experimental results obtained by Az-izinamini and Radziminski [13], a similar relationship has been obtained in the case of top and seat angles with double web angle connections [9]

$$\overline{m} = 0.7474 \, \eta^{-0.4120} \tag{7.35}$$

These relationships are illustrated with their corresponding experimental data in Figs. 7.9–7.11.

Due to the great variety of structural detail which can be adopted for beam-to-column joints, the relationship between joint stiffness and its ultimate moment deserves further investigation and supplementary experimental tests. Notwithstanding, the above equations can be useful for a better comprehension of the inelastic behaviour of the substructure model and, therefore, of semirigid frames.

In particular, taking into account equation (7.31), the use of equations

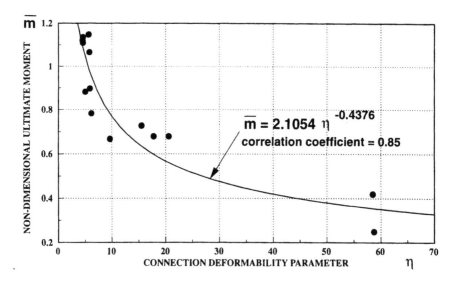

Fig. 7.9 $\overline{m}$–η for extended end plate connections with stiffened column

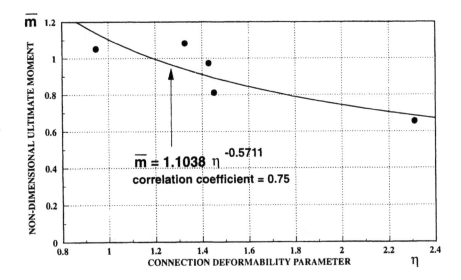

Fig. 7.10 $\overline{m}$–η for extended end plate connections with unstiffened column

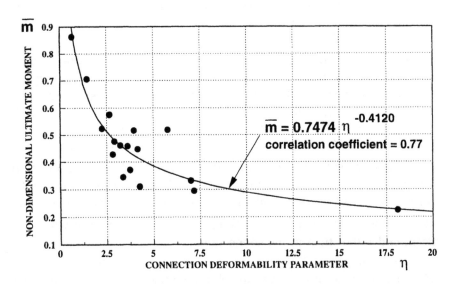

Fig. 7.11 $\overline{m}$–η for top and seat angle with double web angle connections

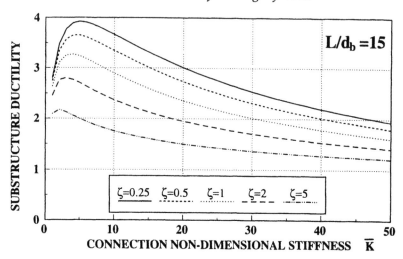

Fig. 7.12   Substructure with partial-strength joints: influence of the connection on the available ductility

(7.33)–(7.35) allows, for the corresponding joint typologies, expression of the ductility of the model through a relationship of the type

$$\mu_k^{(PS)} = \mu_k^{(PS)}\left(\overline{K}, \zeta, \frac{L}{d_b}\right) \tag{7.36}$$

As an example, this relationship is illustrated in Fig. 7.12 with reference to the top and seat angles with double web angle connections for the case $L/d_b = 15$.

The ductility of the substructure has to be derived by considering that, depending on the beam depth-to-span ratio d_b/L and on the connection typology, the existence of a value of the non-dimensional stiffness of the connection leading to the full strength condition has to be recognized. This value indicates two ranges of connection stiffness. The first one corresponds to $M_{uc} < M_{pb}$, i.e. to partial-strength connections, so that the ductility of the substructure is provided by $\mu_k^{(PS)}$ and yielding occurs in the connection rather than in the beam end. The second one leads to the condition $M_{uc} > M_{pb}$, i.e. to full-strength connections, so that the ductility of the substructure is provided by $\mu_k^{(FS)}$, because yielding arises in the beam end rather than in the connection.

As an example, with reference to a beam span-to-depth ratio equal to 15 and to a value of the beam-to-column stiffness ratio $\zeta = 1$, the above behaviour is illustrated in Figs. 7.13–7.15 for $R = 6$ and in Figs. 7.16–7.18 for $R = 2$. In these figures, the value of the non-dimensional ultimate

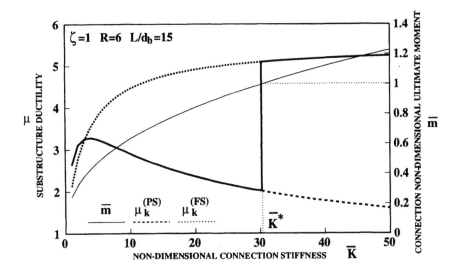

Fig. 7.13   Substructure: connection influence on the available ductility for top and seat angle with double web angles

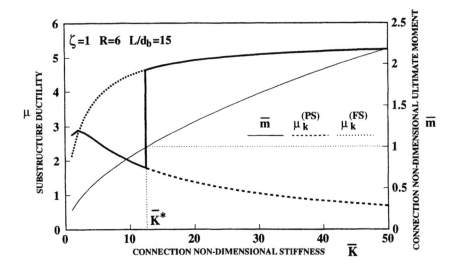

Fig. 7.14   Substructure: connection influence on the available ductility for extended end plate with unstiffened column

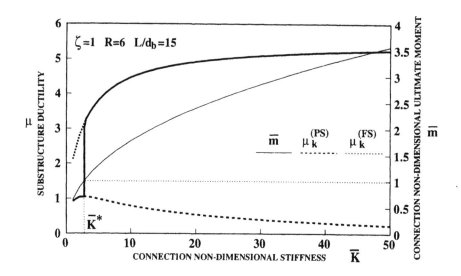

Fig. 7.15   Substructure: connection influence on the available ductility for extended end plate with stiffened column

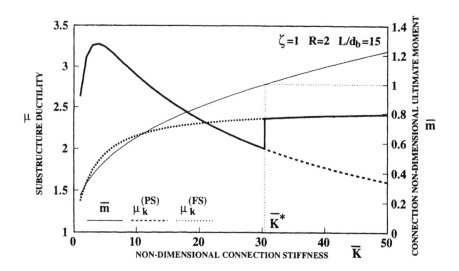

Fig. 7.16   Substructure: connection influence on the available ductility for top and seat angle with double web angles

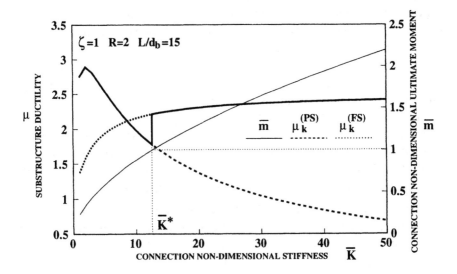

Fig. 7.17 Substructure: connection influence on the available ductility for extended end plate with unstiffened column

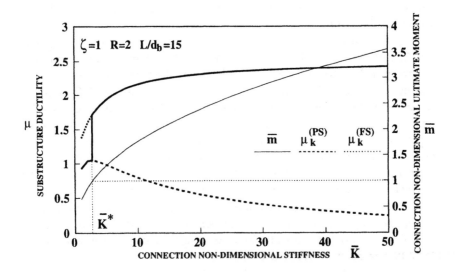

Fig. 7.18 Substructure: connection influence on the available ductility for extended end plate with stiffened column

moment $\overline{m}$ of the connection is given on the right-hand axis. The thin continuous line represents, for the given value of the beam span-to-depth ratio L/d_b, the relationship between the non-dimensional ultimate moment of the connection $\overline{m}$ and its non-dimensional rotational stiffness $\overline{K}$. The thin dotted line, starting from $\overline{m} = 1$ and intercepting the $\overline{m}$ versus $\overline{K}$ curve, allows identification of the value $\overline{K}^*$ of $\overline{K}$ dividing the $\overline{K}$-axis into two ranges: the first range, on the left-hand side, corresponds to the partial-strength condition and the second range corresponds to the full-strength condition. In addition, on the left-hand scale, the value for substructure ductility is provided. The thick dotted line shows, as a function of $\overline{K}$, the ductility $\mu_k^{(FS)}$ of the substructure with full-strength semirigid connections. Moreover, the thick hatched line represents, as a function of $\overline{K}$, the ductility $\mu_k^{(PS)}$ of the substructure with partial-strength semirigid connections. As far as the actual ductility of the substructure is concerned, the validity ranges of the two previous curves which have been identified through the value $\overline{K}^*$ must be accounted for. Consequently, the actual ductility of the substructure is given by the thick continuous line.

It can be seen that for a member behavioural class corresponding to $R = 6$, i.e. for ductile sections (first-class sections), the connection influence always leads to a worsening of the inelastic behaviour, leading to a reduction of the available ductility with respect to full-strength rigid connections. In addition, this worsening seems to be intolerable in the case of partial-strength connections.

On the contrary, for beam sections having a small rotation capacity ($R = 2$) different behaviour patterns are exhibited in the two ranges of $\overline{K}$, corresponding to full-strength and partial-strength connections.

In the range for full-strength connections, the connection deformability always leads to a reduction of the available ductility. Conversely, for the partial-strength range, the connection effects depend on the ratio between its plastic deformation capacity, which increases as the connection stiffness decreases, and the beam rotation capacity. Depending on the connection typology, therefore, a range of connection stiffness can arise in which an improvement of the available ductility is attained.

In order to demonstrate the ability of the analytical model to represent connection influence on the inelastic behaviour of actual frames, a numerical simulation of the static inelastic response of the frame shown in Fig. 7.19 has been performed [10].

In Figures 7.20–7.25, the results of these numerical simulations are compared with those derived from the simplified model of Fig. 7.2, where it has been assumed $\zeta = 0.48$ and $L/d_b = 15$ according to the investigated frame. It can be seen that these results are in good agreement

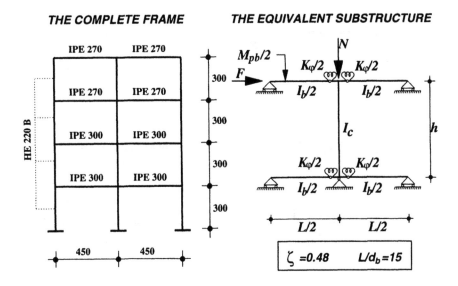

Fig.7.19 Analysed frame

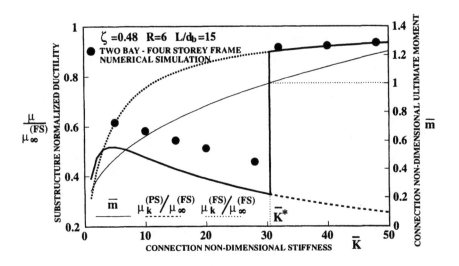

Fig. 7.20 Results of the numerical simulation of the inelastic response of the given frame compared with the substructure model, for top and seat angle with double web angles

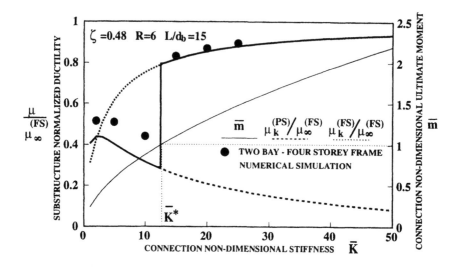

Fig. 7.21   Results of the numerical simulation of the inelastic response of the given frame compared with the substructure model, for extended end plate with unstiffened column

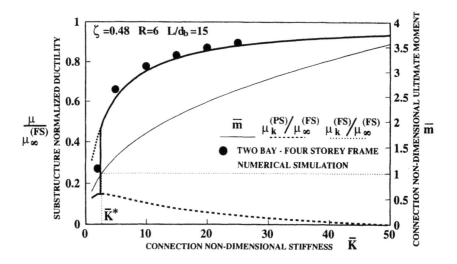

Fig. 7.22   Results of the numerical simulation of the inelastic response of the given frame compared with the substructure model, for extended end plate with stiffened column

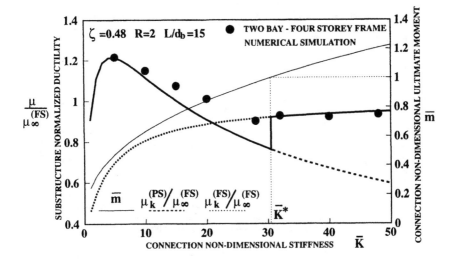

Fig. 7.23 Results of the numerical simulation of the inelastic response of the given frame compared with the substructure model, for top and seat angle with double web angles

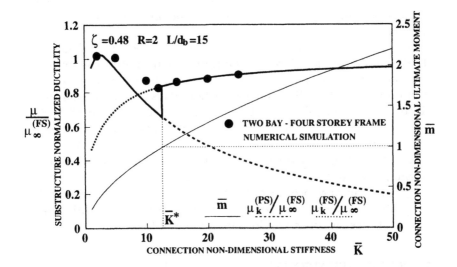

Fig. 7.24 Results of the numerical simulation of the inelastic response of the given frame compared with the substructure model, for extended end plate with unstiffened column

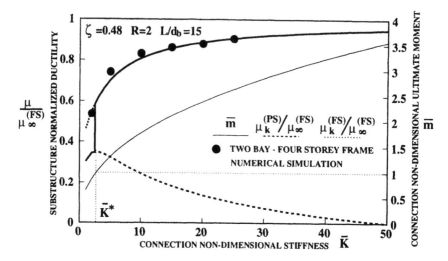

Fig. 7.25   Results of the numerical simulation of the inelastic response of the given frame compared with the substructure model, for extended end plate with stiffened column

with the theoretical curves obtained from the analysis of the simple substructure. Therefore, it can be stated that the simplified model of Fig. 7.2 is able to interpret, to a reasonable approximation, the complex influence of beam-to-column connections on the elastic and inelastic response of steel frames.

These results show that beam-to-column connections influence not only the period of vibration and the frame's sensitivity to second-order effects, but also its ductility. In particular, as far as the available ductility is concerned, the strength of the connection plays a fundamental role. In fact, for partial-strength connections a significant reduction of ductility arises. This reduction increases as the ratio between the beam plastic rotation capacity and the connection plastic deformation capacity increases. In addition, it has been shown that a knowledge of the conditions governing the transition from partial-strength to full-strength behaviour is of primary importance. In order to establish simplified relationships among all parameters characterizing the joint behaviour, additional experimental investigations would therefore be of help.

7.5   SEISMIC BEHAVIOUR

7.5.1 Seismic response of the model as an SDOF system

The analyses presented in the previous sections have pointed out the advantages of investigating the seismic behaviour of semirigid steel frames, with full-strength or partial-strength joints, using a simplified model. In fact, the use of the substructure shown in Fig. 7.2 has led to closed-form solutions which clearly show all the effects due to the beam-to-column connections. In particular, it has been stressed that the joint deformability produces a beneficial increase in the period of vibration, but also a dangerous increase in frame sensitivity to second-order effects. In addition, in the case of full-strength joints, the decrease of frame lateral stiffness is also responsible for a reduction of the available ductility. Regarding the available ductility, a different behaviour has been identified for partial-strength connections. In this case, the connection deformation capacity increases as the stiffness of the connection decreases, but this phenomenon leads to an advantageous increase of the global ductility only for beams having a very small plastic deformation capacity (as can be seen by comparing Fig. 7.20 and Fig. 7.23). In other words, in the case of beams made of ductile sections (first-class sections according to EC3) it is better to adopt full-strength connections so that yielding is forced to occur in the beam ends rather than in the connection, and the beam rotation capacity is intentionally exploited. Conversely, the use of partial-strength connections can prevent complete exploitation of the plastic reserves of the frame, leading to a critical reduction of the global ductility, due to severe local conditions.

Another important advantage, related to the use of a simplified model, is the possibility of predicting seismic behaviour from the great amount of information available for seismic responses of simple degree of freedom (SDOF) systems. In fact, the lateral load versus top displacement curve of the substructure (Figs. 7.5 and 7.8) corresponds to that of an elastic–perfectly plastic SDOF system with geometrical non-linearity, i.e. with second-order effects included.

Therefore, the q-factor of the substructure can be evaluated as

$$q(mu, T, \gamma) = \frac{q_0(\mu, T, \gamma = 0)}{\varphi_\Delta(\mu, \gamma)} \tag{7.37}$$

where $q_0(mu, T)$ is the q-factor of an elastic–perfectly plastic SDOF system, as a function of μ and T which are the available ductility and the period of vibration, respectively, while $\varphi_\Delta(\mu, \gamma)$ is a coefficient which takes into account the influence of second-order effects (Chapter 5) [14–16].

The value of the q-factor, in the absence of second-order effects, can be evaluated from the relationship proposed by Nassar and Krawinkler [17]

$$q_0 = [c(\mu - 1) + 1]^{\frac{1}{c}} \quad \text{with} \quad c = \frac{T}{1 + T} + \frac{0.42}{T} \tag{7.38}$$

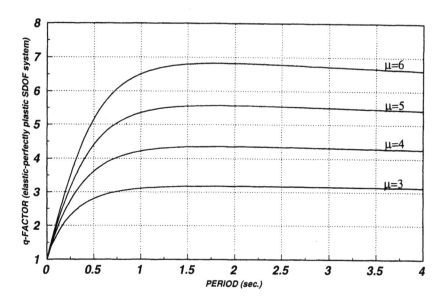

Fig. 7.26 Krawinkler and Nassar formulation for evaluating the *q*-factor of an elastic–perfectly plastic SDOF system in the absence of geometric non-linearity

This relationship is illustrated in Fig. 7.26.

Moreover, the influence of second-order effects can be accounted for through the relationship [14–16]

$$\varphi_\Delta = \frac{1 + \psi_1 \, (\mu - 1)^{\psi_2} \gamma}{1 - \gamma} \qquad (7.39)$$

where the mean values of the coefficient φ_Δ can be obtained through the parameters $\psi_1 = 0.62$ and $\psi_2 = 1.45$.

As a consequence, the ratio between the *q*-factor of the substructure model including the effects of beam-to-column joints and that for the ideal model with full-strength rigid joints is given by

$$\frac{q_k^{(FS)}}{q_\infty^{(FS)}} = \frac{q_{0_k}^{(FS)}}{q_{0_\infty}^{(FS)}} \frac{\varphi_{\Delta_\infty}^{(FS)}}{\varphi_{\Delta_k}^{(FS)}} \qquad (7.40)$$

for full-strength semirigid joints [18], and

$$\frac{q_k^{(PS)}}{q_\infty^{(FS)}} = \frac{q_{0_k}^{(PS)}}{q_{0_\infty}^{(FS)}} \frac{\varphi_{\Delta_\infty}^{(FS)}}{\varphi_{\Delta_k}^{(PS)}} \tag{7.41}$$

for partial-strength semirigid joints [18].

In equations (7.40) and (7.41), $q_\infty^{(FS)}$ is the q-factor of the substructure with full-strength rigid joints and, therefore, it is evaluated through the value $q_{0_\infty}^{(FS)}$ of q_0 provided by equation (7.38) with T_∞ and $\mu_\infty^{(FS)}$ and through the value $\varphi_{\Delta_\infty}^{(FS)}$ of φ_Δ provided by equation (7.39) with γ_∞ and $\mu_\infty^{(FS)}$.

In equation (7.40), $q_k^{(FS)}$ is the q-factor of the substructure with full-strength semirigid joints and, therefore, it is evaluated through the value $q_{0_k}^{(FS)}$ of q_0 provided by equation (7.38) with T_k and $\mu_k^{(FS)}$ and through the value $\varphi_{\Delta_k}^{(FS)}$ of φ_Δ provided by equation (7.39) with γ_k and $\mu_k^{(FS)}$.

Finally, in equation (7.41), $q_k^{(PS)}$ is the q-factor of the substructure with partial-strength semirigid joints. It has to be evaluated as for $q_k^{(FS)}$, but using $\mu_k^{(PS)}$ instead of $\mu_k^{(FS)}$.

It can be seen that $\mu_\infty^{(FS)}$ and $\mu_k^{(FS)}$ can be expressed through equations (7.19) and (7.20) as a function of $\overline{K}$, ζ and R. In addition, by means of equation (7.10), γ_k can be expressed as a function of $\overline{K}$, ζ and γ_∞.

Furthermore, where the value q_0 of the q-factor in the absence of second-order effects is also affected by the period of vibration, as shown by equation (7.38), the influence of connection deformability on the period value should be taken into account through the ratio T_k / T_∞ expressed by equation (7.8). However, in the period range of steel structures ($T > 0.8$ s), this dependence on T can be neglected as shown in Fig. 7.26. With this simplification, the non-dimensional value of the q-factor is given as a function of four parameters [18]

$$\frac{q_k^{(FS)}}{q_\infty^{(FS)}} = f(\overline{K}, \zeta, \gamma_\infty, R) \tag{7.42}$$

In the case of partial-strength connections, a significant increase in the number of parameters involved in the seismic behaviour of the model arises. In fact, taking into account that the ductility $\mu_k^{(PS)}$ can be expressed through equation (7.32), it is easily shown that the non-dimensional value of the q-factor is affected by six parameters [18]

$$\frac{q_k^{(PS)}}{q_\infty^{(FS)}} = f'(\overline{K}, \zeta, \gamma_\infty, R, \overline{m}, L/d_b) \qquad (7.43)$$

Fortunately, this number of parameters can be reduced by taking into account that $\overline{K}, \overline{m}$ and L/d_b are related through equations (7.31) and (7.33–7.35), on the basis of the connection typology.

In addition, relationships (7.33)–(7.35) through equation (7.31) allow the definition for each connection typology a range of $\overline{K}$ in which $\overline{m} \leq 1$, i.e. partial-strength connections are obtained, and another range ($\overline{m} \geq 1$) in which full-strength connections are attained. As a consequence, in the first range the relationship (7.43) has to be considered, while in the second range equation (7.42) is valid.

The resulting behaviour of the substructure is represented, for some values of the parameters governing its response, in Figs. 7.27–7.38 with reference to different connection typologies. In addition, Figs. 7.27–7.32 provide the seismic response of the model for beams having a high rotation capacity ($R = 6$), while Figs. 7.33 to 7.38 illustrate the behaviour for beams having a moderate rotation capacity ($R = 2$).

It can be noted that, in the case of beams having a high rotation capacity, a significant reduction of the q-factor is attained, independently of the connection typology, over the whole range of connection stiffness. In particular, this reduction is critical in the partial strength zone, for connection stiffness values near to that governing the transition from the full-strength to the partial-strength condition. This is due to the loss of ductility which arises when yielding occurs in the connections, rather than in the beam ends. In addition, in the partial strength range, the structure's ability to dissipate the earthquake input energy is partially recovered by decreasing the connection stiffness. This behaviour has to be ascribed to the fact that the rotation capacity of partial-strength connections increases as their stiffness decreases. Conversely, for full-strength connections the reduction of the q-factor is less significant, becoming important only for small values of the connection stiffness.

In the case of beams having a small rotation capacity a similar behaviour is recognized, but the discontinuity due to the transition from the full-strength to the partial-strength condition is strongly reduced. As a consequence, the increased ductility in the partial-strength range can lead to particular situations in which the value of q-factor is greater than the reference value $q_\infty^{(FS)}$, due to the connection influence (this phenomenon can be recognized from Figs. 7.33–7.36 as compared with Figs. 7.27–7.30, respectively). However, for very small values of the connection rotational stiffness, the reduction of the q-factor is always sharp.

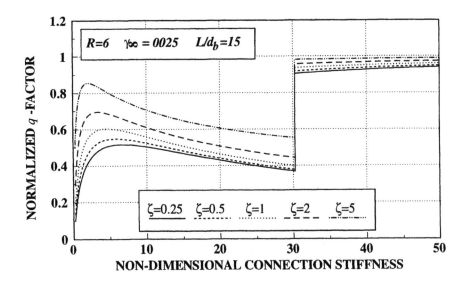

Fig. 7.27 Connection influence on the *q*-factor, for top and seat angle with double web angles

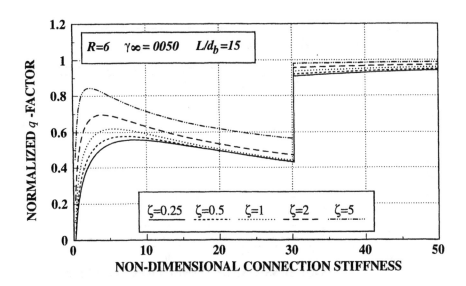

Fig. 7.28 Connection influence on the *q*-factor, for top and seat angle with double web angles

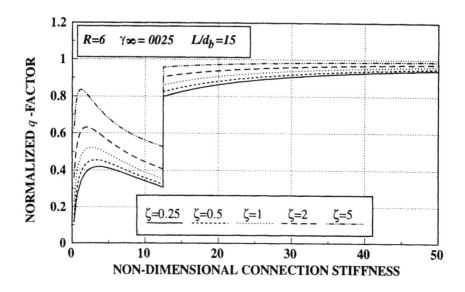

Fig. 7.29   Connection influence on the q-factor, for extended end plate with unstiffened column

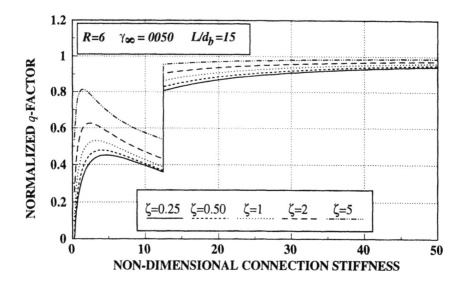

Fig. 7.30   Connection influence on the q-factor, for extended end plate with unstiffened column

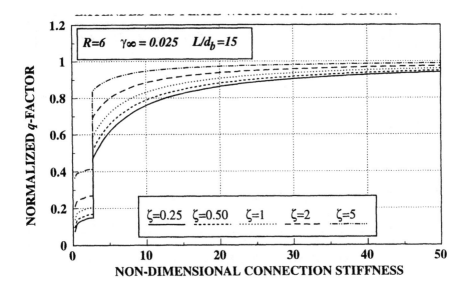

Fig. 7.31 Connection influence on the *q*-factor, for extended end plate with stiffened column

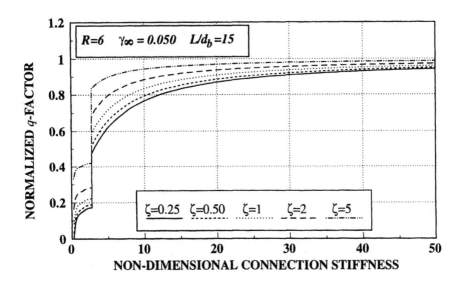

Fig. 7.32 Connection influence on the *q*-factor, for extended end plate with stiffened column

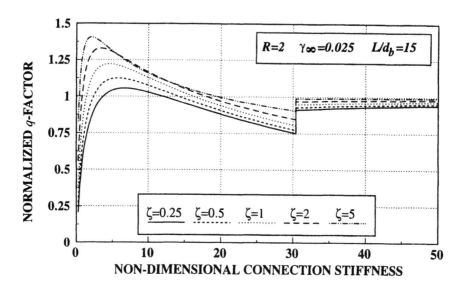

Fig. 7.33   Connection influence on the *q*-factor, for top and seat angle with
double web angles

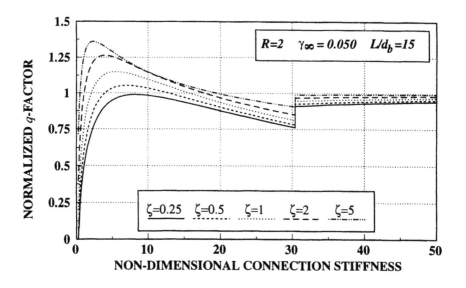

Fig. 7.34   Connection influence on the *q*-factor, for top and seat angle with
double web angles

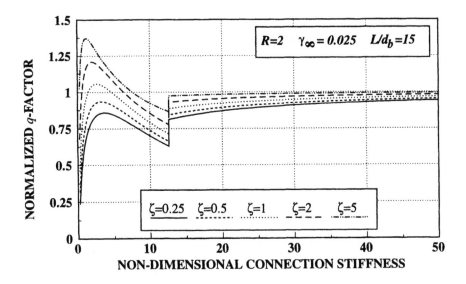

Fig. 7.35 Connection influence on the *q*-factor, for extended end plate with unstiffened column

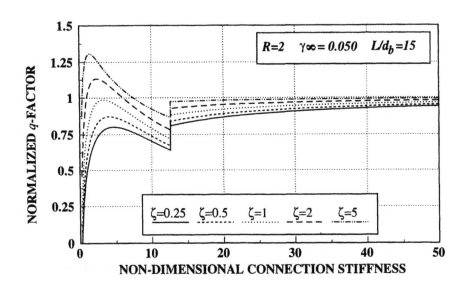

Fig. 7.36 Connection influence on the *q*-factor, for extended end plate with unstiffened column

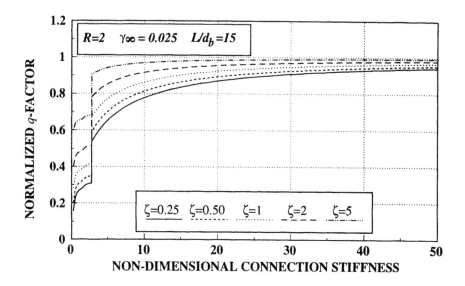

Fig. 7.37   Connection influence on the *q*-factor, for extended end plate with stiffened column

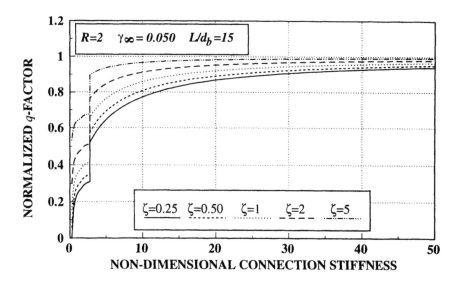

Fig. 7.38   Connection influence on the *q*-factor, for extended end plate with stiffened column

7.5.2 Seismic performance of the subassemblage

In the previous section, the seismic response of an I-shaped subassemblage was investigated from the design point of view and, therefore, the attention was focused on the value of the *q*-factor. In fact, the non-dimensional value of the *q*-factor, expressed by equation (7.43) in the partial-strength range and by equation (7.42) in the full-strength range, allows the establishment of the required amplification or reduction of the design horizontal forces with respect to the reference case represented by the ideal frame with rigid full-strength connections.

Other information can be obtained by analysing the seismic performance of the subassemblage, i.e. its ability to withstand severe earthquakes. This analysis has to be developed taking into account that the ability of a structure to resist destructive earthquakes depends not only on its ductility and energy dissipation capacity, expressed by the *q*-factor, but also on its strength.

A brittle structure might be able to resist a severe ground motion relying on its strength only, because it is a non-dissipative structure ($q = 1$). Its ultimate strength, therefore, has at least to correspond to the following value of the base shear

$$F = M A R(T) \qquad (7.44)$$

where M is the mass of the system, A the peak ground acceleration and $R(T)$ the ordinate of the normalized elastic design response spectrum corresponding to its period T.

As soon as it is realized that the normalized elastic design response spectrum is assigned by seismic codes on the basis of the site soil conditions, it can be recognized that such a structure is able to sustain an earthquake having a peak ground acceleration given by

$$A = \frac{F}{M R(T)} \qquad (7.45)$$

Conversely, a dissipative structure can resist destructive earthquakes by relying on its strength as well as its ductility and energy dissipation capacity ($q > 1$). Therefore, a dissipative structure is able to withstand a peak ground acceleration given by

$$A = \frac{q F_y}{M R(T)} \qquad (7.46)$$

where F_y is the first yielding resistance of the structure.

Therefore, it is clear that, in the case of an I-shaped subassemblage with full-strength semirigid connections, the seismic performance can be

compared with the reference case of full-strength rigid connections by
means of the ratio

$$\frac{A_k^{(FS)}}{A_\infty^{(FS)}} = \frac{F_{y_k}^{(FS)}}{F_{y_\infty}^{(FS)}} \frac{q_k^{(FS)}}{q_\infty^{(FS)}} \frac{R(T_\infty)}{R(T_k)} \tag{7.47}$$

where $F_{y_k}^{(FS)} / F_{y_\infty}^{(FS)} = 1$.

Analogously, in the case of partial-strength semirigid connections, the
comparison with the reference case of full-strength rigid connections
leads to the introduction of the ratio

$$\frac{A_k^{(PS)}}{A_\infty^{(FS)}} = \frac{F_{y_k}^{(PS)}}{F_{y_\infty}^{(FS)}} \frac{q_k^{(PS)}}{q_\infty^{(FS)}} \frac{R(T_\infty)}{R(T_k)} \tag{7.48}$$

where, in this case and with reference to the I-shaped subassemblage,
$F_{y_k}^{(PS)} / F_{y_\infty}^{(FS)} = \overline{m}$.

As steel structures are usually characterized by a period of vibration
located in the softening branch of the elastic design response spectrum,

$$R(T_\infty)/R(T_k) > 1 \tag{7.49}$$

Therefore, from the seismic performance point of view, equations
(7.47) and (7.48) again show the beneficial effect due to the influence of
the connection deformability on the period of vibration.

The resulting behaviour is governed by three parameters as pointed
out by equations (7.47) and (7.48). These are the strength ratio, the q-fac-
tor (combination of ductility and energy dissipation capacity) ratio and
the period ratio. These correspond, respectively, to the first, second and
third term of the right-hand side (where, in the softening branch of the
elastic design response spectrum $R(T_\infty) / R(T_k) = T_k / T_\infty$).

Equations (7.47) and (7.48) also have to be applied according to the
behaviour of the considered connection typology. Therefore, equation
(7.48) has to be used in the range of partial-strength connections and
equation (7.47) in the range of full-strength connections. For some values
of parameters governing the seismic behaviour of the subassemblage, the
resulting response is illustrated in Figs. 7.39–7.44 for beams having high
rotation capacity ($R = 6$) and in Figs. 7.45–7.50 for beams having a small
rotation capacity ($R = 2$).

These figures show that the difference between the seismic performance of the model with full-strength semirigid connections and that
of the same model with full-strength rigid connections is negligible, because the non-dimensional ultimate peak ground acceleration is very

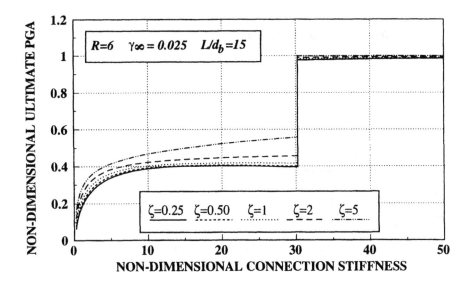

Fig. 7.39   Connection influence on the ultimate peak ground acceleration for top and seat angle with double web angles

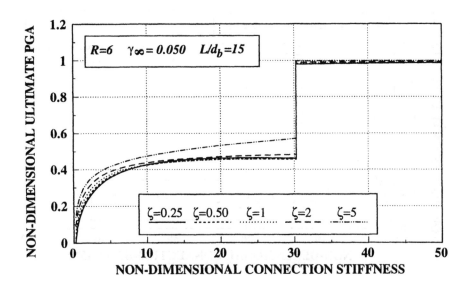

Fig. 7.40   Connection influence on the ultimate peak ground acceleration for top and seat angle with double web angles

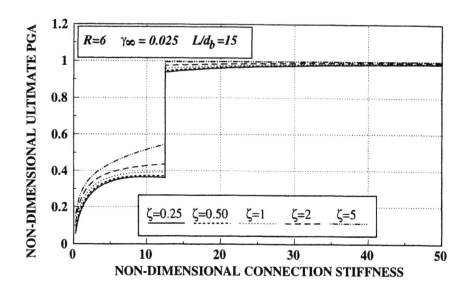

Fig. 7.41 Connection influence on the ultimate peak ground acceleration for extended end plate with unstiffened column

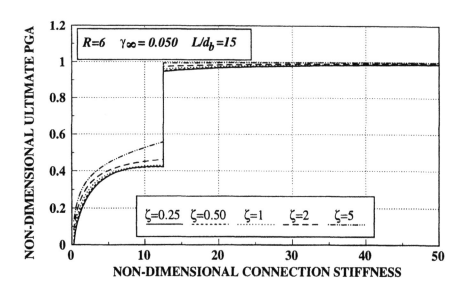

Fig. 7.42 Connection influence on the ultimate peak ground acceleration for extended end plate with unstiffened column

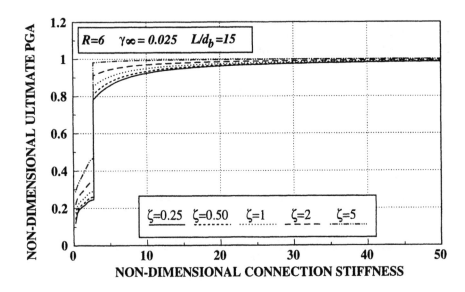

Fig. 7.43 Connection influence on the ultimate peak ground acceleration for extended end plate with stiffened column

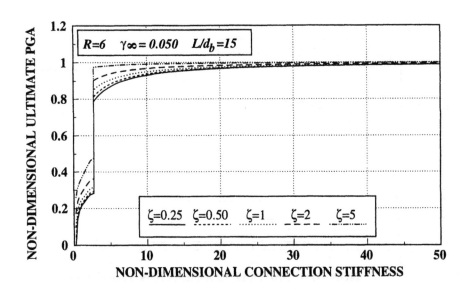

Fig. 7.44 Connection influence on the ultimate peak ground acceleration for extended end plate with stiffened column

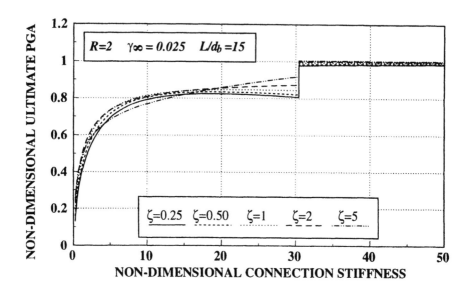

Fig. 7.45 Connection influence on the ultimate peak ground acceleration for top and seat angle with double web angles

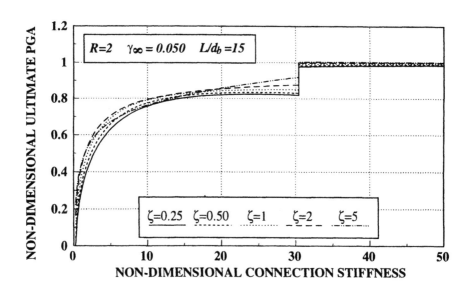

Fig. 7.46 Connection influence on the ultimate peak ground acceleration for top and seat angle with double web angles

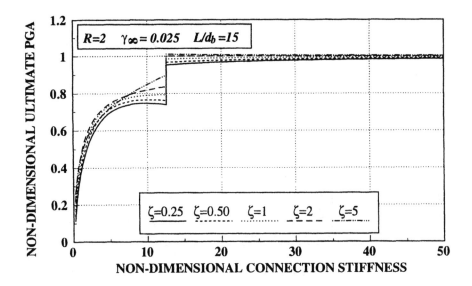

Fig. 7.47 Connection influence on the ultimate peak ground acceleration for extended end plate with unstiffened column

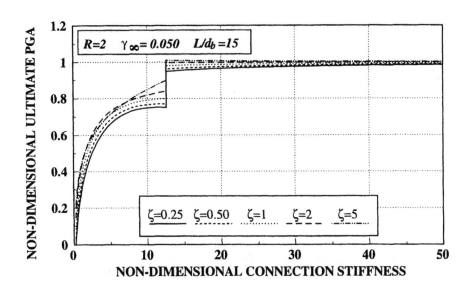

Fig. 7.48 Connection influence on the ultimate peak ground acceleration for extended end plate with unstiffened column

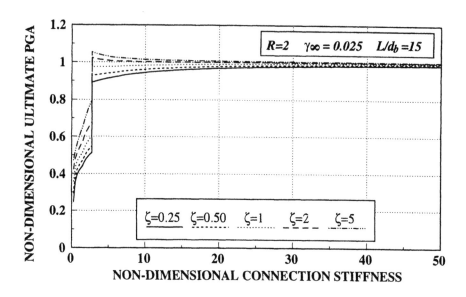

Fig. 7.49   Connection influence on the ultimate peak ground acceleration for
extended end plate with stiffened column

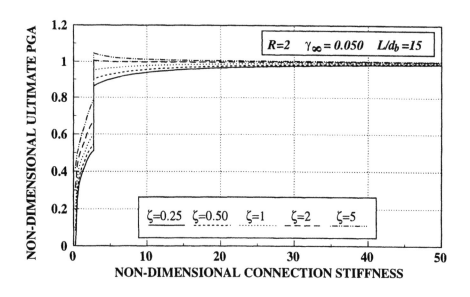

Fig. 7.50   Connection influence on the ultimate peak ground acceleration for
extended end plate with stiffened column

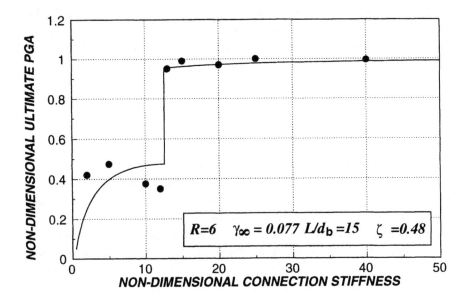

Fig. 7.51 Comparison between the simplified model and the numerical simulation of the dynamic inelastic response of the analysed frame

close to 1 in the full-strength range. On the contrary, comparison between the seismic performance of the model with partial-strength semirigid connections and the same model with full-strength rigid connections shows that there arises a significant decrease in the peak ground acceleration leading to collapse. This reduction increases as the connection stiffness decreases and the beam rotation capacity increases.

In order to investigate the ability of the simplified model to identify the influence of the connection on the seismic behaviour of MR frames, a first comparison between the results coming from application of the model and those obtained by numerical simulations has been carried out for the two-bay four-storey frame shown in Fig. 7.19 [18]. As an example, this comparison is given in Fig. 7.51 for an extended end plate connection with unstiffened column. In this figure, the points represent the results obtained by evaluating, for each value of the connection stiffness, the peak ground acceleration leading to collapse. This value has been computed through dynamic inelastic analyses, which have been repeated for increasing values of the peak ground acceleration up to failure. The collapse condition has been defined as the occurrence of the ultimate plastic rotation at the beam end, in the case of full-strength connections, or in the connection itself, in the case of partial-strength connections. The analyses have been performed using an accelerogram, generated to match the elastic design response spectrum provided in Eurocode 8. This comparison shows reasonable agreement between the curves obtained from the analysis of the simplified model and the numerical simulations of the dynamic inelastic response of the considered frame.

The above results appear encouraging as regards investigation of the complex influence of connection behaviour on the seismic response of actual frames. However, it has to be emphasized that the most important role in the seismic behaviour of full/partial-strength semirigid frames is played by the ratio between the beam plastic rotation capacity and that for the connection. In addition, as the location of the discontinuity characterizing the transition from the partial-strength to the full-strength condition is governed by the connection strength versus stiffness relationship, this aspect also deserves further investigation.

Furthermore, due to the fundamental role exhibited by the beam-to-connection ductility ratio, it seems that the use of partial-strength connections in seismic zones requires some limitations. This will be defined in more detail when new research results become available. At the moment, the use of full-strength connections, as suggested by the modern codes, appears to be preferable.

7.6   REFERENCES

[1] *E. Cosenza, A. De Luca, C. Faella:* 'Nonlinear behaviour of framed structures with semirigid joints', Costruzioni Metalliche, N.4, 1984.

[2] *Commission of the European Communities:* 'Eurocode 8: Earthquake Resistant Design of Structures', draft, October 1993.

[3] *European Convention for Constructional Steelwork (ECCS):* 'European Recommendations for Steel Structures in Seismic Zones', 1988.

[4] *A. Astaneh, N. Nader:* 'Cyclic behaviour of frames with semi-rigid connection', Proceedings of the Second International Workshop 'Connections in Steel Structures II: Behavior, Strength and Design', Pittsburg, Pennsylvania, April, 1991.

[5] *A. Astaneh, N. Nader:* 'Shaking Table Tests of Steel Semi-Rigid Frames and Seismic Design Procedures', 1st COST C1 Workshop, Strasbourg, October, 1992.

[6] *A. Astaneh, N. Nader:* 'Proposed Code Provision for Seismic Design of Steel Semirigid Frames', submitted to AISC Engineering Journal for review and publication, 1992.

[7] *A. Astaneh, J. Shen:* 'Seismic Response Evaluation of an instrumented six story steel building', Earthquake Engineering Research Center, Report UBC/EERC-90/20, University of California at Berkeley, August, 1990.

[8] *C. Faella, V. Piluso, G. Rizzano:* 'Sul Comportamento Sismico Inelastico

dei Telai in Acciaio a Nodi Semirigidi', VI Convegno Nazionale, L'Ingegneria Sismica in Italia, Perugia, 13–15 Ottobre 1993.

[9] *C. Faella, V. Piluso, G. Rizzano:* 'L'Influenza del Comportamento Nodale sulla Risposta Inelastica dei Telai Sismo-Resistenti', Giornate Italiane della Costruzione in Acciaio, C.T.A., Viareggio, 24–27 Ottobre, 1993.

[10] *C. Faella, V. Piluso, G. Rizzano:* 'Connection Influence on the Elastic and Inelastic Behaviour of Steel Frames', International Workshop and Seminar on Behaviour of Steel Structures in Seismic Areas, STESSA 94, Timisoara, Romania, 26 June–1 July, 1994.

[11] *R. Bjorhovde, A. Colson, J. Brozzetti:* 'Classification System for Beam-to-Column Connections', Journal of Structural Engineering, ASCE, Vol. 116, No.11, November 1990.

[12] *K. Weinand:* 'SERICON – Databank on joints in building frames', Proceedings of the 1st COST C1 Workshop, Strasbourg, 28–30 October.

[13] *A. Azizinamini, J.B. Radziminski:* 'Prediction of Moment-Rotation Behaviour of Semi-Rigid Beam-to-Column Connections' in 'Connections in Steel Structures: Behaviour, Strength and Design' Edited by R. Bjorhovde, J. Brozzetti and A. Colson, Elsevier Applied Science, London and New York, 1988.

[14] *E. Cosenza, C. Faella, V. Piluso:* 'Effetto del Degrado Geometrico sul Coefficiente di Struttura', IV Convegno Nazionale, L'Ingegneria Sismica in Italia, Milano, Ottobre, 1989.

[15] *C. Faella, O. Mazzarella, V. Piluso:* 'L'Influenza della non-linearità geometrica sul danneggiamento strutturale sotto azioni sismiche', VI Convegno Nazionale, L'Ingegneria Sismica in Italia, Perugia, 13–15 Ottobre, 1993.

[16] *F.M. Mazzolani, V. Piluso:* 'P-Δ Effect in Seismic Resistant Steel Structures', Structural Stability Research Council, Annual Technical Session & Meeting, Milwaukee, Wisconsin, April, 1993.

[17] *H. Krawinkler, A.A. Nassar:* 'Seismic Design based on Ductility and Cumulative Damage Demands and Capacities', in 'Nonlinear Seismic Analysis and Design of Reinforced Concrete Buildings' edited by P. Fajfar and H. Krawinkler, Elsevier, London, 1992.

[18] *C. Faella, V. Piluso, G. Rizzano:* 'Connection Influence on the Seismic Behaviour of Steel Frames', International Workshop and Seminar on Behaviour of Steel Structures in Seismic Areas, STESSA 94, Timisoara, Romania, 26 June–1 July, 1994.

[19] *F.M. Mazzolani:* 'Influence of Semirigid Connections on the Overall Stability of Steel Frames', State-of-Art Workshop: Connections, Strength and Design of Steel Structures, Cachan, 25–27 May, 1987.

[20] *F.M. Mazzolani:* 'Stability of Steel Frames with Semirigid Joints', published in 'Stability Problems of Steel Structures', Springer Verlag, Wien, New York, 1992.

8

Structural regularity

8.1 BASIC DEFINITIONS

The traditional design philosophy of seismic-resistant structures is based upon the dissipation of the earthquake input energy by means of plastic excursions, leading to structural damage which has to be controlled in order to prevent failure. The ability of a structure to withstand severe earthquakes is strictly related to its capacity to uniformly distribute the structural damage. In the classical design approach the damage concentrations can lead to failure modes which are characterized by a reduced dissipation capacity. Therefore, it is universally recognized that it is necessary to avoid the uncontrolled concentration of the earthquake input energy; an exception is made for cases in which this energy is intentionally directed into specifically designed dissipative elements.

In the case of multistorey buildings, an optimum seismic response is obtained when the structural damage is uniformly distributed among the different storeys of the building. The 'structural regularity' is unanimously considered an indispensable prerequisite to avoid concentration of the earthquake input energy in limited parts of the structure. It leads to a more efficient energy dissipation mechanism and provides a more reliable forecast of the inelastic structural behaviour.

Although the physical meaning of the structural regularity concept is quite intuitive, its quantitative definition is particularly difficult. There are a great number of parameters which influence the energy dissipation mechanism and, therefore, the structural regularity. Moreover, for any given 'degree of regularity', the reliability of the structural design depends upon the structural modelling and the type of structural analysis (static or dynamic, elastic or inelastic).

Structural regularity is conventionally separated into two fundamental types, which can interact:

- regularity of the vertical configuration;
- regularity of the plan configuration.

The term **vertical regularity** is used to indicate the ability of a structure to uniformly distribute the ductility demand or the structural damage between the different storeys of a building. It can be immediately understood that this ability is strictly related to the mass, stiffness and strength distributions along the height of the structure.

The term **plan regularity** is referred to the ability of a building to vibrate separately in two vertical planes without torsional coupling. This feature is considered favourable because it provides both a reduction of the damage concentration in the peripheral elements and it allows the use of simplified assumptions in structural modelling and analysis. This situation can be seen as analogous to vertical regularity because the damage distribution depends upon the in-plan distribution of mass, stiffness and strength.

The distinction between **regular structures** and **irregular structures** has been introduced into modern seismic codes to allow for the extension of results obtained from the analysis of the seismic response of simple degree of freedom (SDOF) systems to real multistorey structures. From the structural damage characterization point of view, the fundamental difference between SDOF and multi degree of freedom (MDOF) systems is that, in the first case, the structural damage can be quantified by means of a single parameter such as the global ductility demand, while, in the second case, a single parameter is not sufficient. This is because different yielding patterns can occur for a given value of the reference parameter. Therefore, the extension to MDOF systems of the all background regarding the seismic inelastic response of SDOF systems can be performed only under the hypothesis of uniformly distributed damage; in fact, only in this case, the structural damage can be quantified by means of a single damage parameter also for MDOF systems. Moreover, it can be concluded that the concept of structural regularity can be interpreted as a synonym of uniform damage.

From the design point of view, a very important aspect is the need to provide simplified design rules for evaluating structural regularity and, as a consequence, for pointing out those structural situations which would lead to concentration of the earthquake input energy and, therefore, a worsening of the seismic inelastic behaviour. In modern seismic codes, this necessity has led to the introduction of much simplification, which consists of the association of structural irregularity with **geometrical irregularity**. The problem itself is very complex and has to be considered as a matter of **structural optimization**. This is particularly cumbersome because such optimization relates to the inelastic performance of structures.

In some cases, simplified procedures are already available which allow the problem to be treated as a whole, taking into account that **geometrical irregularity does not always lead to structural irregularity** [1, 2].

8.2   VERTICAL REGULARITY

With reference to the volume of the building, it is generally accepted that closed and compact shapes are the most favourable so that the ideal configuration is obtained when the building has the shape of a parallelepiped (Fig. 8.1). Irregular configurations are characterized by variation of the boundary lines along the height giving rise to **set-backs** or **off-sets** (Fig. 8.2); they are considered irregular because the corresponding distributions of mass and stiffness are not uniform. Moreover, even if the boundary lines of a building are regular, it has to be considered as irregular, when the vertical distribution of the stiffness is altered. This case is shown in Fig. 8.3 for three buildings in which alteration of the vertical distribution of stiffness could arise due to the absence of cladding on one storey. Structural situations such as these can lead to an undesirable concentration of the earthquake input energy.

It is useful to note that, when the building configuration is characterized by bodies with different heights, the configuration can be regularized by means of the introduction of expansion joints (Fig. 8.4).

Another condition which produces irregular building behaviour is when built on a soil layer with non-uniform geotechnical properties; in such a case the building can still be regularized, by the insertion of expansion joints to isolate the parts lying on the different types of soil.

During an earthquake, the parts separated by an expansion joint can vibrate independently, under their own dynamics, so that collisions can arise. As this 'hammering' phenomenon can seriously undermine the integrity of the building, the expansion joints have to be designed to allow for the peak amplitudes of the oscillations of the parts separated by the joint.

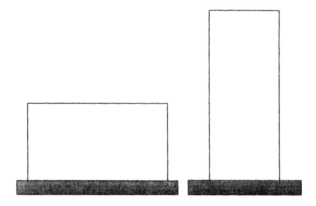

Fig. 8.1   Buildings with closed and compact shapes

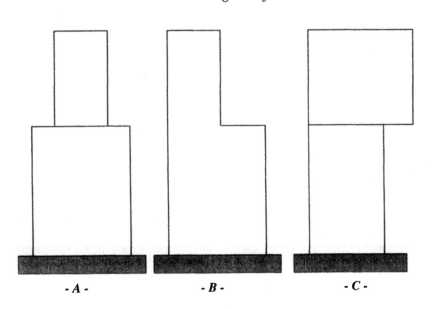

Fig. 8.2   Geometrical configurations with 'set-backs' (A and B) or 'off-sets' (C)

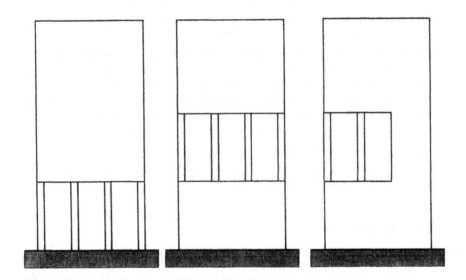

Fig. 8.3   Irregular distributions of stiffness along the height

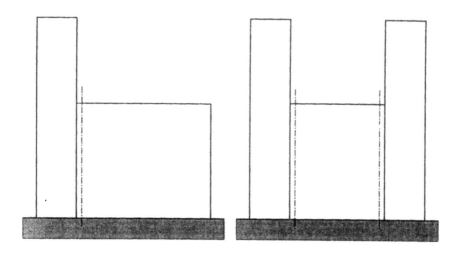

Fig. 8.4   Separation into independent bodies by expansion joints

The type of foundation can also give rise to vertically irregular configurations. This is the case where the ground orography does not allow
all foundations to be built at the same level (Figs. 8.5a and 8.5b). In this
situation the best solution is again the subdivision of the building into
independent parts by the insertion of expansion joints. When this solution cannot be achieved, the columns have to reach the same level of
underground zones (Fig. 8.5c), so that all foundations are located at the
same level. If this solution is particularly expensive, we can adopt the
solution shown in Fig. 8.5d, which is the construction of a rigid reinforced
concrete box, being also the base of the upper structure.

Furthermore, other situations giving rise to irregular configurations
are represented by all cases in which a nonuniform transfer of forces
arises (Fig. 8.6), due to different reasons.

8.2.1 Code provisions

The previous section has pointed out that there is a great variety of
structural situations in which damage concentration can arise due to the
geometrical configuration. It is clear, therefore, that quantitative solutions
cannot be provided which are able to cover all structural situations. For
this reason, modern seismic codes give, in general, qualitative provisions

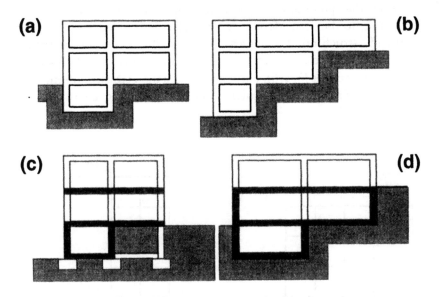

Fig. 8.5 Irregular configurations due to the ground orography and corresponding possible structural solutions

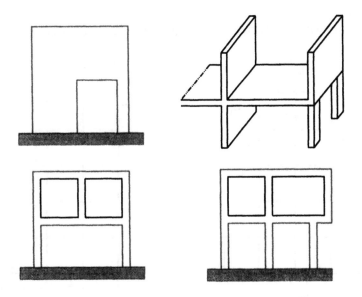

Fig. 8.6 Irregular configurations in which a non-uniform transfer of the forces arises

only, while the quantitative definition of the structural configurations which have to be considered as irregular is limited to the most widespread structural typologies. This is the case for structural configurations characterized by the presence of 'set-backs' or 'off-sets'. In modern seismic codes [3, 4, 5], with reference to the vertical configuration, a building is classified as regular when the following conditions are satisfied (Fig. 8.7).

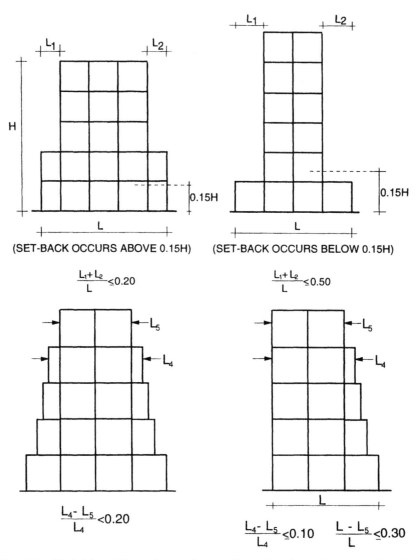

Fig. 8.7 Definition of irregular vertical configurations in modern seismic codes

- The stiffness and mass properties are approximately uniform along the building height.
- In the case of gradually tapering buildings with symmetry about the vertical axis, the extent of the setback at each floor does not exceed 20% of the previous plan dimension.
- The above limit may be increased to 50%, if the setback stops at a level 15% below the top of the building.
- In the case of only one tapered facade, the setback at each floor does not exceed 10% and the overall setback is not greater than 30% of the plan dimension at the first floor.

For structures which are considered irregular, a 25% reduction of the q-factor to be used in the design is recommended.

It can be seen that the restrictions provided by the seismic codes are mainly related to the geometrical configuration of the building and, therefore, they are not able to take into account all factors affecting the inelastic performance of structures. In the following, it will be shown that 'geometrical irregularity', as defined in recent seismic codes, does not necessarily lead to 'structural irregularity' (Section 8.2.4). As a consequence, the penalizing of some geometrical configurations, which is specified in seismic codes, is not justified.

The distinction between regular and irregular buildings has implication on the structural model, the method of analysis and the value of the q-factor.

In the case of plane regular buildings (Section 8.3.1) a plane structural model can be assumed. Regarding the method of analysis and the behaviour factor, two situations can be verified. In the first, the building is also regular in the vertical configuration. In such a case, the reference value of the behaviour factor can be adopted and a simplified modal analysis (or equivalent static analysis) can be performed provided that, according to Eurocode 8 [3], the periods of vibration in both main directions are less than four times the period value corresponding to the beginning of the softening branch of the elastic design response spectrum and, in any case, not greater than 2.0 s. However, in the case of buildings having an irregular vertical configuration, a multimodal response analysis has to be performed and the behaviour factor has to be decreased with respect to its reference value in order to account for the worsening of the structural inelastic behaviour. The reduction of the behaviour factor is, in such a case, assumed equal to 20% in Eurocode 8 [3] and to 25% in the ECCS Recommendations [4].

8.2.2 A general view on research

The research into seismic inelastic behaviour of irregular structures is quite limited when compared with the number of structural situations which should be investigated. A first exploratory analysis on the influence of set-backs on the seismic response of multistorey buildings has been provided by Berg [6], who investigated the dynamic response of a tapered cantilever beam. A wider investigation, although limited to the elastic range, has been performed by Jhaveri [7]. Studies on the influence of set-backs on the inelastic behaviour of steel structures has been carried out by Pekan and Green [8], Varma [9] and by Humar and Wright [10]. More recently, this subject has been investigated by Sedlacek *et al.* [11], who, despite the limitation of their analysis to the elastic range, have demonstrated the incompleteness of the code provisions. The effects related to different stiffness distributions along the height of the building have been examined by Dolce [12–14]. Even though these studies refer to reinforced concrete structures, a significant outcome noted by the author is that, on average, the plastic demand in structures designed by static analysis is lower than that requested for structures designed by dynamic analysis.

For a given ground motion, damage distribution depends upon the distribution of mass, stiffness and strength. This problem has been studied by Kobori and Minai [15], and Kato and Akiyama [16] who, by means of numerical simulations of the seismic response of simplified MDOF models, have determined the 'optimum distributions' which lead to a uniform structural damage. Other authors [17, 18] have examined the seismic response corresponding to given laws of the distribution of mass, stiffness and strength, without looking for the optimum ones.

The above studies have now led to some cases [19–21] in which the definition of simplified procedures, even if they do not require any inelastic analysis, enable evaluation of the q-factor by taking into account all effects which derive from the geometrical configuration of the structure. The use of such procedures could lead, at least for some structural situations such as set-backs, to the overcoming of the distinction between regular and irregular structures.

8.2.3 Behaviour of frames with set-backs

Geometrical definition of the structural schemes

In order to evaluate the possible worsening of the dissipative capacity of steel frames in the presence of set-backs, the seismic inelastic response has been analysed with reference to two series of geometrically irregular frames. The first series derives from a 5-storey 3-bay regular frame, while the second series derives from a 6-storey 5-bay regular frame.

The first series of frames [1], (Fig. 8.8) includes both symmetrical and asymmetrical frames. The study of the inelastic behaviour is limited to

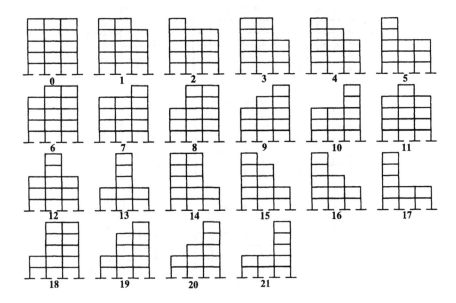

Fig. 8.8 The first series of examined frames

the plan case and, therefore, it does not include torsional effects which arise in asymmetrical schemes due to the irregularity of the plan configuration. The second series of frames [2], (Fig. 8.9) is composed by symmetrical schemes only, which are considered to be regular in plan.

As far as the dimensions of members are concerned, different trends can now be recognized in the frame design, leading to three design levels. The first level is characterized by the column design without any amplification of the bending strength, as following the present Italian code [22]. This type of approach leads, in general, to the design of frames failing with mechanisms other than the global mode. Often, a storey mechanism arises leading to a strong damage concentration.

The second-level design methods are represented by the use of simplified criteria, which have been proposed in order to obtain frames failing in global mode. The design criteria included in the modern codes, such as GNDT [5], ECCS [4] and EC8 [3], belong to this second approach. Unfortunately, as shown in Chapter 2, recent studies [23] have pointed out that the above recommendations very often do not lead to the achievement of the desired result. Therefore a third, more refined, design level has been introduced. The design procedure corresponding to this last level is characterized by the progressive increase of the column sections and on the control, by means of static inelastic analyses, of the

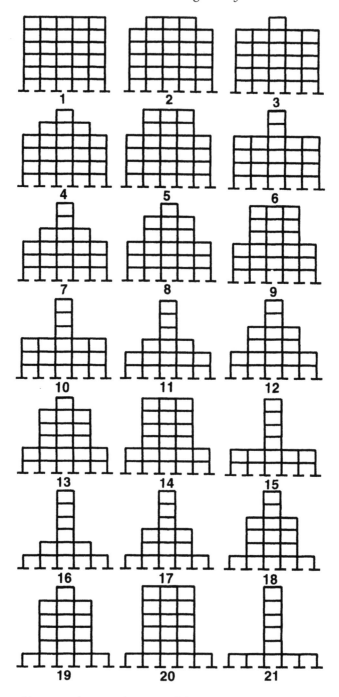

Fig. 8.9   The second series of examined frames

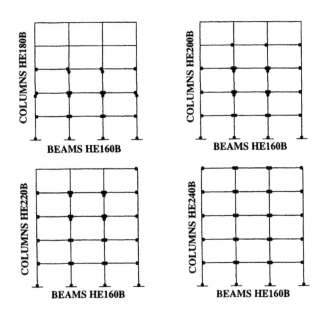

Fig. 8.10 Evolution of the patterns of yielding as the column section increases

corresponding failure mode, until the global-type mechanism is obtained. Even if this procedure leads to the fulfilment of the design goal, it must be stressed that considerable computational effort is required to reach the final design solution. A more sophisticated approach [24, 25], based on the application of the kinematic theorem of plastic collapse, has been described in Chapter 2. This alternative approach leads immediately to the correct design solution without requiring any inelastic analysis. In addition, it allows tapering of the column section along the building height without any additional computational effort.

An analysis has been developed [1, 2, 26] that includes frames designed according to the first and third levels.

As an example, Fig. 8.10 shows the failure modes of the regular frame of the first series obtained by progressively increasing the column section until the third level of design is reached.

With reference to the first design level, each frame has been dimensioned by means of a dynamic elastic analysis including the effects of the first three modes of vibration. The elastic design response spectrum adopted in the analysis is the one proposed by GNDT for stiff-soil site conditions [5] and with a peak ground acceleration equal to 0.35g. The design value of the q-factor is equal to 6 for all frames. The beam and column sections of frames of the first and second series are given in

Table 8.1   Member sections of the two series of frames

FIRST SERIES		
FRAMES	STOREYS 1–2	STOREYS 1–2–3
0	HE200B	HE160B
1–6	HE200B	HE160B
2–7	HE180B	HE160B
3–8	HE200B	HE160B
4–9	HE180B	HE140B
5–10	HE180B	HE140B
11	HE180B	HE160B
12	HE180B	HE140B
13	HE160B	HE140B
14–18	HE200B	HE160B
15–19	HE180B	HE160B
16–20	HE180B	HE140B
17–21	HE160B	HE140B
	IPE 300 BEAMS	
SECOND SERIES		
COLUMNS	FRAMES	
HE200B	1–2–3–4–5–6–7–8–9–12–13–14–19–20	
HE180B	10–11–15–16–17–18	
HE160B	21	
	IPE 300 BEAMS	

Table 8.1. In both series all beams are made of IPE300 and columns of the second series have constant section along the height.

Quantitative evaluation of the geometrical irregularity has been performed by the parameter Φ [24] (Fig. 8.11):

$$\Phi = \frac{1}{1+K} \ (C + KV) \tag{8.1}$$

where the coefficients C and V, which take into account horizontal and vertical setbacks, are provided by

$$C = \frac{1}{N_p} \sum_{i=1}^{N_p} \frac{L_i}{L} \tag{8.2}$$

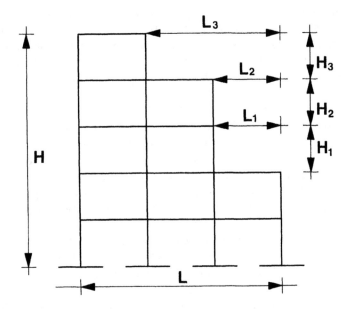

Fig. 8.11   Definition of the geometrical irregularity index

$$V = \frac{1}{N_c} \sum_{j=1}^{N_c} \frac{H_j}{H} \qquad\qquad (8.3)$$

where:

- N_p is the number of storeys with horizontal setbacks;
- N_c is the number of bays where vertical setbacks are present;
- L_i is the length of the horizontal setback at the ith storey or the sum of the lengths of the setbacks, if they are present at both sides of the frame;
- H_j is the height of the vertical setback at the jth bay;
- K is a constant value which can be assumed equal to 2.

The main reason of choice of this parameter, which is not present in the parameters proposed by other codes such as SEAOC [27], is because it does not excessively penalize the configuration when the setback is not abrupt.

Inelastic analyses

It is well known that rigorous evaluation of the q-factor can be performed only by means of a great number of dynamic inelastic analyses, which have to be repeated for different ground motions. The procedure, from the computational point of view, is particularly cumbersome when a large number of structural situations has to be investigated in order to obtain general conclusions. Therefore, even if dynamic inelastic analysis represents the best tool with which to determine the true pattern of damage of a structure under seismic loads, the use of simplified methods for evaluating the q-factor [28, 29] (Chapter 4) allows a quick investigation of the seismic behaviour of structures. In order to obtain a forecast of the pattern of yielding in a structure and, at the same time, obtain a quick evaluation of its seismic behaviour, both static and dynamic inelastic analyses have been performed. In particular, the former have been used for determining all parameters characterizing the $\alpha - \delta$ behavioural curve (Fig. 8.12), where α is the multiplier of horizontal forces and δ is the top sway displacement, as well as for evaluating the q-factor according to the energy method proposed by Como and Lanni [30] (Chapter 4).

From the operative point of view, static inelastic analyses have been performed by the program given in reference 31, while the following computations required for evaluating the q-factor have been carried out by using a specifically developed postprocessor.

The dynamic inelastic analyses have been done by means of the well known DRAIN–2D program [32].

In the case of the 5-storey 3-bay frames designed according to the third-level design criterion, the analyses never suggested, in all the examined cases, a worsening of seismic performance for geometrically irregular schemes with respect to the regular frame from which they are derived [24].

The analysis of the inelastic performance of frames dimensioned according to the first level design criterion is more interesting. With reference to these frames, the q-factor values q_e, computed by the above mentioned energy method, are given in Figs. 8.13 and 8.14 respectively for the first (5-storey 3-bay frames) and the second (6-storey 5-bay frames) series of frames. The most significant results obtained through dynamic analyses are shown, for the first series frames, in Fig. 8.15, where the values of the required cumulated plastic ductility are given for the most damaged storey.

With reference to the frames belonging to the second series, the values of the interstorey drifts are given in Figs. 8.16–8.18. Furthermore, the maximum values of the interstorey drifts, independently of the storey in which it is verified, are shown in Fig. 8.19. The values shown in the above graphical representations are the maximum values obtained from the analyses performed for three different accelerograms, which have been artificially generated from the response spectrum proposed by GNDT [5] for stiff-soil site conditions and a peak ground acceleration equal to $0.35g$. Moreover, for the first series of frames, the collapse mechanisms evaluated by static inelastic analyses are shown in Fig. 8.20; for

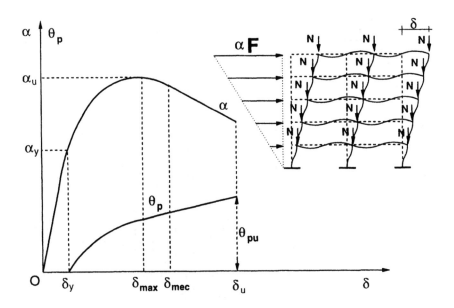

Fig. 8.12 Multiplier of horizontal forces versus top sway displacement curve

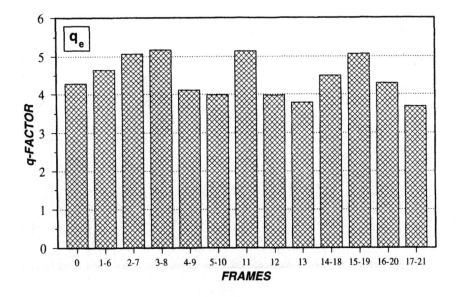

Fig. 8.13 Values of the *q*-factor computed for the first series of frames

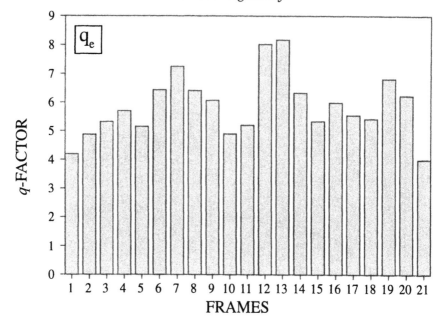

Fig. 8.14 Values of the q-factor computed for the second series of frames

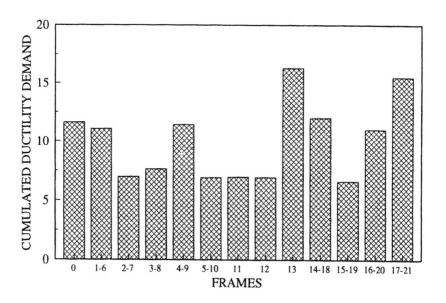

Fig. 8.15 First series of frames: results of dynamic inelastic analyses

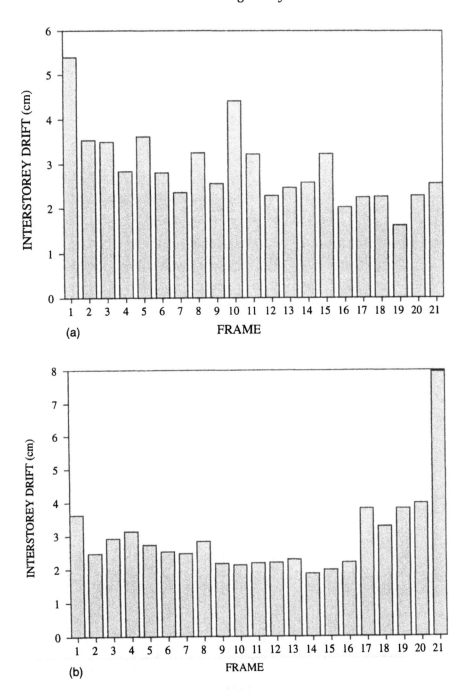

Fig. 8.16 Interstorey drifts computed by means of dynamic inelastic analysis for the second series of frames: (a) first storey; (b) second storey

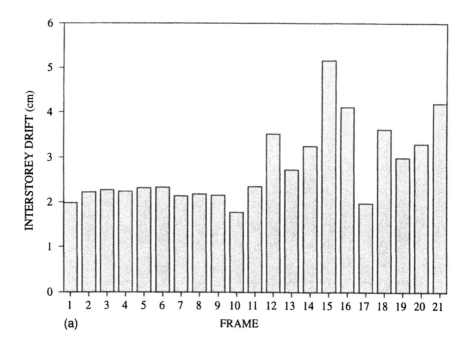

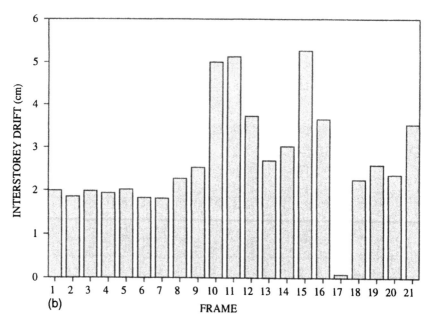

Fig. 8.17   Interstorey drifts computed by means of dynamic inelastic analysis
for the second series of frames: (a) third storey; (b) fourth storey

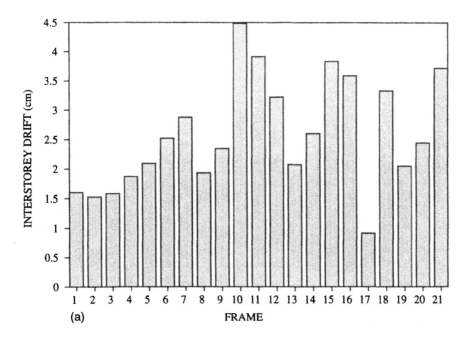

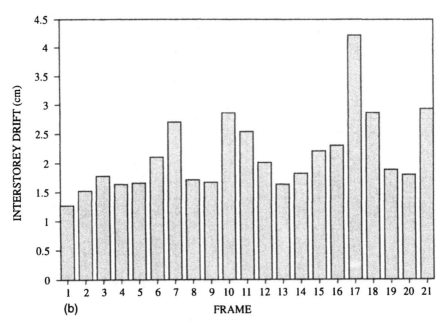

Fig. 8.18   Interstorey drifts computed by means of dynamic inelastic analysis
for the second series of frames: (a) fifth storey; (b) sixth storey

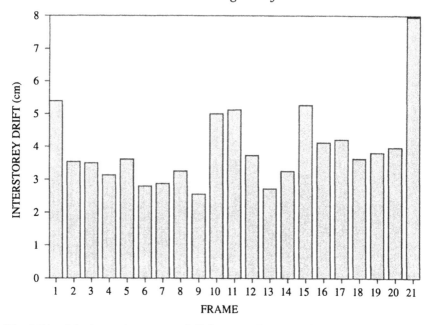

Fig. 8.19   Maximum interstorey drift for second series of frames

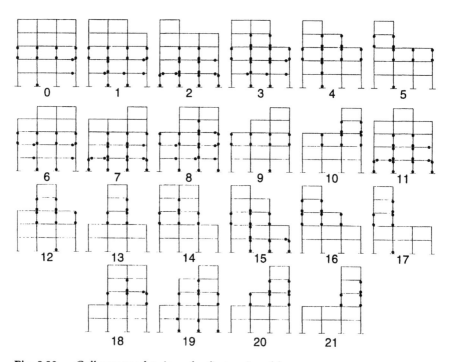

Fig. 8.20   Collapse mechanisms for first series of frames

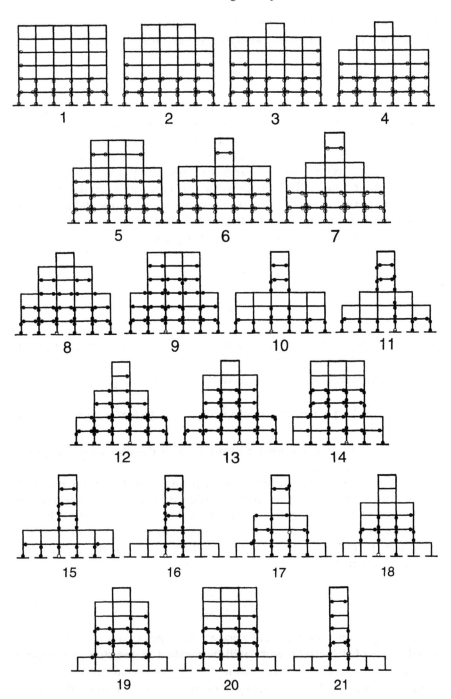

Fig. 8.21 Energy dissipation mechanisms for second series of frames

frames belonging to the second series, the energy dissipation mechanisms derived by dynamic inelastic analyses are given in Fig. 8.21.

Discussion of results

The patterns of yielding which have been established by inelastic analyses have been compared with the forecasts derived from the optimum shear coefficient distribution proposed by Kato and Akiyama [16, 19–21]. This distribution law, which has been derived by dynamic inelastic analysis of MDOF 'shear type' systems and by assuming for the optimum distribution the one which leads to a constant value of the required cumulated plastic ductility at each storey, plays a fundamental role in the 'background' of the Japanese seismic code (Chapter 4).

The shear coefficient of the *i*th storey is given by

$$\alpha_i = \frac{Q_{yi}}{n_s \sum\limits_{j=i} m_j g} \tag{8.4}$$

where Q_{yi} is the yield shear force of the *i*th storey, m_j is the mass of the *j*th storey and n_s is the number of storeys.

The optimum shear coefficient distribution law is given by [16]

$$\frac{\alpha_{opt,i}}{\alpha_1} = 1 + 1.153X - 11.85X^2 + 42.6\,X^3 - 59.5\,X^4 + 30X^5 \tag{8.5}$$

where $X = X_i/H$, X_i is the height of the floor of the *i*th storey and H the total height of the frame.

When the actual shear coefficient distribution differs from the optimum, the structure is incorrectly designed and a concentration of earthquake input energy at a weak storey has to be expected. The unavoidable scatter between the shear coefficient distribution and the optimum distribution can be represented by means of the coefficients, namely damage concentration factors, given by

$$p_i = \frac{\alpha_i}{\alpha_{opti}} \tag{8.6}$$

A damage concentration will arise at the storeys having $p_i < 1$, while a reduction of the plastic work will be verified at the storeys having a coefficient $p_i > 1$. The 'weakest storey' is the one characterized by the minimum value of the coefficient p_i. For the examined frames, the values of the coefficients p_i are provided in Table 8.2 with reference to the first (5-storey 3-bay frames) and the second (6-storey 3-bay frames) series of

Vertical regularity77

Table 8.2 Damage concentration factors of examined frames

Frame	FIRST SERIES				SECOND SERIES				
	p_2	p_3	p_4	p_5	p_2	p_3	p_4	p_5	p_6
0	1.13	0.73	0.89	1.31	–	–	–	–	–
1	1.15	0.76	0.99	1.37	1.10	1.26	1.43	1.81	2.65
2	1.18	1.08	1.55	2.25	1.11	1.31	1.55	2.12	2.73
3	1.18	0.81	0.86	1.28	1.13	1.37	1.71	2.65	3.78
4	1.21	0.81	0.99	1.44	1.15	1.44	1.89	2.31	3.53
5	1.25	0.89	0.90	1.32	1.13	1.36	1.67	1.67	2.56
6	1.15	0.76	0.99	1.37	1.18	1.55	2.26	2.18	3.20
7	1.18	1.08	1.55	2.25	1.22	1.69	1.91	1.98	2.91
8	1.18	0.81	0.86	1.28	1.18	1.55	1.51	2.23	3.20
9	1.21	0.81	0.99	1.44	1.15	1.44	1.27	1.60	2.34
10	1.25	0.89	0.90	1.32	1.27	1.89	1.42	1.79	2.62
11	1.18	1.08	1.55	2.25	1.33	1.49	1.26	1.59	2.33
12	1.25	0.89	0.90	1.32	1.27	1.26	1.72	1.79	2.62
13	1.36	0.82	1.00	1.48	1.22	1.12	1.37	2.03	2.91
14	1.21	0.65	0.80	1.18	1.18	1.02	1.16	1.47	2.15
15	1.25	0.96	1.31	1.91	1.42	0.97	1.10	1.39	2.04
16	1.29	0.75	0.81	1.20	1.04	0.83	0.95	1.19	1.75
17	1.36	1.83	1.03	1.59	0.95	1.31	1.10	1.39	2.04
18	1.21	0.65	0.80	1.18	0.89	1.12	1.53	1.59	2.33
19	1.25	0.96	1.31	1.91	0.84	1.01	1.23	1.83	2.62
20	1.29	0.75	0.81	1.20	0.81	0.93	1.06	1.33	1.96
21	1.36	1.83	1.03	1.59	0.60	0.70	0.79	1.00	1.46

frames (where $p_i = 1$ in all cases). The comparison between these values and the collapse mechanisms indicated by inelastic analyses (Figs. 8.20 and 8.21) shows that in all the examined cases there is a good agreement between the storeys in which the maximum damage occurs and the storeys in which the coefficient p_i attains its minimum value. This result suggests that the optimum shear coefficient distribution, previously described, could be a powerful tool with which to forecast the pattern of damage [1, 2].

In order to show the influence of geometrical irregularity on the seismic behaviour of framed structures, the values of the q-factor q_e, computed by means of the energy methods, have to be examined as a function of the irregularity index Φ. As this examination has to be independent of the design value q_d of the q-factor, values of the ratio q_e/q_d have been used. In recent seismic codes [3–5] for a period value $T > T_2$ (where T_2 is the period value corresponding to the beginning of the softening branch of the spectrum), the elastic design response spectrum is given by

$$S_{a,e}(T) = A\,R_e(T) = \frac{A\,\beta_o}{T/T_2} \tag{8.7}$$

where A is the peak ground acceleration and β_o is the maximum spectral amplification, while the inelastic design response spectrum is given by

$$S_{a,d}(T) = \frac{A\,\beta_o}{\bar{q}\,(T/T_2)^{2/3}} \tag{8.8}$$

The design value of the structural coefficient can be therefore expressed as (Chapter 1)

$$q_d = \frac{S_{a,e}(T)}{S_{a,d}(T)} = \bar{q}\,(T/T_2)^{-1/3} \tag{8.9}$$

where $\bar{q}$ is the behaviour factor.

For each frame the ratio $q^* = q_e/q_d$ has been computed by assuming, according to GNDT code [5], the values $\bar{q} = 6$ and $T_2 = 0.35$s.

The values obtained have been successively normalized by considering the ratio between q^* and q_o^*, where q_o^* is the value of q^* for Φ = 0. The results are illustrated in Fig. 8.22 as a function of the irregularity index Φ. In the same figure the points shown by solid squares refer to the 6-storey 5-bay frames (second series), while the ones denoted by a cross correspond to the 5-storey 3-bay frames (first series). These have been dimensioned with the same criteria but with columns characterized by a variation of the section along the height. Examination of the corresponding points in Fig. 8.22 shows that this column section variation, which produces a variation of the shear stiffness and strength, seems to have more significant effects than those related to the geometrical configuration.

Moreover, the same figure shows that for the first series of frames only, and for values of the geometrical irregularity index Φ greater than 0.25, the geometrical irregularity leads to the structural irregularity. In this case only (Φ > 0.25), a reduction of the design value of the q-factor has to be adopted reaching a maximum reduction equal to 20% for Φ ≥ 0.6. However, for frames with constant column sections the corre-

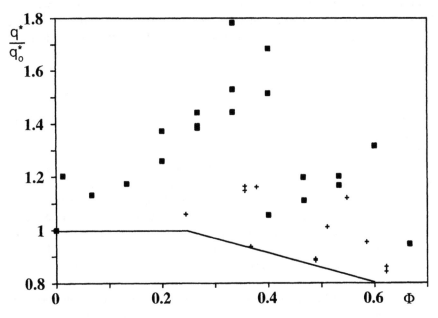

Fig. 8.22 Normalized values of the q-factor for the first and second series of frames

sponding points, which are denoted by means of squares, are all above 1; an exception is frame number 21 which has a high degree of irregularity ($\Phi = 0.67$) and shows a small reduction of about 5%.

8.2.4 Conclusions about vertical irregularity

The analyses just described have demonstrated that the recommendations provided by modern seismic codes on the subject of vertical irregularity unfairly penalize structural configurations characterized by the presence of set-backs. The criteria adopted for the classification in regular and irregular structures seem to be unsatisfactory, because the geometrical irregularity does not always produce structural irregularity. As a consequence, it seems to be particularly important in the generalization of the structural irregularity concept to introduce all causes which produce a non-uniform damage distribution under seismic loads.

The introduction of more rational irregularity parameters, which take into account mass, stiffness and strength distributions, could allow a correct evaluation of the structural irregularity and the definition of the range in which it actually produces a worsening of the seismic behaviour. In the case of frames with set-backs, the introduction of simplified procedures which allow evaluation of the q-factor by taking into account the effects arising from its geometrical configuration could enable, at least for some specific cases (symmetrical schemes with set-backs), elimination of the unjustified distinction between regular and irregular frames.

8.3   PLAN REGULARITY

It has already been pointed out that the concept of plan regularity is related to the ability of a building to vibrate separately in two vertical planes without torsional coupling. Any irregularity and discontinuity in the horizontal layout could cause serious damage concentrations. In fact, it is well known that asymmetric plan-wise buildings undergo translational as well as torsional motions during seismic action and, therefore, they are subjected to a nonuniform plan distribution of damage. Symmetric forms are desirable, because the ability to predict the response of a building is considerably greater for simple symmetric layouts than for complex ones. The ideal shape is undoubtedly the squared one (Fig. 8.23a). As this shape is not always feasible, the design has to aim at a similar shape; rectangular shapes are desirable when they are not too lengthened (Fig. 8.23b). Horizontal building layouts with re-entrant corners provide an unsatisfactory seismic behaviour leading to damage concentrations. When these shapes are unavoidable, it is desirable to use

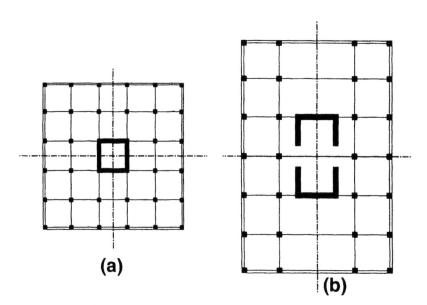

(a)   **(b)**

Fig. 8.23   Simple symmetric layouts

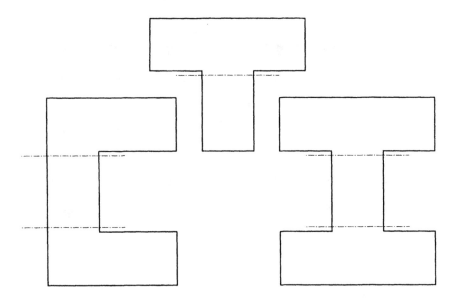

Fig. 8.24　Independent parts divided by expansion joints

expansion joints to divide the building into independent parts, which provide a regular seismic behaviour (Fig. 8.24). The expansion joints have to continue through the foundations if these are on different levels or on soil layers with different mechanical properties. This solution is not necessary when the foundations are on a homogeneous soil layer at the same level.

8.3.1 Code provisions

In order to correctly implement simplified computational methods and to establish the behaviour factor q, a distinction has to be made between regular and irregular buildings. Regarding the plan configuration, according to Eurocode 8 [3], a building is referred to as regular when the following conditions are satisfied (Fig. 8.25).

- The building has a significant symmetry of structure and mass with regard to at least two orthogonal axes. When re-entrant corners or recesses exist, their dimensions do not exceed 25% of the external size of the building in the corresponding direction.
- Under the seismic forces at any storey the maximum storey displacement in the direction of the seismic forces does not exceed the average storey displacement by more than 20%.

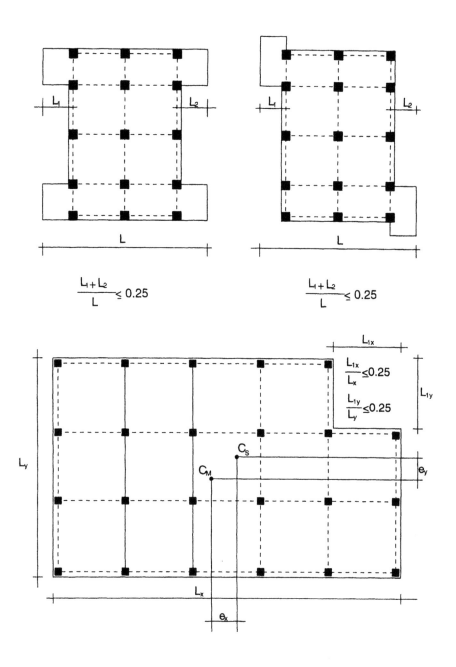

Fig. 8.25   Conditions for plan regularity

- The in-plane stiffness of the floors is large enough, in comparison with the lateral stiffness of the vertical structural elements, that a rigid floor diaphragm behaviour can be assumed.

According to the ECCS Recommendations, the following additional requirement has to be fulfilled:

- The distance (as measured perpendicularly to the direction of the seismic action) between the centre of gravity of the mass and the centre of stiffness does not exceed, at each floor, 15% of the 'torsional stiffness distance' (ρ_{s_x} or ρ_{s_y}), defined as the square root of the ratio between the torsional stiffness K_Θ and the translational stiffness (K_x or K_y) of the building at the considered floor:

$$\frac{e_{s_x}}{\rho_{s_x}} \leq 0.15 \quad \text{and} \quad \frac{e_{s_y}}{\rho_{s_y}} \leq 0.15 \tag{8.10}$$

The computed mass at each storey of the building has to be moved from its nominal position by $\pm 0.05\, L$ [3, 4] (where L is the largest plan size orthogonal to the seismic action); the selected sign is the one giving the most unfavourable effect in the heaviest loaded members. This additional eccentricity is usually referred to as **accidental eccentricity**, because it takes into account uncertainties in the location of the centre of mass and centre of stiffness, a distribution of the live loads which differs from the design assumption and a possible phase shift of seismic horizontal accelerations (rotational component of the input ground motion).

The seismic effects and other action effects should be determined on the basis of a model of the structure as a whole. Three different types of structural analysis are allowed in ECCS Recommendations and EC8 [3, 4]:

- direct dynamic linear or nonlinear analysis;
- response spectrum modal elastic analysis;
- simplified modal analysis (or equivalent-static elastic analysis).

It has already been pointed out that the equivalent-static analysis is allowed in the case of buildings complying with the structural regularity requirements (in vertical and plan configuration); otherwise a complete multimodal analysis has to be performed. In the case of buildings having an irregular configuration both in plan and in elevation, a spatial model has to be considered in the structural analyses. The method to be used is multimodal response analysis. In addition, in order to account for the damage concentration effects due to plan irregularity, a reduced value of the behaviour factor has to be considered.

In the case of buildings having an irregular plan configuration, but

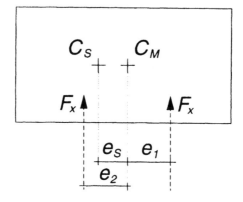

Fig. 8.26   Shifting of the seismic action

with a regular vertical configuration, the reference value of the behaviour factor can be used and, in addition, a simplified analysis of the torsional effects is allowed. In fact, the simplified modal analysis (or equivalent static analysis) is allowed provided that at each storey of the building the horizontal force equivalent to the seismic action is shifted from its nominal position (centre of the masses) by quantities e_1 and e_2, as it is illustrated in Fig. 8.26 where: C_M is the centre of masses at the storey and C_S is the centre of stiffness of the storey.

The values of e_1 and e_2 are generally provided in the form

$$e_1 = \Delta'e_s + e_a$$

$$e_2 = \Delta''e_s - e_a \tag{8.11}$$

where e_a is the previously mentioned **accidental eccentricity**, while $\Delta'e_s$ and $\Delta''e_s$ are additional terms which provide the **dynamic eccentricity**, where e_d is $e_d = e_s + \Delta'e_s$ and $e_d = e_s + \Delta''e_s$ in the first and second cases respectively (where e_s is the **static eccentricity**, i.e. the distance between C_S and C_M).

Following ECCS Recommendations it has to be assumed that

$$e_a = 0.05\ L$$

$$\Delta'e_s = 0.5\ e_s \tag{8.12}$$

$$\Delta''e_s = 0$$

Eurocode 8 [3] gives the following values

$$e_a = 0.05L$$

$$\Delta''e_s = 0$$

(8.13)

while $\Delta'e_s$ is obtained by the minimum value provided by the relationships

$$\Delta'e_s = 0.1\,(\,L + B\,)\,\sqrt{10e_s/L}\,\le 0.1\,(\,L + B\,)$$

$$\Delta''e_s = \frac{1}{2\,e_s}\left[\,(\rho_M^2 - e_s^2 - \rho_s^2\,) + \sqrt{(\rho_M^2 + e_s^2 - \rho_s^2)^2 + 4\,e_s^2\rho_s^2}\,\right]$$

(8.14)

where B is the plan size parallel to the earthquake direction, ρ_M the mass radius of gyration and ρ_s the stiffness radius of gyration.

In Table 8.3, the values of e_a, $\Delta'e_s$ and $\Delta''e_s$ adopted by other codes are given.

The difference between the Mexico 77 and Mexico 87 codes [33, 34] is that the last one requires, in addition, that strength and stiffness eccentricities (e_R and e_s respectively) must have the same sign. Moreover, the fulfilment of the following limitations is requested in the Mexico 87 code:

$$e_R > e_s - 0.2L, \quad \text{for} \quad q^* \ge 3$$

$$e_R > e_s - 0.1_L, \quad \text{for} \quad q^* < 3$$

(8.15)

where q^* is a parameter related to the behaviour factor q, which is coincident with it in the case of medium- and long-period structures [60, 61].

The purpose of the above limitations is to increase the strength of the stiff side of the structural system.

Table 8.3   Provisions adopted by different seismic codes

Code	e_a	$\Delta'e_s$	$\Delta''e_s$
Canada 90	0.10 L	0.50 e_s	-0.50 e_s
Mexico 77	0.10 L	0.50 e_s	0
UBC79	0.05 L	0	0

The Uniform Building Code (UBC79) [35] requires, in addition, that design forces of the vertical-resisting system cannot be less than those which the system should sustain in a symmetric building. This recommendation governs the strength of the vertical bracings belonging to the stiff side of the building.

In Section 8.3.6, the inelastic response of one-storey structural systems, representing 'plan-irregular vertical-regular' buildings, dimensioned according to the above different design criteria will be examined.

8.3.2 A general view on research

Buildings characterized by plan irregularity undergo translational as well as torsional motions during seismic excitation. In particular, a non-uniform plan distribution of actions arises from deck rotations, leading to an increase of the structural element damage with respect to that which elements experience if located in a symmetric system.

The undesirable effects of a building's torsional response were pointed out over half a century ago; in 1938 Ayre [36] proved that, in general, the centre of stiffness and the centre of mass are not coincident, so that horizontal translational motions are accompanied by torsional motions even if torsional ground motion is not present. As a consequence, the fundamental definition of **torsional coupling** was introduced. The importance of this phenomenon was enphasized by Housner and Outinen [37], who demonstrated that the torsional moment, which arises in eccentric systems under seismic loads, is greater than that computed by means of an equivalent static analysis. Successively, this result has led to the introduction of the concept of **dynamic eccentricity** and the dangers which arise from asymmetrical structural configurations were clarified.

The problem of torsional coupling in real structures is complicated by interaction with effects arising from vertical irregularity. For this reason, the study of the torsional response is now limited to simplified models which represent one-storey buildings, whose behaviour is believed to be representative also for multistorey buildings regular along their height. The parameters governing the elastic dynamic response of these one-storey models have been identified as the translational uncoupled period, the ratio between the radius of gyration of the stiffness and that of the mass, the stiffness to mass eccentricity ratio and the damping coefficient. Therefore, the elastic behaviour depends on parameters which globally describe the mass and stiffness distribution, while it is not influenced by the number, the position and the stiffness of single bracing elements. The influence of the above parameters on the elastic torsional response of eccentric systems has been widely investigated [36–46]. Many authors have shown that the code provisions regarding the dynamic eccentricity should be reviewed. However, the results obtained cannot be directly inserted into seismic codes because significant excursions in the plastic range have to be expected under severe earthquakes; therefore, research interest is now devoted to investigation of the inelastic torsional response.

In the inelastic range, the torsional response produces a damage concentration in some bracing elements, leading to the collapse of the structure as a whole, as seen by the examination of damage resulting from recent earthquakes [47, 48]. For this reason, great effort has been spent in analysing the seismic inelastic response of eccentric schemes, but the obtained results are often in contradiction, being influenced by the model adopted in the analysis.

Complete characterization of the inelastic response, even for a simple one-storey model, requires a high number of parameters to be added to those describing the elastic response. In fact, the inelastic response is also influenced by the position and the constitutive law of the single bracing elements, so that systems having the same elastic torsional response can exhibit completely different inelastic behaviours. Therefore, it is clear that a complete parametric analysis would be very costly, without providing effective design indications. In order to overcome the above difficulty, global parameters defining the plan-wise strength distribution have been successfully used. The concept of 'strength eccentricity' has been introduced in order to characterize the non-uniform strength plan distribution [49–54]. The inelastic torsional response is clearly related to this parameter [49]. Goel and Chopra [52–54] have pointed out that, in the case of systems with a symmetric strength distribution, the largest plastic excursion has to be expected on the stiff side of the structural scheme, while it is experienced by the flexible side when the strength eccentricity is equal to the stiffness eccentricity.

The available analyses have shown the roles of the parameters influencing the problem, but they have not allowed the definition of general design rules. The main design goal is the reduction of peak ductility demand and structural damage, and can be attained by two different design strategies. The first is based on the reduction of plastic work by giving the asymmetric system a total strength larger than that corresponding to the equivalent torsionally uncoupled system. The second strategy consists of a carefully planned distribution of the yield strength of the bracing elements, and can be obtained by using plan-position-dependent design levels [55]. The use of such a design criterion leads to a significant improvement of the inelastic response compared to that of systems designed with the same total strength, but with elements having a constant value of design level [56].

8.3.3 A simplified model for torsionally coupled systems

The torsional coupling which characterizes the inelastic response of plan-irregular structures can be investigated by a simplified model having only two degrees of freedom. The model [56] represents the behaviour of a rigid deck of mass M and mass radius of gyration ρ_M, supported by lateral load-resisting elements located along the x as well as y direction. The resisting elements are considered massless and able to withstand forces acting only in their plane. The system, which is subjected to translational ground motion in the x direction, is assumed to be symmetric about the y axis (Fig. 8.27). The two degrees of freedom are the x displacement $x(t)$ of the mass centre C_M and the deck rotation $\Phi(t)$ around the vertical axis.

The translational stiffness of the system is given by

$$K_x = \sum_i k_{x_i} \tag{8.16}$$

and

$$K_y = \sum_j k_{y_j} \tag{8.17}$$

along the x and y directions, respectively, where k_{xi} is the stiffness of the ith element along the x direction and k_{yj} is the stiffness of the jth element along the y direction.

The stiffness centre C_s is located at a distance e_s from the mass centre C_M, where the reference system origin is placed. This is given by

$$e_s = \frac{\displaystyle\sum_i k_{x_i} y_i}{K_x} \tag{8.18}$$

where y_i is the distance of the ith element (along the x direction) from the x axis.

The torsional stiffness of the system is provided by the relationship

$$K_\Theta = \sum_i k_{x_i} (y_i - e_s)^2 + \sum_j k_{y_j} x_j^2 \tag{8.19}$$

where x_j is the distance of the jth element (along the y direction) from the y axis. The stiffness radius of gyration is defined by

$$\rho_s = \sqrt{K_\Theta / K_x} \tag{8.20}$$

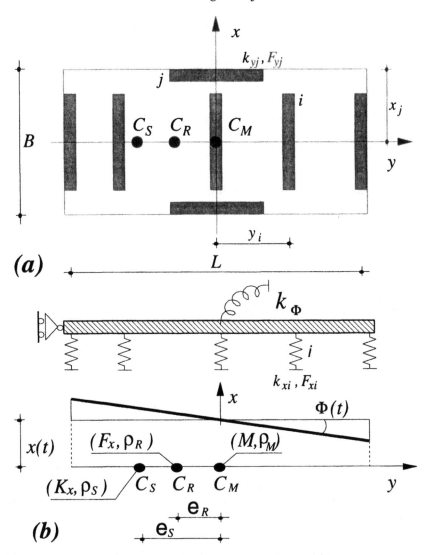

Fig. 8.27 Simplified model for torsionally coupled systems [56]

The motion equations for the undamped system are given by

$$[M] \{\ddot{u}\} + [K] \{u\} = -[M] \{\ddot{u}_g\} \tag{8.21}$$

where:

$$[M] = \begin{bmatrix} M & 0 \\ 0 & M\rho_M^2 \end{bmatrix} \quad [K] = \begin{bmatrix} K_x & K_x e_s \\ K_x e_s & K_x (\rho_s^2 + e_s^2) \end{bmatrix} \tag{8.22}$$

are the mass and stiffness matrix respectively, and

$$\{\ddot{u}\} = \left\{ \begin{matrix} \ddot{x} \\ \ddot{\Phi} \end{matrix} \right\} \qquad \{u\} = \left\{ \begin{matrix} x \\ \Phi \end{matrix} \right\} \qquad \{\ddot{u}_g\} = \left\{ \begin{matrix} \ddot{x}_g \\ 0 \end{matrix} \right\} \qquad (8.23)$$

where $\ddot{x}_g$ is the ground acceleration.

By introducing the non-dimensional parameters

$$\bar{\rho}_M = \frac{\rho_M}{L}, \qquad \bar{\rho}_s = \frac{\rho_s}{L}, \qquad \bar{e}_s = \frac{e_s}{L} \qquad (8.24)$$

the equations of motion can be written as:

$$\begin{bmatrix} 1 & 0 \\ 0 & 1 \end{bmatrix} \left\{ \begin{matrix} \ddot{x} \\ \rho_M \ddot{\Phi} \end{matrix} \right\} + \omega_x^2 \begin{bmatrix} 1 & \bar{e}_s/\bar{\rho}_M \\ \bar{e}_s/\bar{\rho}_M & (\bar{\rho}_s/\bar{\rho}_M)^2 + (\bar{e}_s/\bar{\rho}_M)^2 \end{bmatrix} \left\{ \begin{matrix} x \\ \rho_M \Phi \end{matrix} \right\} = \left\{ \begin{matrix} -\ddot{x}_g \\ 0 \end{matrix} \right\} \qquad (8.25)$$

Therefore, the undamped elastic response of the system depends on the **uncoupled translational frequency**

$$\omega_x = \sqrt{K_x/M} \qquad (8.26)$$

and on the non-dimensional parameters $\bar{e}s/\bar{\rho}_M$ and $\bar{\rho}_s/\bar{\rho}_M.$ In damped systems the damping ratio v has to be added to the above parameters.

8.3.4 Inelastic response parameters

It has been pointed out earlier that the characterization of the elastic response of torsionally coupled systems can be achieved through the uncoupled translational period $T = 2\pi/\omega_x$, the normalized stiffness eccentricity $\bar{e}_s/\bar{\rho}_M$, the normalized stiffness radius of gyration $\bar{\rho}_s/\bar{\rho}_M$ and the damping ratio v. In addition to the above parameters, the inelastic response also depends on the force–displacement relationship and on the location of each resisting element. In the following [56], an elastic–perfectly plastic behaviour is assumed for resisting elements along the ground motion direction, while resisting elements along the y axis (orthogonal to the seismic action) are supposed to behave elastically as they undergo small deformations due to the symmetry about the y axis. As a consequence, the torsional stiffness arising from these elements can be represented by introducing an elastic rotational spring of stiffness k_ϕ, which is related to the global torsional stiffness K_Θ by means of a parameter γ defined as

$$\gamma = \frac{k_\varphi}{K_\Theta} = \frac{\sum\limits_j k_{x_j} \, x_j^2}{K_\Theta} \qquad (8.27)$$

Using F_{xi} to denote the yield force of the ith element, the total strength capacity of the system is defined as

$$F_x = \sum_i F_{x_i} \qquad (8.28)$$

According to Tso and Sadek [49], the plan distribution of the strength can be characterized by means of the distance e_R between the centre of the strength C_R and the centre of mass C_M, which defines the strength eccentricity e_R as:

$$e_R = \frac{\sum\limits_i F_{x_i} y_i}{F_x} \qquad (8.29)$$

Moreover, the strength radius of gyration $\rho_{R'}$ computed with respect to $e_{R'}$ can be considered [56] to be

$$\rho_R = \left[\frac{\sum\limits_i F_{x_i} \, (y_i - e_R)^2}{F_x} \right]^{1/2} \qquad (8.30)$$

which represents a measure of the strength centrifugation. The corresponding non-dimensional parameters are given by

$$\bar{e}_R = \frac{e_R}{L} \qquad \bar{\rho}_R = \frac{\rho_R}{L} \qquad (8.31)$$

The total strength capacity of the system F_x can be evaluated by reducing, according to a factor α, the maximum force F_x^e sustained by the system considered as indefinitely elastic and subjected to a given ground motion. The force F_x^e can be derived either from an elastic dynamic analysis of an equivalent SDOF system (characterized by a period equal to the first period of the coupled system) or from the elastic response of the actual torsionally coupled system. Clearly, the above procedures lead to different values of F_x, because in the second case the force F_x is provided by the sum of peak elastic forces $F_{x_i}^e$ acting on the resisting elements which are not reached simultaneously. A wide parametric analysis of the inelastic response of the above simplified model has been performed in [56] starting from one-storey models designed according to the second procedure. As a consequence, the total strength of the system is given by

$$F_X = \sum_i F_{x_i} = \frac{F_x^e}{\alpha} \qquad (8.32)$$

Regarding the plan distribution of the element strengths F_{x_i}, which strongly influences the damage distribution, two different design criteria can be adopted. The first design criterion is based on the use of a constant value $\alpha_i = \alpha$ of the design level of the elements, so that the inelastic parameters $\bar{e}_R$ and $\bar{\rho}_R$ are uniquely defined. The second one is characterized by the use of different local design levels α_i. As a consequence, different values of the inelastic parameters $\bar{e}_R$ and $\bar{\rho}_R$ correspond at each distribution of the local design levels α_i. As the number of the resisting elements does not provide a significant influence on the maximum value of the required ductility [52–54], reference can be made to the simple system with only three resisting elements. In this case, relations (8.23)–(8.25) uniquely provide the yield forces F_{x_i} for given values of α, $\bar{e}_R$ and $\bar{\rho}_R$. In the following, the strength plan distribution, defined by means of the values of $\bar{e}_R$ and $\bar{\rho}_R$, that produces in-plan uniform damage, is investigated.

8.3.5 Inelastic response optimization

As already pointed out, the inelastic response of torsionally coupled systems depends on the design levels of the resisting elements. An extensive parametric analysis has been carried out [56]. The first period of the analysed systems is included in the range 0.30–0.80 s. The non-dimensional radius of gyration of the masses is $\bar{\rho}_M = 0.35$. With reference to the stiffness eccentricity, two cases have been considered which correspond to the non-dimensional values $\bar{e}_s = -0.10$ and $\bar{e}_s = -0.20$. Moreover, in order to consider both torsionally flexible and torsionally stiff systems, two values of the non-dimensional stiffness radius of gyration, $\bar{\rho}_s = 0.35$ and $\bar{\rho}_s = 0.40$, have been included in the analysis. The first one can be used for representing torsionally flexible systems, while the second is typical of torsionally stiff elements. Finally, the parameter γ has been assumed equal to 0.40, while the value $\alpha = 4$ has been adopted for the global design level. The authors have investigated the inelastic response for two different earthquake accelerograms recorded at

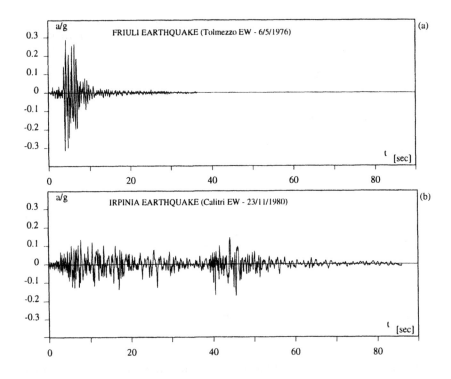

Fig. 8.28   Tolmezzo and Calitri accelerograms

Tolmezzo (Friuli earthquake, EW – 06/05/1976) and Calitri (Campania earthquake, EW – 23/11/1980) (Fig. 8.28), which can be considered particularly significant as their durations and frequency contents are completely different [57] (Fig. 8.29). For given values of the parameters characterizing the elastic response and for a given value of the total strength capacity of the system, a suitable choice of the inelastic parameters $\bar{e}_R$ and $\bar{\rho}_R$ allows the plan distribution of the damage to be made uniform and, as a consequence, a reduction of the peak ductility demand on the heaviest damaged element to be obtained. As an example, with reference to the Tolmezzo ground motion, the required hysteretic ductility is given in Fig. 8.30, for the three elements of systems characterized by $T_1 = 0.4$ s, $\bar{e}_s = -0.10$ and $\bar{\rho}_s/\bar{\rho}_M = 1.43$ and with a design level $\alpha = 4$, as a function of the inelastic parameters $\bar{e}_R$ and $\bar{\rho}_R$. The required ductility in element 1 on the stiff side, which experiences the largest inelasticity, decreases as the strength eccentricity $\bar{e}_R$ decreases. This

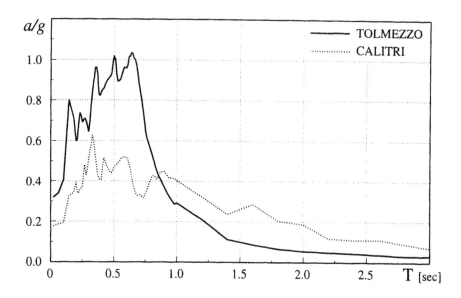

Fig. 8.29   Tolmezzo and Calitri response spectra

reduction of element 1 damage is accompanied by an increase of the ductility demand in element 3, while the required ductility in the central element (element 2) is not significantly affected by the value of $\bar{e}_R$. As a consequence, a range of $\bar{e}_R$ values for which an almost uniform plan distribution of the damage is attained, can be pointed out. Moreover, the same figure shows that plastic excursions of both central and stiff side elements are more severe than those evaluated for the SDOF system having the same period ($T = 0.4$ s). Furthermore, it can be noted that the possibility of attaining a uniform damage plan distribution is not significantly affected by $\bar{\rho}_R$. As a consequence, the authors [56] have paid particular attention to the influence of $\bar{e}_R$. Similar conclusions can be obtained from the analysis of Fig. 8.31 where, by varying T_1 and $\bar{\rho}_s$, the Park–Ang's damage index values D_i [56] of models characterized by $\bar{e}_s = -0.10$ are given. It can be noted that the damage index D_1 of the element 1 is greater than D_3 for element 3 and, in general, those of the corresponding SDOF system. The values referring to element 2 are not shown as they always fall between D_1 and D_3. A measure of the non-uniform damage plan distribution is provided by the parameter

$$R = \frac{D_{\max} - D_{\min}}{D_{\max}} \qquad (8.33)$$

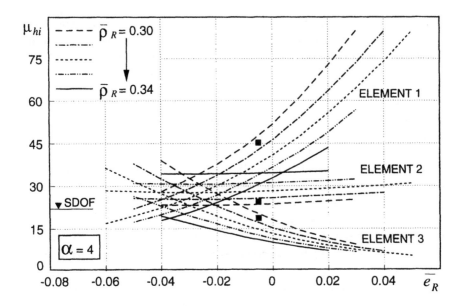

Fig. 8.30 Required hysteretic ductility μ_{hi} of the resisting elements as a function of the inelastic parameters $\bar{e}_R$ and $\bar{\rho}_R$ [56]

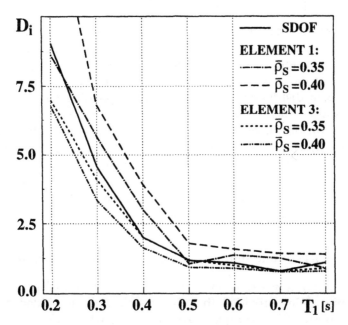

Fig. 8.31 Damage spectra of exterior elements for the model subjected to Tolmezzo earthquake and comparison with the corresponding SDOF system [56]

where D_{max} and D_{min} are the maximum and minimum value assumed by index D_i for the resisting elements. In Fig. 8.32 the influence of the strength eccentricity $\bar{e}_R$ on parameter R is evidenced. In the same figure the response of the corresponding reference models (designed with equal design levels for all resisting elements) is represented by means of a square symbol. The above representations show that the parameter R achieves a minimum value for values of $\bar{e}_R$ which are different from the value of $\bar{e}_R$ corresponding to the reference model. This result confirms that a more uniform damage distribution can be obtained by means of a proper value of $\bar{e}_R$, which can be obtained by using different design levels for the resisting elements. The minimum value of R is attained for $\bar{e}_R$ varying between 0 and $\bar{e}_s$ and often close to $\bar{e}_s/2$. The most significant parameter for evaluating inelastic response is represented by the maximum damage D_{max} among the resisting elements, which is examined as a function of $\bar{e}_R$ for the same cases previously considered (Fig. 8.33). The value of D_{max}, as well as in SDOF systems, is strongly influenced by the period and the earthquake excitation (Fig. 8.34). The comparison between Fig. 8.33 and Fig. 8.34 shows that in the first case the values of D_{max} are more sensitive to the variation of $\bar{e}_R$, having a greater slope. However, independently of the earthquake motion and of the values of T_1, $\bar{\rho}_s$ and $\bar{e}_s$, the curves show a minimum value of D_{max} for small values of the strength eccentricity $\bar{e}_R$. In many cases, these values are close to $\bar{e}_s/2$, which locates the centre of strength in an intermediate position between the centre of mass and the centre of stiffness. Therefore, the reduction of the maximum damage is associated with a more uniform damage plan distribution, because the corresponding values of $\bar{e}_R$ are close to those which minimize the parameter R. The same figures point out that (in torsionally stiff systems $\bar{\rho}_s/\bar{\rho}_M = 1.143$) the damage reduction, which can be obtained by using different design levels for each resisting element, can be more significant than the one which can be attained in torsionally flexible systems ($\bar{\rho}_s/\bar{\rho}_M = 1$), as shown by the comparison with the corresponding reference models (equal design levels) denoted by the square symbols.

8.3.6 Response of one-storey buildings designed according to codes

In Section 8.3.1, it was shown that the equivalent static analysis is allowed by seismic codes provided that the seismic force is shifted from its nominal position by two quantities, e_1 and e_2, which take into account amplification of the torsional effects due to the dynamic behaviour and the accidental error which can occur in the evaluation of the centre of mass and/or centre of stiffness. As a consequence, two structural analyses

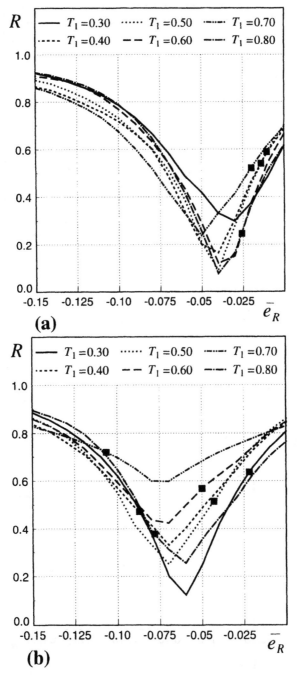

Fig. 8.32 Variation of the scatter R for (a) torsionally stiff and (b) torsionally flexible systems characterized by $\bar{e}_s = -0.10$ and subjected to Calitri ground motion [56]

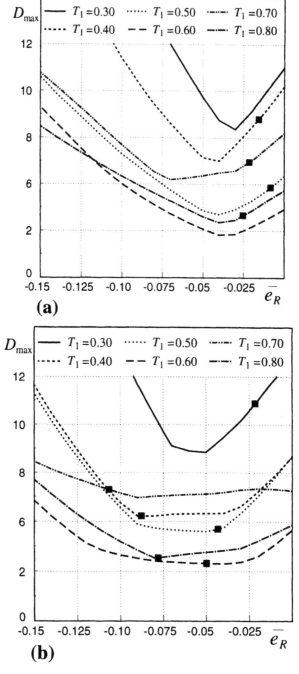

Fig. 8.33 Variation of the maximum damage D_{max} for (a) torsionally stiff and (b) torsionally flexible systems characterized by $\bar{e}_s = -0.10$ and subjected to Calitri ground motion [56]

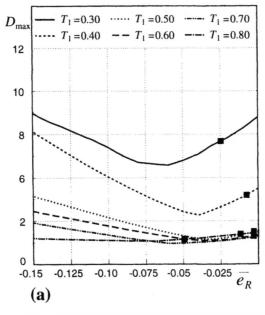

(a)

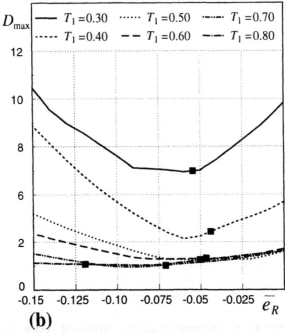

(b)

Fig. 8.34   Variation of the maximum damage D_{max} for (a) torsionally stiff and (b) torsionally flexible systems characterized by $\overline{e}_s = -0.10$ and subjected to Tolmezzo ground motion [56]

have to be performed by assuming a design eccentricity e_d of the seismic force given by

$$e_{d_1} = e_s + e_1$$

$$e_{d_2} = e_s - e_2 \qquad\qquad (8.34)$$

The design force of the *i*th resisting element can be expressed as:

$$F_{x_i} = F_x \frac{k_{x_i}}{K_x}\left[1 + \frac{\bar{e}_d}{\bar{\rho}_s^2}\left(\frac{y_i}{L} - \bar{e}_s \right)\right] \qquad\qquad (8.35)$$

where $\bar{e}_d = e_d/L$ is the non-dimensional design eccentricity, which has to be computed by assuming for e_d the value given by equation (8.34) which provides the largest value of F_{x_i}. The application of equation (8.35) with the severest condition provided by equation (8.34) leads to resisting element design forces given by

$$F_{x_i} = \tau_i F_x \qquad\qquad (8.36)$$

where:

$$\tau_i = \frac{k_{x_i}}{K_x}\left[1 + \frac{\bar{e}_d}{\bar{\rho}_s^2}\left(\frac{y_i}{L} - \bar{e}_s \right)\right] \qquad\qquad (8.37)$$

with

$$O_s = \sum_i \tau_i > 1 \qquad\qquad (8.38)$$

Therefore, an overstrength arises whose magnitude depends on the considered seismic code. As soon as the design forces F_{x_i} of the resisting elements have been defined, the corresponding values of $\bar{e}_R$ and $\bar{\rho}_R$ can be evaluated. One-storey systems with three resisting elements have been designed [56, 58] according to different seismic codes [59–62] and the corresponding inelastic response under seismic loads has been examined. In order to obtain a comparison among different codes, which is independent of the overstrength factor O_s, the authors have considered for each seismic code a seismic action scaled by F_x/O_s, where the non-scaled seismic force F_x has been computed by assuming for all codes a design level equal to 4 (i.e. $F_y/M A = 4$). In addition, the accidental eccentricity

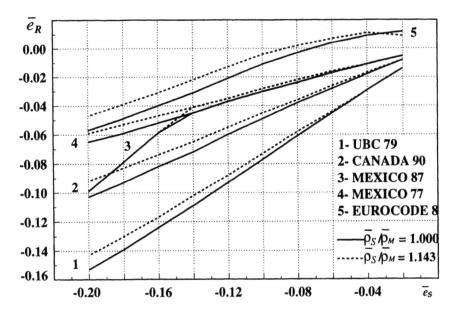

Fig. 8.35 Non-dimensional strength eccentricity $\bar{e}_R$ of one-storey models designed according to different seismic codes [58]

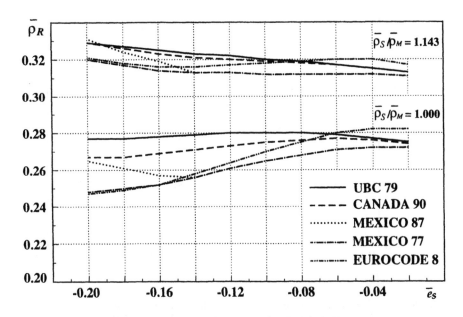

Fig. 8.36 Non-dimensional strength radius of gyration of one-storey models designed according to different seismic codes [58]

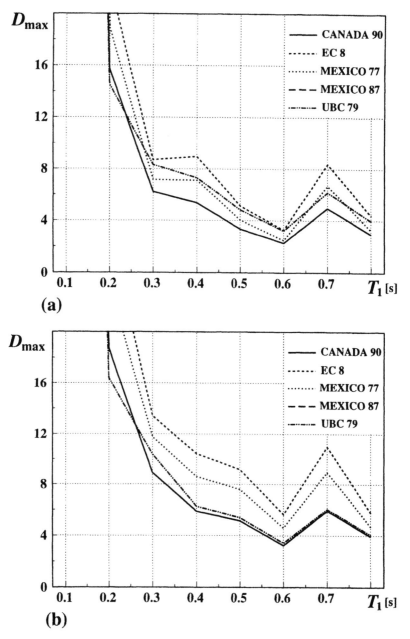

Fig. 8.37 Maximum damage D_{max} of one-storey models with $\bar{e}_s = -0.10$, characterized by (a) $\bar{\rho}_s / \bar{\rho}_M = 1.143$ (torsionally stiff) and (b) $\bar{\rho}_s / \bar{\rho}_M = 1.000$ (torsionally flexible) and subjected to Calitri ground motion [58]

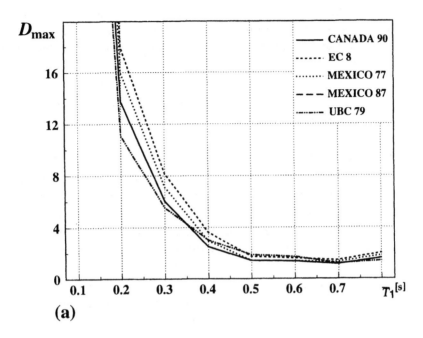

(a)

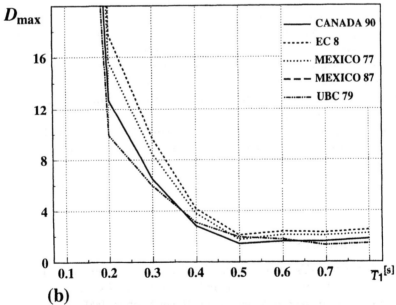

(b)

Fig. 8.38   Maximum damage D_{max} of one-storey models with $\bar{e}_s = -0.10$, characterized by (a) $\bar{\rho}_s / \bar{\rho}_M = 1.143$ (torsionally stiff) and (b) $\bar{\rho}_s / \bar{\rho}_M = 1.000$ (torsionally flexible) and subjected to Tolmezzo ground motion [58]

has not been taken into account since uncertainties that are intended to be covered by this term are not introduced in the dynamic analysis.

In Figs. 8.35 and 8.36 the values of $\bar{e}_R$ and $\bar{\rho}_R$ derived from the application of different seismic codes are presented. The influence of $\bar{\rho}_s/\bar{\rho}_M$ on the value of $\bar{e}_R$ is not particularly significant. The Canadian code leads to the values of $\bar{e}_R$ closest to $\bar{e}_s/2$, while the UBC79 code provides the maximum values of $|\bar{e}_R|$, thus leading to the minimum distance between the centre of stiffness and the centre of strength. The EC8 code provides the minimum values of $|\bar{e}_R|$ while, finally, both Mexican codes provide intermediate values. The parameter $\bar{\rho}_R$ is not particularly influenced by the code provisions, while it is significantly affected by the ratio $\bar{\rho}_s/\bar{\rho}_M$. In Figs. 8.37 and 8.38, with reference to the accelerograms of Calitri and Tolmezzo, the maximum damage between the resisting elements is presented as a function of the uncoupled translational period T_1 for systems characterized by $\bar{e}_s = -0.10$. With reference to the Calitri accelerogram, the Canadian code provides the minimum values of the damage, while the EC8 code provides the maximum values. Figure 8.37b shows that, in the case of torsionally flexible systems, the UBC79 code provides damage values close to those provided by the Canadian code, while, in the case of torsionally stiff systems (Fig. 8.37a), the damage to systems designed according to the UBC79 code is significantly larger than the one obtained by using the Canadian design criteria. Mexican codes, which in the examined cases provide equal results, lead to intermediate damage values. For the Tolmezzo record (Fig. 8.38), the influence of the design criterion is less significant, but the best performances are still obtained by systems designed according to the Canada 90 code or, in some cases, the UBC79 code. The EC8 provisions always lead to the worst behaviour. As the Canadian code provides the best inelastic response, it is most important to emphasize that it leads to a value of $\bar{e}_R$ close to $\bar{e}_s/2$, which has been previously recognized as an optimum condition.

8.4   REFERENCES

[1] *C.A. Guerra, F.M. Mazzolani, V. Piluso:* 'On the Seismic Behaviour of Irregular Steel Frames', 9th ECEE (European Conference on Earthquake Engineering), Moscow, 11–16 September 1990.
[2] *C.A. Guerra, F.M. Mazzolani, V. Piluso:* 'Sul Comportamento Sismico di Strutture Intelaiate in Acciaio in presenza di "Set-Backs"', V Convegno Nazionale, L'Ingegneria Sismica in Italia, Palermo, 29 Settembre–2 Ottobre, 1991.

[3] *Commission of the European Communities:* 'Eurocode 8: European Code for Seismic Regions', draft of October 1993.

[4] *European Convention for Constructional Steelwork (ECCS):* 'European Recommendations for Steel Structures in Seismic Zones', 1988.

[5] *Gruppo Nazionale per la Difesa dai Terremoti (GNDT):* 'Norme Tecniche per le Costruzioni in zone sismiche', 1984.

[6] *G.V. Berg:* 'Earthquake Stresses in Tall Buildings with Setback', Proceedings of the Second Symposium on Earthquake Engineering, University of Roorkee, India, 1962.

[7] *D.P. Jhaveri:* 'Earthquake Forces in Tall Buildings with Set-Back', Ph.D Thesis, University of Michigan, USA, 1967.

[8] *O.A. Pekan, R. Green:* 'Inelastic Structures with Set-backs', 5th World Conference on Earthquake Engineering, Rome, Italy, 1973.

[9] *S.K. Varma:* 'Dynamic Behaviour Of Setback Structures', M.Sc. Thesis, Sir George Williams University, Montreal, Canada, 1974.

[10] *J.L. Humar, E.W. Wright:* 'Earthquake Response of Steel Framed Multistorey Buildings with Setbacks', Earthquake Engineering and Structural Dynamics, N.5, pp. 15–39, 1977.

[11] *G. Sedlacek, J. Kuck, S. Kook:* 'Parameter Study by Simplified Dynamic Analysis according to EC8', Lehrstuhl fur Stahlban, TH Aachen, April, 1989.

[12] *M. Dolce, A. Simonini:* 'The Influence of Structural Regularity on the Seismic Behaviour of Buildings', 8th European Conference on Earthquake Engineering, Lisbon, 1986.

[13] *M. Dolce:* 'Non-Linear Response of Buildings vs. Vertical Regularity Requirements of Seismic Codes: a Parametric Study', 9th World Conference on Earthquake Engineering, Tokyo, August, 1988.

[14] *M. Dolce, A. Masciotta:* 'Il Comportamento Sismico di Edifici in C.A. caratterizzati da variazioni in Elevazione della Rigidezza dei Pilastri', Atti dell'Istituto di Scienza e Tecnica delle Costruzioni, N.2, Novembre, 1988.

[15] *T. Kobori, R. Minai:* 'Evaluation of Seismic Loads', Draft on Seismic Loads, N.2, Architectural Institute of Japan, 1974.

[16] *B. Kato, H. Akiyama:* 'Energy Concentration of Multi-Storey Buildings', 7th World Conference on Earthquake Engineering, Istanbul, 1980.

[17] *D. Capecchi, C. Rega, F. Vestroni:* 'A Study of the Effect of Stiffness Distribution on the Non-Linear Seismic Response of Multi-Degree-of-Freedom Systems', Engineering Structures, Vol. 2, 1980.

[18] *G. Anagnostides:* 'Optimum Distribution of Stiffness in Multistorey Steel Building Frames Subjected to Earthquake Excitation', European Earthquake Engineering, N.2, 1988.

[19] *H. Akiyama:* 'Earthquake Resistant Limit State Design for Buildings', University of Tokyo Press, 1985.

[20] *B. Kato, H. Akiyama:* 'Earthquake Resistant Design of Steel Buildings', 6th World Conference on Earthquake Engineering, 1977.

[21] *H. Akiyama:* 'Earthquake Resistant Design Based on Energy Concept', 9th World Conference on Earthquake Engineering, Tokyo, August, 1988.

[22] *Ministero dei Lavori Pubblici, D.M.24 Gennaio 1986:* 'Norme Tecniche relative alle Costruzioni Sismiche', Gazzetta Ufficiale N.108, 12 Maggio 1986.

[23] *R. Landolfo, F.M. Mazzolani:* 'The Consequences of the Design Criteria on the Seismic Behaviour of Steel Frames', 9th European Conference on Earthquake Engineering, Moscow, 11–16 September, 1990.

[24] *F.M. Mazzolani, V. Piluso:* 'Failure Mode and Ductility Control of Seismic Resistant MR Frames', XIV Congresso C.T.A. Viareggio, 24–27 Ottobre 1993.

[25] *F.M. Mazzolani, V. Piluso:* 'Dimensionamento a Collasso dei Telai Sismo-Resistenti in Acciaio', VI Convegno Nazionale, L'Ingegneria Sismica in Italia, Perugia, 13–15 Ottobre 1993.

[26] *C.A. Guerra, F.M. Mazzolani, V. Piluso:* 'Le Conseguenze delle Irregolarità Strutturali: Problematiche e Prospettive', IV Convegno Nazionale, L'Ingegneria Sismica in Italia, Milano, 5–7 Ottobre 1989.

[27] *Structural Engineers Association of California (SEAOC):* 'Aseismic Design', 1958.

[28] *C.A. Guerra, F.M. Mazzolani, V. Piluso:* 'Esame Critico sulle Metodologie di Valutazione del Fattore di struttura nelle Strutture Intelaiate in acciaio', IV Convegno Nazionale, L'Ingegneria Sismica in Italia, Milano, 5–7 Ottobre, 1989.

[29] *C.A. Guerra, F.M. Mazzolani, V. Piluso:* 'Evaluation of the q-factor in steel framed structures: State of art', Ingegneria Sismica, N.2, 1990.

[30] *M. Como, G. Lanni:* 'Aseismic Toughness of Structures', Meccanica, N.18, pp. 107–114, 1983.

[31] *E. Cosenza, A. De Luca, C. Faella:* 'Analisi Inelastica di Strutture Intelaiate a Nodi Semirigidi', Costruzioni Metalliche, N.4, 1984.

[32] *G.H.Powell:* 'DRAIN – 2D User's Guide', Earthquake Engineering Research Center, University of California, Berkeley, September, 1973.

[33] *National University of Mexico:* 'Design manual for earthquake construction regulation for the Federal District of Mexico', No. 406, 1977.

[34] *R. Gomez, F. Garcia-Ranz:* 'The Mexico earthquake of September 19, 1985', Complementary technical norms for earthquake resistant design, 1987 edition, Earthquake Spectra, Vol. 4, 1988, pp. 441–459.

[35] *UBC:* 'Uniform Building Code – Earthquake Regulations', International Conference of Building Officials, Whittier, California, 1979.

[36] *R.S. Ayre:* 'Interconnection of translational and torsional vibrations in buildings', Bulletin of the Seismological Society of America, vol. 28, No. 2, April 1938, pp. 89–130.

[37] *G.W. Housner, H. Outinen:* 'The effect of torsional oscillations on earthquake stresses', Bulletin of the Seismological Society of America, vol. 48, No. 2, July 1958, pp. 221–229.

[38] *R. Hejal, A.K. Chopra:* 'Earthquake response of torsionally-coupled buildings', Report No. UCB/EERC 87/20, Earthquake Engineering Research Center, University of California, Berkeley, California, December 1987.

[39] *P. D'Andria, R. Ramasco:* 'L'eccentricità convenzionale delle azioni sismiche orizzontali negli edifici multipiano dissimmetrici', Giornale del Genio Civile, Luglio, Settembre 1980.

[40] *K.M. Dempsey, H.M. Irvine:* 'Envelopes of maximum seismic response for a partially symmetric single storey building model', Earthquake Engineering and Structural Dynamics, vol. 7, 1979, pp. 161–180.

[41] *M. De Stefano, G. Faella, R. Ramasco:* 'Eccentricità delle azioni sismiche orizzontali negli edifici non simmetrici', Ingegneria Sismica, No. 1, 1987.

[42] *M. De Stefano, G. Faella, R. Ramasco:* 'Eccentricità dinamica delle azioni sismiche orizzontali sugli edifici multipiano', Atti III Convegno Nazionale 'L'Ingegneria Sismica in Italia', Roma, Settembre–Ottobre 1987.

[43] *M. De Stefano, G. Faella, R. Ramasco:* 'Ammissibilità del calcolo statico per gli edifici multipiano dissimmetrici', Volume dedicato al Prof. F. Jossa in occasione del cinquantenario della Facoltà di Architettura di Napoli, Giannini Editore, 1988.

[44] *G.C. Hart, R.M. Di Julio, M. Lew:* 'Torsional response of high-rise buildings', Journal of the Structural Division, ASCE, Vol. 101, No. ST2, February 1975, pp. 397–416.

[45] *C.L. Kan, A.K. Chopra:* 'Elastic earthquake analysis of torsionally coupled buildings', International Journal of Earthquake Engineering and Structural Dynamics, Vol. 5, No. 4, October 1977, pp. 395–412.

[46] *S.Y. Kung, D.A. Pecknold:* 'Seismic response of torsionally coupled single-storey structures', Proc. VIII World Conference on Earthquake Engineering, San Francisco, USA, 1984, pp. 235–242.

[47] *P. Mazilu, H. Sandi, D. Teodorescu:* 'Analysis of torsional oscillations', Proc. V World Conference on Earthquake Engineering, Rome, Italy, 1973, pp. 153–162.

[48] *A. Rutenberg, T.I. Hsu, W.K. Tso:* 'Response spectrum techniques for asymmetric buildings', Earthquake Engineering and Structural Dynamics, Vol. 6, 1978, pp. 427–435.

[49] *W.K. Tso, V. Meng:* 'Torsional provisions in building codes', Canadian Journal of Civil Engineering, Vol. 9, 1982, pp. 38–46.

[50] *R. Meli:* 'Evaluation of performance of concrete buildings damaged by the September 19, 1985 Mexico earthquake', The Mexico Earthquake of 1985, Proc. Int. Conf., ASCE, Mexico City, Mexico, 1986, pp. 308–327.

[51] *L. Esteva:* 'Earthquake engineering research and practice in Mexico after the 1985 earthquake', Bulletin of the New Zealand National Society for Earthquake Engineering, Vol. 20, No. 3, September 1987, pp. 159–200.

[52] *W.K. Tso, A.W. Sadek:* 'Inelastic seismic response of simple eccentric structures', Earthquake Engineering and Structural Dynamics, vol. 13, 1985, pp. 255–269.

[53] *W.K. Tso, Y. Bozorgnia:* 'Effective eccentricity for inelastic seismic response of buildings', Earthquake Engineering and Structural Dynamics, Vol. 14, 1986, pp. 413–427.

[54] *W.K. Tso, H. Ying:* 'Additional seismic inelastic deformation caused by structural asymmetry', Earthquake Engineering and Structural Dynamics, Vol. 19, 1990, pp. 243–258.

[55] *R.K. Goel, A.K. Chopra:* 'Inelastic seismic response of one-storey, asymmetric-plan systems: effects of stiffness and strength distribution', Earthquake Engineering and Structural Dynamics, Vol. 19, 1990, pp. 949–970.

[56] *R.K. Goel, A.K. Chopra:* 'Inelastic seismic response of one-storey, asymmetric-plan systems', Report No. UCB/EERC 90/14, Earthquake Engineering Research Center, University of California, Berkeley, California, 1990.

[57] *R.K. Goel, A.K. Chopra:* 'Inelastic seismic response of one-storey, asymmetric-plan systems: effects of system parameters and yielding', Earthquake Engineering and Structural Dynamics, Vol. 20, 1991, pp. 201–222.

[58] *M. De Stefano, G. Faella, R. Ramasco:* 'Inelastic Spectra for Eccentric Systems', 10th World Conference on Earthquake Engineering, Madrid, July, 1992.

[59] *M. De Stefano, G. Faella, R. Ramasco:* 'Inelastic Response and Design Criteria of Plan-wise Asymmetric Systems', Earthquake Engineering and Structural Dynamics, Vol. 22, March, 1993.

[60] *E. Cosenza, G. Manfredi, R. Ramasco:* 'L'uso dei funzionali di danneggiamento in Ingegneria Sismica: confronto fra diverse metodologie' Rapporto dell'Istituto di Ingegneria Civile No. 25, Università di Salerno, Luglio 1991.

[61] *M. De Stefano, G. Faella, R. Ramasco:* 'Inelastic Response of code-designed asymmetric systems', European Earthquake Engineering, Vol. 2, 1993.

[62] *Associate Committee on the National Building Code:* 'National Building Code of Canada', National Research Council of Canada, Ottawa, Ontario, 1985.

9

Influence of random material variability

9.1 INTRODUCTION

Seismic design is usually based upon the ability of structures to sustain plastic deformations so that earthquake input energy is dissipated through hysteretic behaviour of the material. In any case, the damage due to plastic deformations has to be limited in order to prevent collapse.

As this design criterion involves the plastic redistribution capacity of the structure, the safety of a structure against the severest design earthquake is guaranteed only if local and global ductility demands are compatible with the geometrical and mechanical properties of the structure and its members.

The available global ductility is strictly related to the collapse mechanism and it assumes the maximum value when the global-type mechanism occurs. For this reason, modern seismic codes such as ECCS Recommendations and EC8 provide simplified design criteria in order to control the failure mode [1].

Under this point of view, randomness of the yield strength of members plays a very important role, because it affects the plastic hinge formation process and, as a consequence, the ultimate behaviour of structures, leading to an energy dissipation capacity and to an available ductility different from the predicted ones. Even if the structure is designed to fail in global mode on the basis of the nominal yield strength, an increase in the coefficient of variation of the yield strength of members can be responsible for the formation of local failure modes. These reduce both ductility and energy dissipation capacity and, therefore, the structure is no longer able to resist the severest design earthquake.

The problem under examination is also strictly related to that of 'structural regularity'. Modern seismic codes classify structures into two categories, regular and irregular structures, on the basis of their geometrical configuration (Chapter 8). In the case of irregular structures, damage concentration is expected so that the worsening of the inelastic behaviour is taken into account by means of a reduction of the q-factor which has to be used in design. As previously pointed out, the concept of structural regularity allows the definition of structural situations for which the seismic response of simple degree of freedom (SDOF) systems can be used to approximately represent the situation for multi degree of freedom (MDOF) structures.

It has been observed that the use of results regarding the seismic response of SDOF systems for predicting the seismic behaviour of real MDOF structures requires the hypothesis of uniform damage distribution. In addition, it can be stated that even if a structure is designed in such a way that this very ambitious pattern of damage is assured on the basis of the nominal yield strength, the actual damage distribution can be very different due to the randomness of yield strength in structural members. From this point of view the randomness of yield strength can be considered as a type of 'irregularity'.

9.2 RANDOM MATERIAL VARIABILITY

9.2.1 Analysis of experimental data

The first step in the analysis of the impact of fluctuating material properties on structural inelastic behaviour is the generation of random yield stresses, as in a real structure. To this end a preliminary statistical analysis of available experimental data has been performed [2, 3], where data collected in recent years by the Institute of 'Tecnica delle Costruzioni' of Naples University (as part of its testing and quality certification activity) have been analysed. The statistical analysis has been developed for three grades of steel: Fe360, Fe430 and Fe510, type B, C and D.

The analysed random variables are the yield stress, the ultimate tensile strength and the yield ratio. The yield ratio, which is defined as the ratio between the yield strength and the ultimate tensile strength, is an indirect variable and is not shown in inspection certificates.

The influence of the plate thickness has been also considered.

Figures 9.1, 9.2 and 9.4 show, for Fe360 steel, the influence of thickness on yield stress, ultimate strength and yield ratio, respectively. In Fig. 9.3 the yield stress versus ultimate strength relationship is also shown. The values of the correlation coefficient r and of the variance v are also given. Figures 9.5–9.12 show the same relationships for Fe430 and Fe510 steels. For all the examined steels the influence of thickness on the yield stress is very important, even if the corresponding correlation coefficient is not particularly high. With reference to the influence of thickness on ultimate strength, in the case of Fe360 and Fe430 steels it can be stated that these variables can be considered statistically independent as the corresponding correlation coefficient is close to zero. The same is not true for Fe510 steel.

From the structural point of view, the dependence of the yield ratio

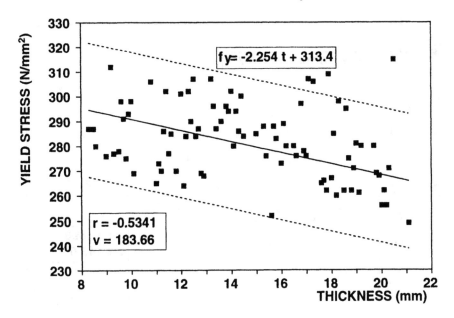

Fig. 9.1   Yield stress versus thickness regression, for Fe360 steel

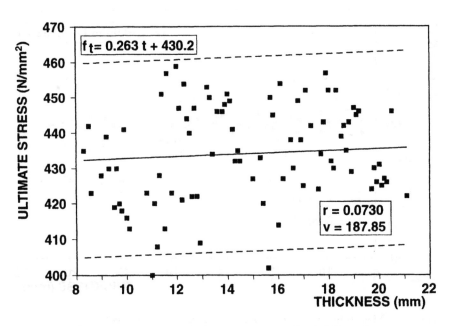

Fig. 9.2   Ultimate stress versus thickness regression, for Fe360 steel

Influence of random material variability

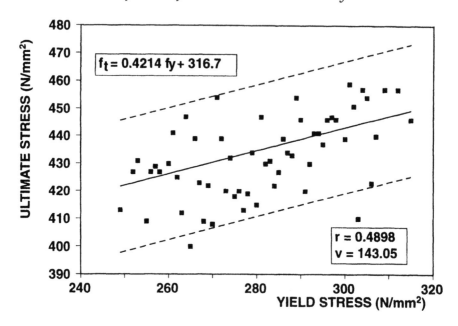

Fig. 9.3   Ultimate stress versus yield stress regression, for Fe360 steel

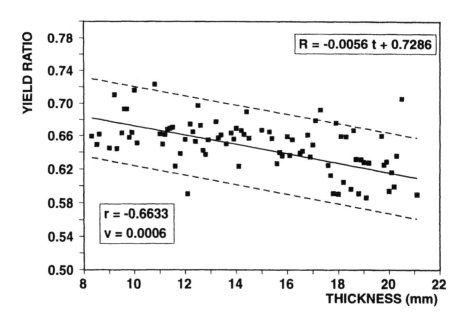

Fig. 9.4   Yield ratio versus thickness regression, for Fe360 steel

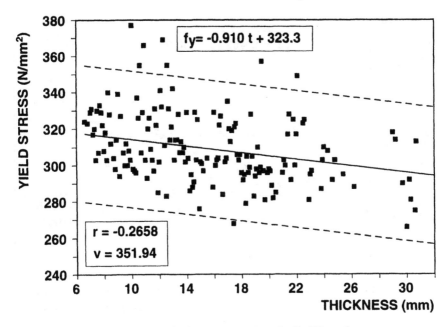

Fig. 9.5   Yield stress versus thickness regression, for Fe430 steel

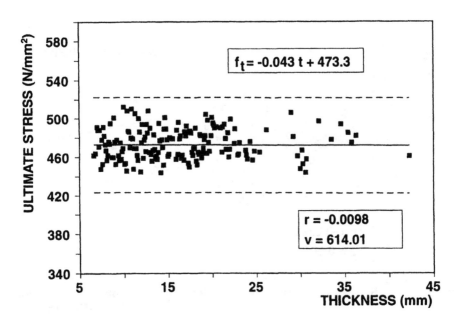

Fig. 9.6   Ultimate stress versus thickness regression, for Fe430 steel

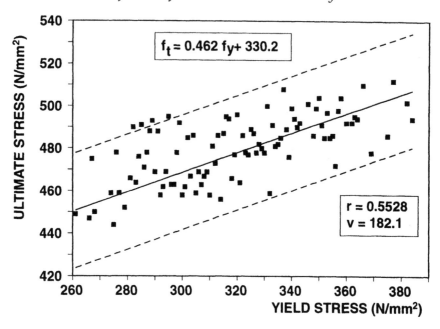

Fig. 9.7   Ultimate stress versus yield stress regression, for Fe430 steel

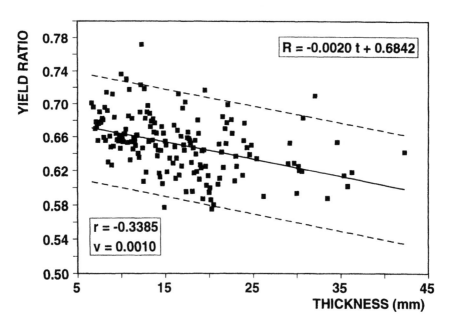

Fig. 9.8   Yield ratio versus thickness regression, for Fe430 steel

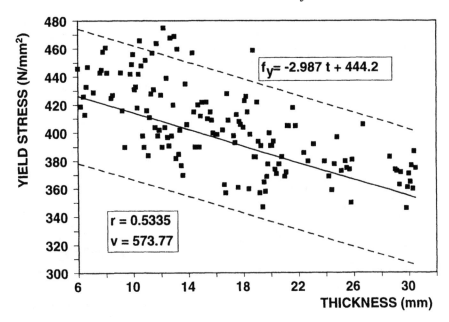

Fig. 9.9 Yield stress versus thickness regression, for Fe510 steel

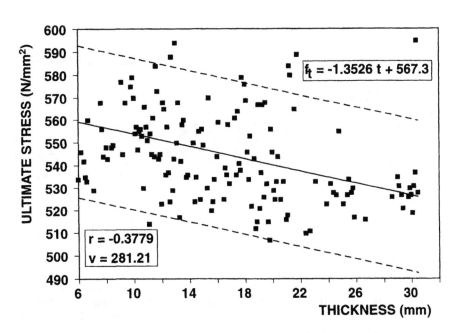

Fig. 9.10 Ultimate stress versus thickness regression, for Fe510 steel

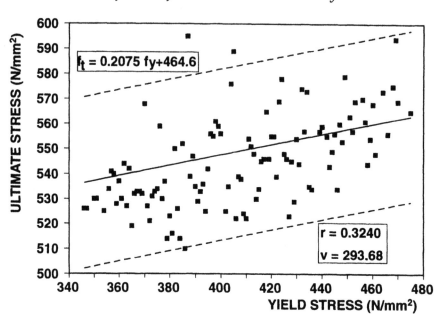

Fig. 9.11   Ultimate stress versus yield stress regression, for Fe510 steel

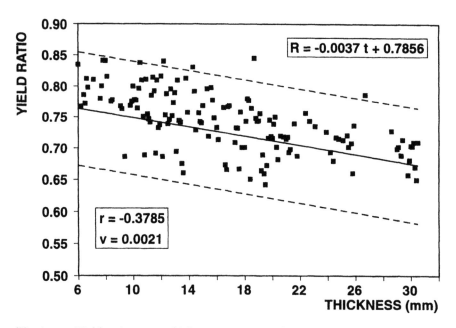

Fig. 9.12   Yield ratio versus thickness regression, for Fe510 steel

upon the thickness is very important, because it produces an increase of the plastic strength reserve as the thickness increases. The values obtained for the yield ratio are all compatible with the EC3 provisions, which require the condition $f_t/f_y > 1.2$ when the plastic behaviour of structures is taken into account in the design [4].

The dependence of yield strength upon plate thickness plays an important role in the investigation of the inelastic behaviour of frames with random variability of member yield strength. In order to point out the importance of the above statement a very common structural typology for seismic-resistant structures can be considered: the strong-column weak-beam frame. This typology leads to columns having an inertia greater than for the beam. Usually, the thickness of beam flanges is less than for column flanges. As a consequence, the yield strength of beams is greater than for columns. Due to this condition the formation of plastic hinges in the beams rather than in the columns can be undermined, leading to local failure modes which cannot provide the required ductility.

Therefore, the dependence of yield strength upon the thickness has to be taken into account in order to prevent premature collapse. For these reasons, particular attention has been paid to the yield strength versus thickness relationship. Moreover, data representations on normal probability paper have been performed for different thickness classes. One of these is given, in Fig. 9.13 for Fe360 steel with thicknesses ranging between 10 and 14 mm.

It is well known that normal distributions are represented by a straight line in the normal standard variable u versus random variable (f_y) plane. Data are shown using their cumulated probability Φ and the relationship $\Phi = \Phi(u)$, which is provided on the ordinate axis. Additionally, the 5% confidence interval is also shown on the figure and the most important statistical features are given in the table. The complete series of these representations is given elsewhere [2], where reference is also made to the ultimate tensile strength.

9.2.2 Discussion on the statistical features obtained

Examination of the above data representations shows that, for each steel, the yield stress decreases as the thickness of plates increases. This feature is already well known, but the data analysed have not provided a high correlation coefficient and the scatter is non-negligible. The influence of the thickness on the ultimate tensile strength is lower than for yield stress and in some cases it is negligible.

From the complete statistical analysis available [2], values for the coefficient of variation of yield stress are found to be larger than those obtained for the ultimate strength. This feature is so strong that in some

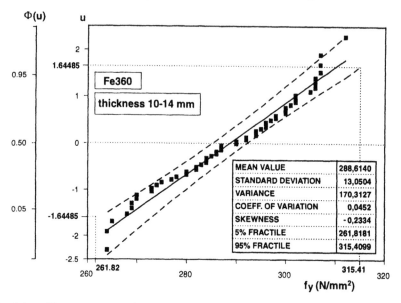

Fig. 9.13   Data representation on normal probability graph paper and corresponding to 5% confidence interval

cases the coefficient of variation of yield stress is twice that for tensile strength. This is a very important result because an increase in the coefficient of variation of yield strength of the members will induce the structural system to collapse in local failure modes, even if designed to collapse in a global mechanism.

The magnitude of the scatters in the values of yield stress, ultimate strength and yield ratio is, in general, dependent on the manufacturer, as emphasized elsewhere [5].

Different distribution laws have been examined [2] by the skewness test. In some cases the normal distribution has been recognized as the optimum one, while in other cases the log–normal distribution should be adopted. In spite of this, the choice of the distribution law does not present a significant influence on the values computed for the 5% and 95% fractiles.

It is important to realize that the coefficient of variation of the yield stress is very often greater than 0.05, while an improvement of the controls in steel production could provide a value of 0.025 [6]. Therefore, due to random yield stresses, structural steels do not always match the expectations coming from plastic design results. In seismic design recommendations the material specifications regarding the upper bound of the yield stress are usually absent and the lower limits only are specified [1, 7]; this gap must be eliminated and an appropriate coefficient of variation for the design stress should be introduced.

9.3 ANALYSIS METHODOLOGY

9.3.1 Parameters considered

The inelastic behaviour of steel frames with random material variability has already been investigated by several authors [6, 8, 9]. A statistical simulation of the load-bearing capacity of a 6-storey 3-bay frame has been carried out and the influence of random material variability shown for different values of the coefficient of variation of the material yield stress [6, 8]. On the basis of the results obtained, the advantages of a more severe quality control procedure, which produces a reduction of the variability range of the material mechanical properties, are demonstrated.

Similar conclusions have been obtained, based upon the analysis of a simple portal frame [9].

A full knowledge of the influence of material property randomness on the inelastic response of steel frames under seismic loads requires a great number of numerical simulations. As a consequence of the practical difficulties, published studies refer to single examples.

In order to analyse the influence of the randomness of member yield strength on the failure of frames, the inelastic behaviour of frames has been investigated [10], starting with a frame designed on the basis of the nominal value of the yield stress, so that the corresponding collapse mechanism is of global type. This choice is justified on the basis that the 'weak-beam strong-column' design philosophy represents the case in which the plastic hinge formation process has the most decisive influence on the ultimate behaviour [11–14]. In this way the random variability of material properties can lead to significant effects on the design criteria, which state the column-to-beam strength ratio.

The ability of a member or a structural system to undergo deformations beyond its first yielding strain is usually referred as ductility. In modern seismic codes, knowledge of the structural ductility plays a very important role in the design of structures able to resist destructive earthquakes. The following parameters have been considered in order to characterize the inelastic response of the examined structure:

- the top displacement corresponding to first yielding, δ_y;
- the multiplier of horizontal forces corresponding to first yielding, α_y;
- the ultimate multiplier of horizontal forces (maximum value), α_u;
- the plastic redistribution parameter, α_u/α_y;
- the ultimate displacement at 95% of α_u (5% strength degradation), δ_1;
- the ultimate displacement at 90% of α_u (10% strength degradation), δ_2;
- the ductility ratio computed as δ_1/δ_y (5% strength degradation), μ_1;
- the ductility ratio computed as δ_2/δ_y (10% strength degradation), μ_2;
- the ductility ratio computed by assuming δ_2 as ultimate displacement

and by characterizing the yield state by means of the top displacement δ_y^*, which the indefinitely elastic structure exhibits under the horizontal forces corresponding to the ultimate multiplier α_u, denoted as μ_2^*.

Figure 9.14 illustrates the definition of the various parameters on a typical diagram relating the multiplier of horizontal forces versus the top displacement.

As discussed in Chapter 2, the use of different definitions of the ductility ratio is justified by the consideration that there is no general agreement on its evaluation [15]. In addition, it is possible to clarify the sensitivity of the ductility ratio definition to the effects due to random variability of yield strength.

Furthermore it is convenient to note that collapse is usually defined as the ultimate limit situation in which the maximum rotational capacity at the yielded end of a member is reached. The ultimate displacement should therefore be computed for this condition, and the available ductility should be evaluated as the ratio between this ultimate displacement and that which produces the first yielding. This consideration leads to the consequent admission of the necessity to take into account the effect of randomness in yield stress on the rotational capacity of members. Considering the experimental relation provided in Chapter 3 for the evaluation of rotational capacity of beam-columns [16], it can be observed

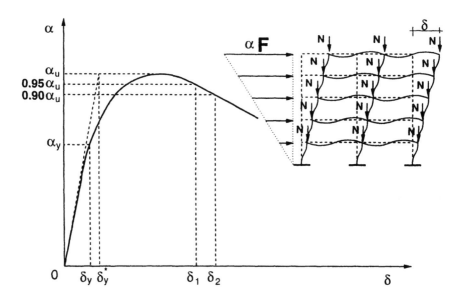

Fig. 9.14   Multiplier of horizontal forces versus top sway displacement behavioural curve: definition of the examined parameters

that, when the true yield stress is less than the nominal value, the formation of the first plastic hinge can develop for a displacement less than that predicted, but modification of the rotational capacity can lead to a higher value of ultimate displacement. The ductility ratio is influenced by both these two phenomena.

However, due to approximations in the available relationships for computing the rotation capacity of steel members, and taking into account that there is no general agreement about its definition [16, 17], the phenomena previously described are difficult to quantify. In any case it has been ensured that values of the ductility ratios, previously defined as derived by the inelastic analysis, are compatible with the available rotational capacity which has been estimated with the relationship provided elsewhere [16]. In this way it is possible to state that the values obtained for the ductility ratio are not only conventional values, but also values that the structure is really able to withstand.

The main problem to be faced in the seismic design of structures is the evaluation of an IDRS starting from an LEDRS. This step requires the definition of a coefficient, namely the q-factor, which takes into account the plastic redistribution capacity, available ductility and dissipative capacities of the structure. Evaluation of the q-factor can be achieved by performing a large number of dynamic inelastic analyses, which have to be repeated for different ground motions. Rigorous computation of the q-factor is therefore particularly cumbersome, so that the use of simplified methods proposed by many authors is justified [18], as discussed in Chapter 4.

As the q-factor represents the most synthetic parameter for evaluating the inelastic performance of seismic-resistant structures, its evaluation with simplified methods, such as the energy method [19], can be considered as a functional tool. As mentioned in Chapter 4, according to the above method, practical evaluation of the q-factor can be performed by considering the work done by a system of equivalent horizontal forces, statically applied and distributed according to a combination of a selected number of vibration modes. The q-factor is evaluated as

$$q = \left(\frac{W_u}{W_y} \right)^{1/2} \tag{9.1}$$

where W_y is the elastic strain energy stored by the system when first yielding occurs and W_u is the total energy stored and dissipated up to failure.

The member sections have to be chosen among the standard shapes and, therefore, an overstrength is obtained. As a consequence, the first yielding corresponding to α_y is attained under horizontal forces $\alpha_y F_d$ greater than the design ones ($F_d = M A R_e(T)/q_d$) provided by the codes, where $\alpha_y \geq 1$. The structure is therefore able to withstand the severest design earthquake provided that $\alpha_y q$ is greater than q_d (where q_d is the design value of the q-factor). For this reason, the 'modified' q-factor

$$q^* = \alpha_y \; q = \left(\frac{W_u}{W_1} \right)^{1/2} \tag{9.2}$$

which takes into account this overstrength effect due to sizing, has to be added to the previous list of parameters characterizing the curve relating the multiplier of horizontal forces to the top displacement (where W_1 is the elastic strain energy stored by the system under the design horizontal forces, i.e. for $\alpha = 1$).

Therefore, parameters to be added to the previous list are:

- q_1^*, computed assuming δ_1 is the ultimate condition;
- q_2^*, computed assuming δ_2 is the ultimate condition.

Evaluation of the random properties of these q-factors provides new information, at present unavailable in the technical literature [6, 8, 9].

9.3.2 Simulation method

In the last 15 years, the problem of evaluation of structural safety by means of probabilistic approaches, as an alternative to the traditional methods of limit-state analysis, has led to substantial theoretical developments due to progress in structural reliability theory. The application of these procedures for evaluating structural reliability is in general very cumbersome so that, in spite of this theoretical progress, the use of structural reliability theory as a design tool is not widespread.

The analysis of structural system reliability can be based on either static or kinematic methods [20]; however, in both cases the numerical procedures are difficult to apply even to very simple structural systems. In particular, the required numerical procedures are inefficient or even unavailable for problems involving geometrical and/or mechanical non-linearities, or for problems where the evaluation of damage accumulation is required. Therefore, the use of numerical simulation seems to be the best solution for complex structural systems [21].

This investigation deals with a particular aspect of structural safety, because in the evaluation of the probability of failure the randomness of loads is neglected and the attention is focused on the parameters which are usually considered intrinsic to the structural behaviour. The usual numerical procedure is Monte Carlo simulation in which the generation of the random yield strength of frame members and the inelastic structural analysis are carried out in sequence [22]. The first step is represented by the generation of numbers uniformly distributed between

0 and 1. Each random number corresponds to a random value of the yield strength, for a given probability distribution law of the structural steels. The transformation of random numbers into random values of yield strength satisfying a given probability distribution law has been performed by the Box and Müller method [22].

In order to simplify the analysis, the geographical distribution of yield strength along the cross-section has been neglected, and just the mean value for a given thickness has been assumed. In addition, because the flexural behaviour of double-T sections is mainly governed by the stress state of flanges, the yield strength of flanges can be taken as the yield strength of the member.

As the examined frame is made of HE160B beams and HE240B columns (Fig. 9.15), two different normal distribution laws have been used by considering both thicknesses of beam and column flanges. Each value of random yield stress is assigned to a member, so that any beam and column are assumed to be independent.

The analysed frame is composed of 35 members, for which 100 series of 35 random-yield stresses have been generated, to give 100 different frames with the same geometrical configuration and loading conditions.

It should be understood that the randomness of the plastic section modulus can be neglected, because the corresponding coefficient of variation is very small when compared with that for yield stress [23]. The

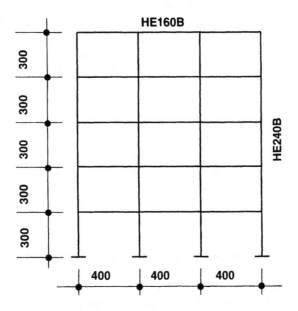

Fig. 9.15 The examined frame

coefficient of variation of the plastic moment of the section can therefore be considered coincident with yield stress.

Random yield stress generation and inelastic structural analysis are carried out in sequence. As a final result, the relationship between the multiplier of horizontal forces and the top displacement is available for each frame. The results of static inelastic analyses have been collected and are shown in reference 24, which gives the required information for each frame. The statistical characterization of the inelastic response of the structure requires a second statistical analysis for the parameters characterizing the α–δ curve and the values of the q-factor.

This method should therefore be considered as a Hybrid Monte Carlo simulation [25]. In it the numerical simulation is used to obtain a statistical sample of the structural response, whose dimension is specified in order to obtain sufficient accuracy in the estimation of the 'central values' of the variables which constitute the object of the investigation. Successively, mathematical models are investigated in order to define the probability distribution functions.

9.3.3 Discussion of results

The main results of the hybrid numerical simulation procedure just described are represented by the 5% fractiles, which have to be directly compared with the values obtained by assuming a constant value of the yield strength equal to the nominal one.

The representation on normal probability graph paper of the values assumed by the ultimate multiplier is given in Fig. 9.16. For this parameter the response is still of the Gaussian type; all values of the numerical simulation, except one, fall within the 5% confidence interval and, in addition, the skewness test is satisfied. This means that the absolute value of the skewness coefficient is less than $(6/n)^{1/2}$, where $n = 100$. The 5% fractile (equal to 3.02) is 15% greater than that computed on the basis of nominal yield strength (equal to 2.63).

The same type of representation is given in Fig. 9.17 for the plastic redistribution parameter α_u/α_y. In this case the distribution cannot be considered as normal, as can be seen from the figure, and the skewness test is not satisfied. The 5% fractile (equal to 1.87), which in this case assumes an indicative meaning due to the non-normal response, presents a 9% decrease with respect to the nominal value.

In Fig. 9.18 the representation on normal probability graph paper is given for the global ductility computed assuming an ultimate displacement corresponding to a 5% degradation. The same representation is given in Fig. 9.19 for the q-factor values computed by assuming the same ultimate condition.

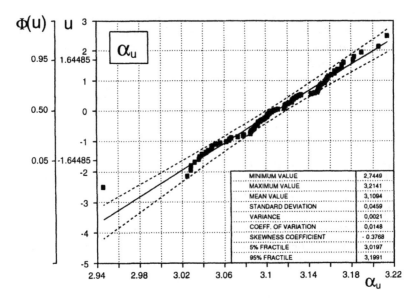

Fig. 9.16 Representation on normal probability graph paper of the inelastic random response: ultimate multiplier

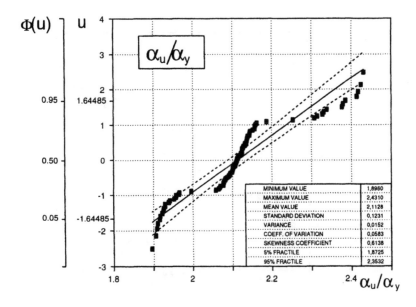

Fig. 9.17 Representation on normal probability graph paper of the inelastic random response: plastic redistribution coefficient

From the design point of view the most interesting results are represented by the 5% fractiles. These have to be compared with the values obtained from the inelastic analysis which has been carried out on the basis of the nominal yield strength of members. These values are 9.59 and 7.94 for the global ductility and the q-factor, respectively.

The comparison shows that, due to yield strength random variability, the 5% fractile of global ductility, equal to 8.44 (Fig. 9.18), presents a 12% reduction with respect to the value estimated on the basis of nominal strength (equal to 9.59). In addition, by using the representation on normal probability graph paper and the 5% confidence interval, it can be stated that the global ductility of the actual structure can be less than that computed for the nominal yield strength with 50% probability, which is non-negligible (Fig. 9.18). However, this value is not rigorous from a statistical point of view because, as shown by the same representation, the normal distribution law does not provide a rigorous statistical interpretation of the global ductility values. This is also confirmed by the skewness test which is not satisfied. As a consequence, from the statistical point of view, the possibility of interpreting the results by distribution laws other than the nominal one should be investigated. This need derives from the non-linearity of the problem under examination, so that although the input (randomness of yield strength) is

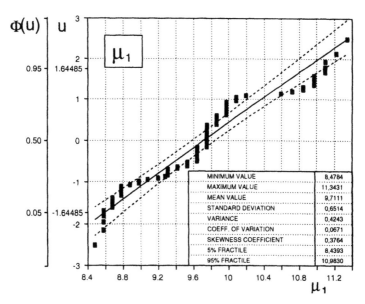

Fig. 9.18   Representation on normal probability graph paper of the inelastic random response: global ductility corresponding to a 5% strength

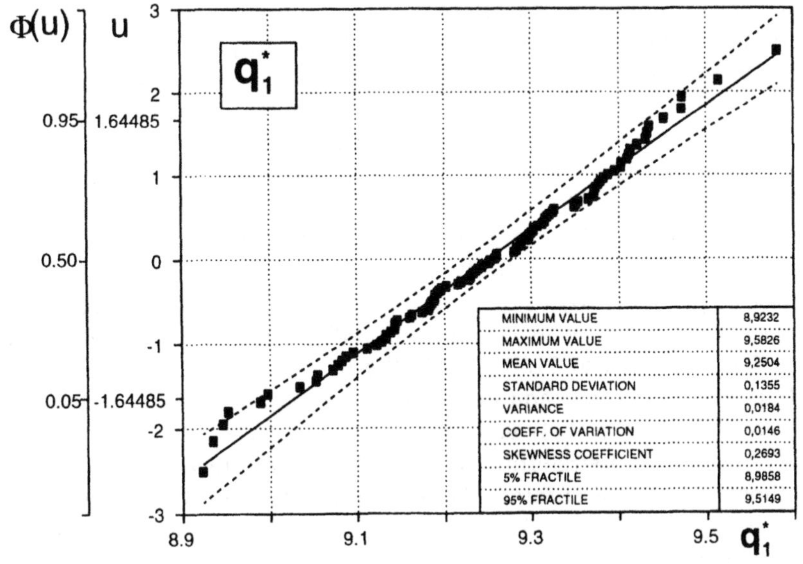
Φ(u)
u
3
2
0.95 1.64485
1
0.50 0
-1
0.05 -1.64485
-2
-3
8.9
9.1
9.3
9.5
q*1
MINIMUM VALUE 8,9232
MAXIMUM VALUE 9,5826
MEAN VALUE 9,2504
STANDARD DEVIATION 0,1355
VARIANCE 0,0184
COEFF. OF VARIATION 0,0146
SKEWNESS COEFFICIENT 0,2693
5% FRACTILE 8,9858
95% FRACTILE 9,5149

Fig. 9.19 Representation on normal probability graph paper of the inelastic random response: *q*-factor corresponding to a 5% strength degradation

described by a normal law, the output (random inelastic response of the structure) can be statistically distributed according to a different law. Notwithstanding, it can be assumed that the use of distribution laws different from the normal one, for analysing the random inelastic response of the system, does not produce significant variation of the above results from the practical point of view.

Regarding the *q*-factor, it can be seen that the corresponding 5% fractile, equal to 8.99 (Fig. 9.19), is 13% greater than that computed on the basis of nominal yield strength (equal to 7.94). In spite of the fact that values of the *q*-factor are computed by a simplified method, this result is particularly interesting both because it is favourable from the seimic behaviour point of view, contrary to global ductility, and because evaluations of this type are absent from the available technical literature [6, 8, 9].

Due to the reduction of the available ductility provided by the random variability of member yield strength, this result might seem unexpected. However, it is easy to recognize that it is in perfect agreement with the interpretation, already proved by other authors [19], which considers the *q*-factor as a measure of the 'seismic toughness' of the structure and, therefore, as a combination of strength and ductility. In fact, while the available ductility is less than the 'nominal' one with 50% probability, the nominal strength or characteristic strength, due to its definition, can be less than the actual strength with 95% probability. Therefore, the first effect (random reduction of available ductility) is largely balanced by the second effect (random increase of strength), which produces an increase both of the first yielding multiplier of horizontal forces and of the ultimate multiplier. This last effect has been already pointed out in the examination of Fig. 9.16.

With reference to 10% degradation as the ultimate condition, statistical representations of global ductility and the q-factor are given in Figs 9.20 and 9.21 respectively. By examining the 5% fractiles, it can be stated that global ductility presents an 11% reduction (5% fractile equal to 10.38 and nominal value equal to 11.71), while the increase of the q-factor is 13% (5% fractile equal to 9.95 and nominal value equal to 8.77). Therefore the percentage differences between nominal values (computed on the basis of the nominal yield strength) and 5% fractiles are practically coincident with those obtained assuming 5% degradation as the collapse condition.

It is interesting to note that, as in the case where 10% degradation is assumed as the ultimate condition, the actual global ductility is less than the nominal value (equal to 11.71) with about 50% probability (Fig. 9.20). Therefore, in the case under examination, it can be concluded that the effects of random variability of yield strength are independent of the assumed collapse condition.

Finally, it is important to realize that by examining the collapse mechanisms of the 100 generated frames, which have been derived from a nominal one designed in order to obtain a collapse mechanism of global type, it can be seen that the inelastic analyses have led to different failure modes in about 10 cases, although in these cases the failure mode is very similar to the global mode [24]. Local failure modes have not been shown by the inelastic analyses so that the slopes of the softening branch of the α–δ curve are practically constant [24].

9.4   INFLUENCE OF THE COEFFICIENT OF VARIATION

The previous section described the randomness of the parameters characterizing the inelastic response of frames, due to the random variability of the yield strength of the members. It was shown that the statistical parameters describing the inelastic random response can be studied by means of the hybrid Monte Carlo simulation, leading to useful results. However, general conclusions cannot be obtained because the previous analysis applies to a single structural scheme.

In order to grasp all parameters affecting the random inelastic response of steel framed structures, a parametric analysis is required. However, the computational effort is so cumbersome that a compromise has to be found between the necessity to point out the influence of the

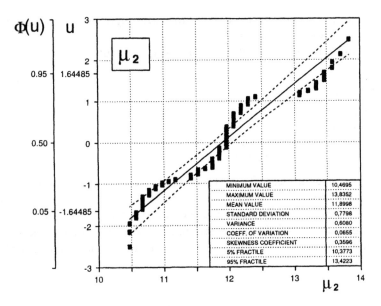

Fig. 9.20 Representation on normal probability graph paper of the inelastic random response: global ductility corresponding to a 10% strength degradation

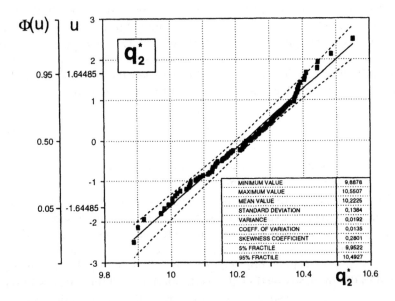

Fig. 9.21 Representation on normal probability graph paper of the inelastic random response: q-factor corresponding to a 10% strength degradation

main parameters and the need to reduce the high computational time required by Monte Carlo simulation.

The following aspects have been investigated [26]:

- the influence of the coefficient of variation (COV) of the yield strength of the structural steel;
- the dependence of the yield strength upon the flange thickness of the frame members;
- the influence of the column-to-beam strength ratio;
- the influence of the structural scheme.

With reference to the COV of the material yield strength, the following values have been considered: 0.025, 0.050, 0.075, 0.100 and 0.150. They cover the whole variability range of standard structural steels.

The dependence of the yield strength upon the flange thickness of frame members has been introduced by means of the linear variation law

$$f_{y,m}(t_f) = -\alpha\, t_f + \beta \tag{9.3}$$

where $f_{y,m}$ is the average value of the yield stress and t_f is the flange thickness.

For inelastic behaviour, the importance of the dependence of the yield stress upon the thickness increases as the value of α increases. The values $\alpha = 2.987$ N/mm^3 and $\beta = 444.2$ N/mm^2, which correspond to Fe510 steel, have been adopted because they correspond to the case of maximum influence (Section 9.2). The ratio between the flexural strength of columns and beams has been introduced by the parameter

$$\rho = \frac{Z_{p,c}}{Z_{p,b}} \tag{9.4}$$

where $Z_{p,c}$ is the plastic section modulus of columns and $Z_{p,b}$ is the plastic section modulus of beams.

This parameter obviously has a decisive influence on the collapse mechanism, which is expected on the basis of the nominal yield strength of members. The influence of this parameter should be investigated until a global type collapse mechanism is attained. However, as the random response of steel frames nominally failing in global mode has been analysed in the previous section, it was decided to limit the analysis initially to the values $\rho = 1$ and $\rho = 1.64$ (to reduce the computational effort). These values, for beams of HE200B shape, correspond to columns of HE200B and HE240B shapes, respectively.

Finally, with reference to the possible influence of the structural scheme, two different types of frame, each with two bays, have been considered: a three-storey frame (FRAME 1) and a six-storey frame (FRAME 2). In both cases the length of the bays is 450 cm and the interstorey height is 300 cm. These frames are shown in Fig. 9.22, together with the corresponding nominal failure modes, i.e. the failure modes determined by assuming a constant value of the member yield strength equal to the nominal value.

Each case of the parametric analysis is defined by the values assumed by the COV and ρ for a given structural scheme, leading to a total of 20 studied cases. For each COV value of material yield strength, 100 frames with a random Gaussian distribution of member yield strength have been generated using the procedure described in Section 9.3.2. The entire parametric analysis required 2000 numerical simulations of the structural inelastic response.

The random frame generation, the inelastic analysis and the evaluation of the parameters characterizing the structural inelastic response have been carried out by means of three specially designed computer programs, which work in sequence.

The characterization of the inelastic response was performed using the parameters described in the previous section. For the available global ductility, however, on the basis of the discussion in Chapter 2 where

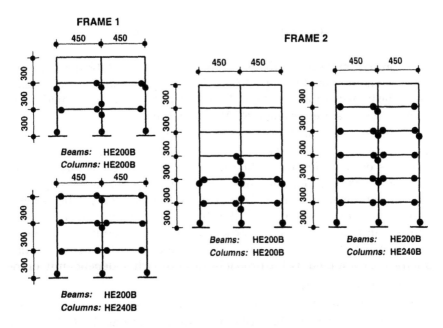

Fig. 9.22   Analysed frames and corresponding nominal failure mode

difficulties related to the quantitative evaluation of the global ductility were described, the introduction of the parameter μ_2^* has been considered useful. This parameter represents the global ductility computed as the ratio between the ultimate displacement corresponding to a 10% strength degradation and that which the frame, assumed indefinitely elastic, should exhibit under the action of horizontal forces corresponding to the ultimate multiplier (Section 9.3.1).

For all cases of the parametric analysis the random response is shown on normal probability graph paper, and has been performed for all parameters which describe structural inelastic behaviour. The confidence interval with a 5% confidence level has also been derived. Use of the confidence interval allows the estimation of fractiles including the scatters due to the fact that, unavoidably, we can only provide an estimate of the values for the average and the standard deviation.

The comparison between the results of the probabilistic analysis, obtained through the Monte Carlo simulation, and those obtained through deterministic analysis seems to be particularly interesting.

The results of probabilistic analyses are mainly represented by the values of the 5% fractiles, while the results of the deterministic analysis correspond to the values obtained through the inelastic response simulation performed by assuming a constant value (for members with the same flange thickness) of the yield stress given by

$$f_y(t_f) = f_{y,k}(t_f) = f_{y,m}(t_f)(1 - 2\,COV) \tag{9.5}$$

A comparison between the probabilistic analysis and the deterministic approach is presented in Figs. 9.23–9.29 for all parameters characterizing the inelastic structural response. As an example, reference is made here only to the six-storey frame (FRAME 2) with $\rho = 1$.

It can be seen that the prediction of the ultimate multiplier of horizontal forces and of the q-factor, by means of the deterministic approach, is always on the safe side leading to values which are less than the corresponding 5% fractiles computed by using the probabilistic approach. Conversely, deterministic analysis is not safe when the plastic redistribution capacity and global ductility have to be computed. In fact, the above comparison shows, in particular, that the deterministic value of the available global ductility is always greater than the corresponding 5% fractile computed by the probabilistic approach, independently of the adopted definition (μ_1, μ_2 and μ_2^*).

Finally, it is worth pointing out that the scatters between the results of deterministic and probabilistic analyses increase as the COV of the material yield strength increases.

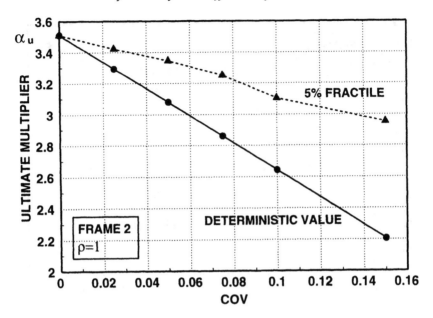

Fig. 9.23 Comparison between probabilistic analysis and deterministic analysis for the ultimate multiplier

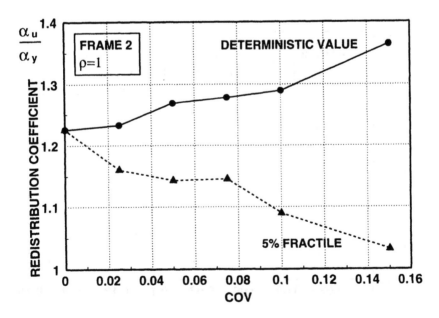

Fig. 9.24 Comparison between probabilistic analysis and deterministic analysis for the plastic redistribution coefficient

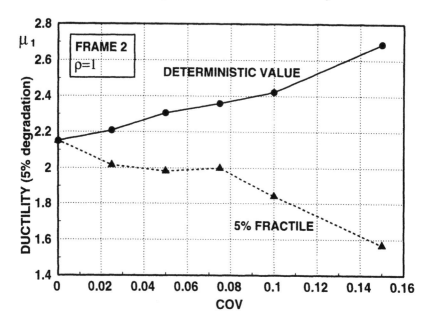

Fig. 9.25 Comparison between probabilistic analysis and deterministic analysis for the global ductility μ_1 corresponding to a 5% strength degradation

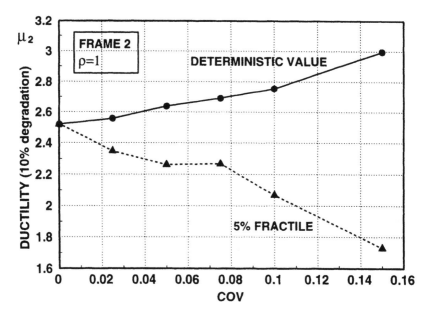

Fig. 9.26 Comparison between probabilistic analysis and deterministic analysis for the global ductility μ_2 corresponding to a 10% strength degradation

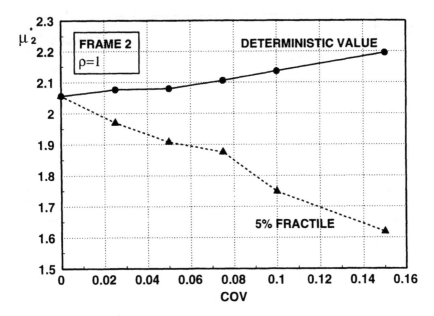

Fig. 9.27 Comparison between probabilistic analysis and deterministic analysis for the global ductility μ_2^*

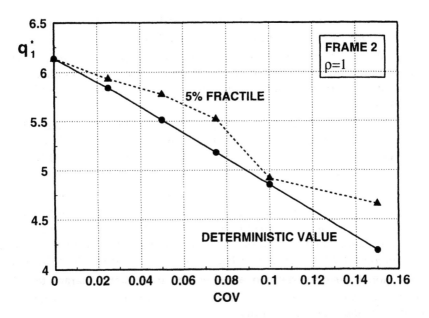

Fig. 9.28 Comparison between probabilistic analysis and deterministic analysis for the q-factor corresponding to a 5% strength degradation

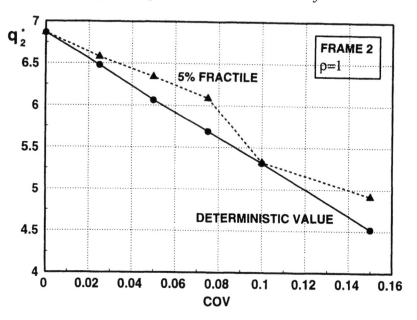

Fig. 9.29   Comparison between probabilistic analysis and deterministic analysis
for the *q*-factor corresponding to a 10% strength degradation

The same results have also been obtained for $\rho = 1.64$ for FRAME 2,
and for both cases $\rho = 1$ and $\rho = 1.64$ for FRAME 1.

Another interesting result comes from a comparison between the different definitions of available global ductility in terms of stability, from
the statistical point of view, expressed by the corresponding coefficient
of variation.

In Fig. 9.30, for FRAME 2 with $\rho = 1$ and $\rho = 1.64$, the values of the
coefficient of variation corresponding to the different definitions of the
available global ductility are given as a function of the COV of the material yield strength. From this representation, it can be observed that
global ductility defined as μ_2^* is the most stable parameter, from the statistical point of view, leading to small values of the corresponding
coefficient of variation. Moreover, it is more stable than the material yield
strength itself, its coefficient of variation being less than for the material
(represented by the bisectrix in Fig. 9.30). Due to the fact that the available global ductility defined as μ_2^* is the most stable parameter, this
parameter can be suggested as the most suitable to provide a measure
of the available global ductility.

This kind of result, i.e. a coefficient of variation less than that of the
material yield strength, which is very important in order to define sig-

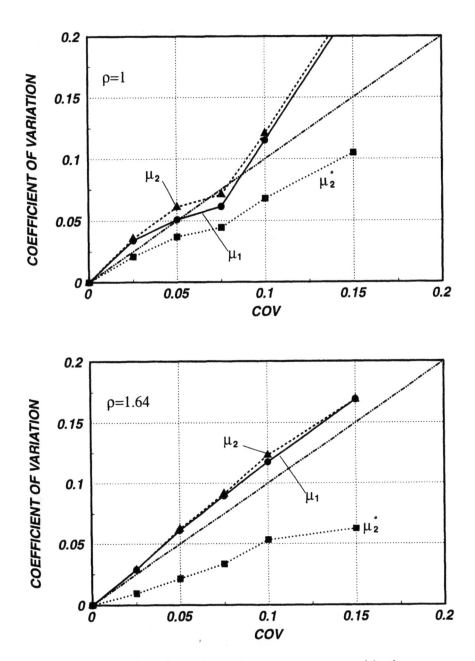

Fig. 9.30   Statistical stability of the various parameters assumed for the
quantitative evaluation of the global ductility for FRAME 2

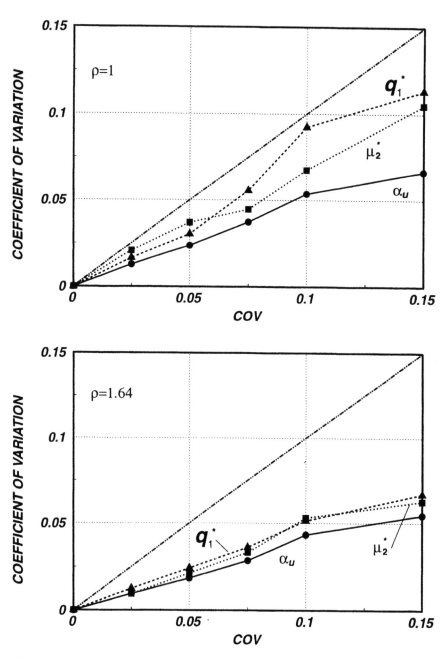

Fig. 9.31   Statistical stability of the main parameters for the characterization of
the inelastic structural responce for FRAME 2

nificant parameters for the structural inelastic response characterization, has also been obtained for the 'modified' q-factor q^*, computed by the energy approach, and for the ultimate multiplier (Fig. 9.31).

A useful representation of the 5% fractiles for parameters characterizing the inelastic response can be obtained by normalizing them with the corresponding value computed for COV = 0. For each parameter the ratio φ between the 5% fractile calculated for COV $\neq$ 0 and the value computed for COV = 0 has been considered. In Figs. 9.32–9.38, the relationship φ versus COV is illustrated for all parameters. It can be seen that the column-to-beam strength ratio ρ always plays an important role, while the influence of the structural scheme is not so significant and, in any case, it is dependent upon the column-to-beam strength ratio.

Additionally, the ratio φ can be interpreted as a 'correction factor' which can be used to predict the 5% fractiles starting from values computed by a deterministic analysis (COV = 0). Furthermore, they are able to characterize the influence of the randomness in material yield strength on the inelastic behaviour of steel framed structures. In particular, it is most important to be aware, from the seismic point of view, that the increase of the coefficient of variation of the member yield strength can produce a significant reduction of the available global ductility. This reduction can reach about 40%, which is far from negligible.

These results have confirmed that the effects of randomness in material yield strength increase as the COV increases. In addition they have pointed out that the importance of this influence depends also upon both the column-to-beam strength ratio and the structural scheme.

The increase of the coefficient of variation of material yield strength can produce a significant reduction of the available global ductility, which is non-negligible from the seismic design point of view.

Finally, it has shown that the use of the probabilistic approach represents a useful tool with which to define stable parameters for characterizing the inelastic structural response of seismic-resistant steel frames.

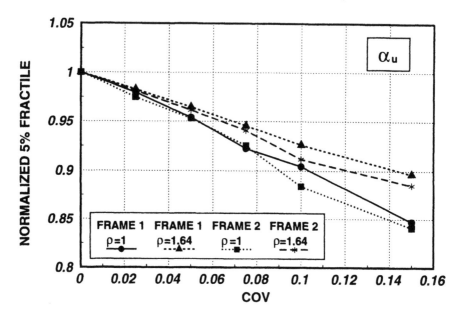

Fig. 9.32   'Correction factor' for the ultimate multiplier

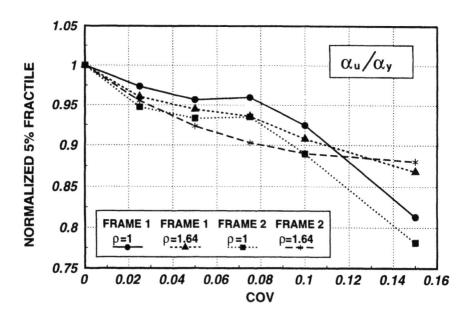

Fig. 9.33   'Correction factor' for the plastic redistribution coefficient

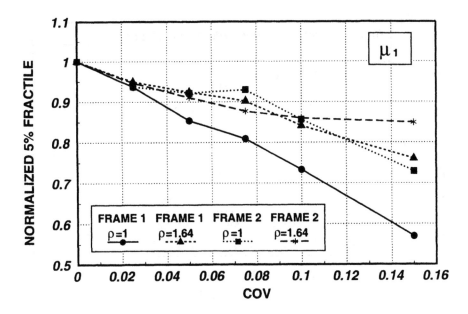

Fig. 9.34 'Correction factor' for the global ductility corresponding to a 5% strength degradation

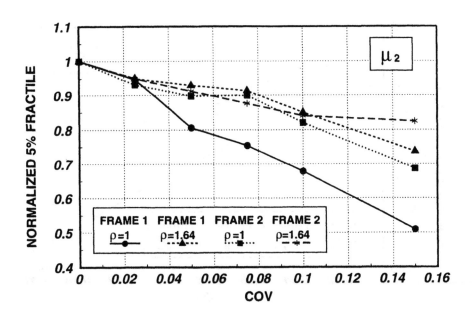

Fig. 9.35 'Correction factor' for the global ductility corresponding to a 10% strength degradation

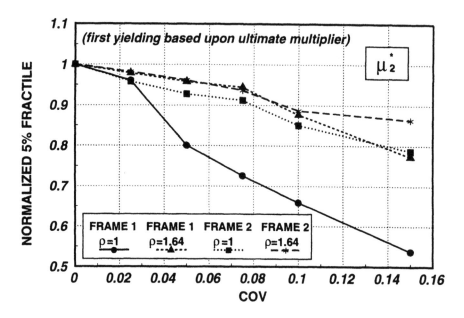

Fig. 9.36   'Correction factor' for the global ductility μ_2^* corresponding to a 10% strength degradation

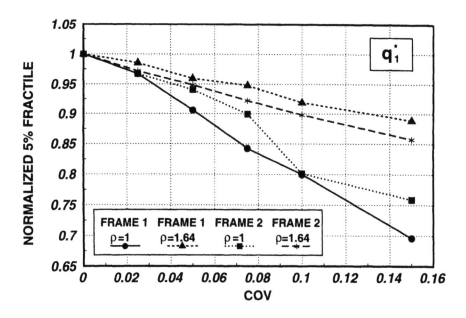

Fig. 9.37   'Correction factor' for the q-factor corresponding to a 5% strength degradation

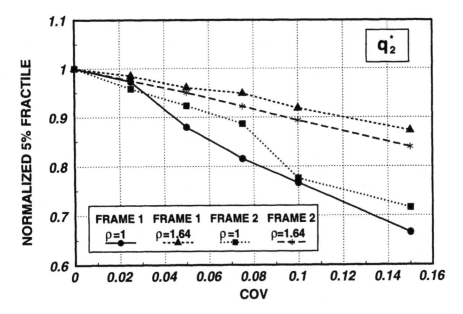

Fig. 9.38   'Correction factor' for the *q*-factor corresponding to a 10% strength degradation

9.5   REFERENCES

[1] *Commission of the European Communities:* 'Eurocode 8: European Code for Seismic Regions', 1988.

[2] *F.M. Mazzolani, E. Mele, V. Piluso:* 'Statistical Features of Mechanical Properties of Structural Steels', ECCS Document TC13.26.90, 1990.

[3] *F.M. Mazzolani, E. Mele, V. Piluso:* 'Caratterizzazione Statistica degli Acciai da Carpenteria per il Controllo della Duttilità Strutturale', Giornate Italiane della Costruzione in Acciaio, C.T.A., Ottobre 1991.

[4] *Commission of the European Communities:* 'Eurocode 3: Design of Steel Structures', 1988.

[5] *H. Yamanouchi, B. Kato, H. Aoki:* 'Statistical Features of Mechanical Properties of Current Japanese Steels', [TC13.17.90]

[6] *H. Kuwamura, B. Kato:* 'Effect of Randomness in Structural Members' Yield Strength on the Structural Systems' Ductility', Journal of Constructional Steel Research, N.13, 1989.

[7] *ECCS (European Convention for Constructional Steelwork):* 'European Recommendations for Steel Structures in Seismic Zones', 1988.

[8] *H. Kuwamura, M. Sasaki:* 'Control of Random Yield-Strength for Mechanism-Based Seismic Design', Journal of Structural Engineering, ASCE, Vol. 116, pp. 98–110, 1990.

[9] *A.S. Elnashai, M. Chryssanthopoulos:* 'Effect of Random Material Vari-

ability on Seismic Design Parameters of Steel Frames', Earthquake Engineering and Structural Dynamics (under review), ECCS Document TC13.7.90, 1990.

[10] *F.M. Mazzolani, E. Mele, V. Piluso:* 'Analisi Statistica del Comportamento Inelastico di Telai in Acciaio con Resistenza Casuale', V Convegno Nazionale, L'Ingegneria Sismica in Italia, Palermo, 29 Settembre–2 Ottobre 1991.

[11] *J.C. Anderson, R.P. Gupta:* 'Earthquake Resistant Design of Unbraced Frames', Journal of Structural Division, ASCE, 98 (11), pp. 2523–2539, 1972.

[12] *R.W. Clough, J. Penzien:* 'Dynamics of Structures', McGraw Hill Book Company, International Student Edition, 6th printing, Singapore, pp. 597–602, 1982.

[13] *R. Park, T. Paulay:* 'Reinforced Concrete Structures', John Wiley and Sons, New York, pp. 545–609, 1975.

[14] *W.R. Walpole, R. Shepherd:* 'Elasto-Plastic Seismic Response of Reinforced Concrete Frames', Journal of Structural Division, ASCE, 95 (10), pp. 2031–2055, 1969.

[15] *G.H. Powell, R. Allahabadi:* 'Seismic Damage Prediction by Deterministic Methods: Concepts and Procedures', Earthquake Engineering and Structural Dynamics, Vol. 16, pp. 719–734, 1988.

[16] *I. Mitani, M. Makino:* 'Post Local Buckling Behaviour and Plastic Rotation Capacity of Steel Beam-Columns', 7th World Conference on Earthquake Engineering, Istanbul, 1980.

[17] *B. Kato, H. Akiyama:* 'Ductility of Members and Frames Subjected to Buckling', ASCE, May, New York, 1981.

[18] *F.M. Mazzolani, C.A. Guerra, V. Piluso:* 'Evaluation of the q-Factor in Steel Framed Structures: State-of-Art', Ingegneria Sismica, N.2, 1990.

[19] *M. Como, G. Lanni:* 'Aseismic Toughness of Structures', Meccanica, N18, pp. 107–114, 1983.

[20] *A.M. Nafday, R.B. Corotis, J.L. Cohon:* 'System Reliability of Rigid Plastic Frames', Reliability and Risk Analysis in Civil Engineering, Vol. I, N.C. Lind, ed., of Waterloo, Ontario, Canada, pp.119–126, 1987.

[21] *F. Casciati, L. Faravelli:* 'Elasto-Plastic Analysis of Random Structures by Simulation Methods', Simulation of Systems '79, North-Holland, Amsterdam, The Netherlands, pp. 497–508, 1980.

[22] *Reuven Y. Rubistein:* 'Simulation and the Monte Carlo Method', Wiley Series in Probability and Mathematical Statistics, 1981.

[23] *H. Kuwamura, T.V. Galambos:* 'Reliability Analysis of Steel Building Structures Under Earthquakes', Structural Engineering Report 86–01, University of Minnesota, Minneapolis, May, 1986.

[24] *F.M. Mazzolani, E. Mele, V. Piluso:* 'On the Effect of Randomness of Yield Strength in Steel Framed Structures Under Seismic Loads', ECCS Document TC13.01.91, 1991.

[25] *G. Augusti, A. Baratta, F. Casciati:* 'Probabilistic Methods in Structural Engineering', Chapman and Hall, London, New York, 1984.

[26] *F.M. Mazzolani, E. Mele, V. Piluso:* 'The Seismic Behaviour of Steel Frames with Random Material Variability', X World Conference on Earthquake Engineering, Madrid, July, 1992.

10

Influence of claddings

10.1 INTRODUCTION

The design method known as stressed skin design was proposed for steel structures during the 1970s and is within the field of activity of ECCS. The principle consisted of considering the capacity of claddings made of light-gauge steel panels to contribute to the behaviour of the structure as a whole, acting as a diaphragm, i.e. with a 'skin-effect'.

It is well known that the interaction between cladding panels and a frame structure represents difficulties when forecasting the seismic behaviour of steel structures. The stiffening effect of the panels is always present and influences the behaviour of the structure (even if it is generally neglected due to the difficulties of taking it into account).

Although cladding panels made using modern steel technology are very light, they participate in the seismic response of the structure and can suffer considerable damage when absorbing the earthquake input energy.

The computation of this contribution is important to either preserve the panels themselves from damage, or to assess the actual bearing capacity of the whole structural system, with an important saving in structural steel weight. Therefore, a seismic-resistant 'all-steel' structure should be designed taking into account the presence of cladding panels by using a reliable methodology and on the basis of suitable design criteria.

This chapter considers the diaphragm action of trapezoidal sheet panels, with the purpose of evaluating the load carrying capacity of the whole structure including the 'skin effect' of claddings.

The degree of interaction between cladding panels and the main structure can be used in order to classify structures [1].

a. The main structure is designed to resist vertical and seismic loads, while the panels contribute to the serviceability limit state only, when checking maximum sways and storey drifts.

b. The whole structure is designed so that panels and frames together have to resist vertical and seismic loads.

c. The frame has the task of resisting vertical loads only, while horizontal forces due to earthquake or wind are supported by cladding panels.

As a consequence, the choice of connections should be made according to the above design criteria. Referring to beam-to-column joints, case **a** requires rigid connections and case **b** can accept semi-rigid connections. As the bracing effect can be guaranteed by claddings, pin-ended connections can also be used in case **c**. This provides maximum economy, by reducing the structural weight and also the manufacturing and erecting costs. In spite of these advantages, examples of c-type structures are not yet found in seismic zones, probably due to lack of specific knowledge and experience.

As a first contribution to this development, the seismic behaviour of pin-jointed structures with bracing panels has been examined [2] as an economical structural solution for buildings in low-seismicity zones. The important influence of the connecting system on the dynamic behaviour of the structure has also been pointed out [3], by considering that the transfer of shear forces from the structure to the panels is guaranteed by means of different technological systems which use bolts, rivets, screws or welds.

The use of collaborating cladding panels is also a useful tool with which to upgrade an existing building as required by the transfer of a given geographic area to a new seismic zone.

Finally it is of interest that corrugated sheet panels of the third-generation type, such as TRP200, have already been used as floor bracings in industrial one-storey buildings constructed in seismic zones [4].

10.2 DIAPHRAGM CONNECTING SYSTEM

The greatest diaphragm strength and stiffness is obtained when each panel of sheeting is fastened to all four edge members. Therefore, this method of fastening has to be preferred, whenever possible. In particular, this fastening method requires that all edge members are at the same level.

When purlins pass over rafters so that the members are not at the same level, the use of 'shear connectors' located above the rafters is suggested, as shown in Fig. 10.1, to bring the members up to the same level and allow fastening on all four sides. (This is the case for roofs, rather than for trapezoidal sheet panels as used for cladding in multistorey frames.)

The typical diaphragm shown in Fig. 10.1 is applicable to sheeting which spans parallel to the direction of the shear force. Such a panel is constituted by the following components:

- individual lengths of profiled steel sheeting or decking;
- purlins or secondary supporting members normal to the direction of the span of the sheeting;
- rafters or main beams supporting the purlins;
- seam fasteners between individual sheet widths;
- fasteners between sheeting and purlins;
- fasteners between sheeting and shear connectors.

The above assemblage is typical of roofs. For multistorey moment-re-sisting frames infilled by trapezoidal sheet panels, the connecting system is always simplified because the edge members are always at the same level. Usually, the width of the diaphragm is greater than the height, so that the corrugations run vertically in the direction of the shorter span (Fig. 10.2).

The connecting system has to be able to transfer the shear forces from the frame to the diaphragm, so that the sheeting is in a state of pure shear. As the corrugations are vertically oriented, the ends of the sheeting require lateral support from the primary beams and would normally be

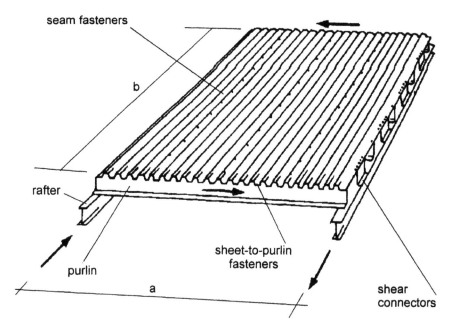

Fig. 10.1   Typical assemblage of a light-gauge steel diaphragm for roofs

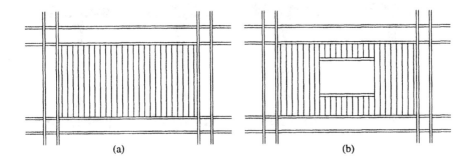

Fig. 10.2 Trapezoidal sheet infill panels for multistorey frames

fixed to them. Unfortunately, a direct fixing cannot be adopted, because considerable relative vertical displacements are likely to occur between the primary beams above and below the panel. As a consequence, a direct fixing could lead to vertical loads on the diaphragm which can cause the failure of either the diaphragm or its connections.

In order to avoid this problem, the diaphragm can be connected to the supporting structure at its corners only, relying on longitudinal members spanning between the columns to support the corrugated infill panel [5]. As this solution seems to be needlessly expensive, alternative connecting systems have been proposed [6, 7]. These alternative solutions are based on the incorporation of additional elements able to provide a degree of flexibility at the top and bottom of the panel (Fig. 10.3).

The top and bottom edge members are perpendicular to the direction of the sheet corrugations, playing the role of purlins; the side members play the role of the rafters. Shear connectors are not present and the sheet is directly fastened to the side members. In the following section, design expressions for evaluating the shear strength and the shear stiffness of corrugated sheet panels will be given with reference to the specific case of infilled MR frames, where the shear force is always perpendicular to the direction of the sheet corrugations (Fig. 10.2). A more detailed description of the European method for evaluating the shear strength and the shear stiffness of trapezoidal sheet panels, including the case for roofs, is provided elsewhere [8–10].

10.3 PREDICTION OF MONOTONIC BEHAVIOUR

The prediction of the shear behaviour of cladding panels made of trapezoidal sheets is carried out mainly by analytical methods or by the finite element method.

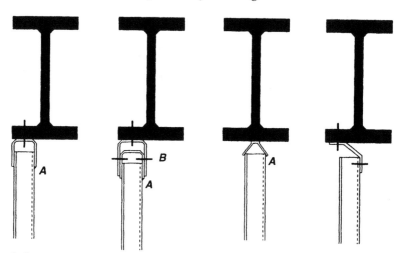

A=intermittent welds (one side only) B=bolt in vertically slotted hole

Fig. 10.3   Connecting systems for trapezoidal sheet infill panels

In this section the analytical approach is described, while the numerical approach based on the finite element method is examined in the next section.

Considerable research effort has been devoted to the provision of reliable analytical methods for predicting the ultimate shear resistance and the shear stiffness of trapezoidal sheet panels. This work has led to different methods for predicting the shear diaphragm behaviour under monotonic loads: the SDI method [11], the Triservice method [12], the West Virginia University method [13] and the European method [14].

Only the European method is described here. A comparison of the features of the different methods is presented in reference [15].

The analytical method suggested by the European Recommendations for the stressed skin design of steel structures is based on the extensive research activity performed in this field by Bryan and Davies [8–10, 16, 17], who developed design expressions for evaluating the shear strength and the shear stiffness of diaphragms made of trapezoidal sheets.

The design expressions given in this section refer to the direction perpendicular to the corrugations, because this is the typical case for multistorey MR frames infilled by light-gauge steel panels. It is seen that the dimension a is always the length of the shear panel perpendicular to the corrugations, and that the dimension b is always the depth of the shear panel parallel to the corrugations.

Design expressions are given below for evaluating the shear strength of a single diaphragm panel. They correspond to different failure modes so that the ultimate resistance of the panel is governed by the weakest component.

a. Tearing of a line of seam fasteners

$$V_{ult} = \frac{a}{b}\left(n_s\, F_s + \frac{\beta_1}{\beta_3} n_p\, F_p\right) \tag{10.1}$$

where:

- n_s is the number of seam fasteners per side lap excluding the supporting edge members (acting as purlins);
- F_s is the design strength of individual seam fasteners;
- n_p is the number of purlins ($n_p = 2$ in this specific case);
- F_p is the design strength of individual sheet-to-purlin fasteners;
- β_1 is a factor accounting for the number of sheet-to-purlin fasteners for sheet width. It depends on the force distribution in the sheet-to-purlin fasteners along the member. Its values are given in Table 10.1 for the two cases of sheeting (seam fasteners in the crests) and decking (seam fasteners in the troughs);
- β_3 is a coefficient given by

$$\beta_3 = n_f - \frac{1}{n_f} \tag{10.2}$$

for sheeting (where n_f is the number of sheet-to-purlin fasteners for sheet width), while $\beta_3 = 1.0$ for decking.

b. Tearing of a line of fasteners connecting the sheet to the edge members (side fasteners)

$$V_{ult} = \frac{a}{b}\, n_{sc}\, F_{sc} \tag{10.3}$$

where n_{sc} is the number of side fasteners, and F_{sc} is the design strength of individual side fasteners. Equation (10.3) is valid provided that all four sides of the panel are fastened.

c. Failure in the top and bottom edge fasteners
In order to avoid the brittle failure of the sheet-to-purlin fasteners a 40% reserve of safety is allowed. Therefore, it should be checked that

$$\frac{0.6\, a\, F_p}{p} \geq V^* \tag{10.4}$$

where V^* is the design shear capacity of the diaphragms given as the

Influence of claddings

Table 10.1 Values of the factor β_1 for sheeting and decking as a function of the total number of fasteners per sheet width n_f

n_f	Factor β_1	
	sheeting	decking
2	0.13	1.0
3	0.30	1.0
4	0.44	1.04
5	0.58	1.13
6	0.71	1.22
7	0.84	1.33
8	0.97	1.45
9	1.10	1.56
10	1.23	1.68

CASE 1: SHEETING (seams at crests)

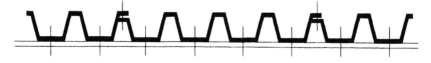

CASE 2: DECKING (seams in troughs)

smallest of the values provided by equations (10.1) and (10.3), and p is the pitch of the sheet-to-purlin fasteners.

d. Shear buckling of the sheeting
In order to avoid the possibility of failure due to the shear buckling of the sheeting a 25% reserve of safety is allowed in the expressions for the capacity against shear buckling. It should be checked that

$$\frac{28.8a}{b^2} D_x^{1/4} D_y^{3/4} > V^* \qquad (10.5)$$

where

$$D_x = \frac{E\,t^3\,d}{12\,(1 - v^2)\,u} \tag{10.6}$$

$$D_y = \frac{E\,I}{d} \tag{10.7}$$

and where

- t is the net sheet thickness excluding galvanizing and coating;
- d is the pitch of the corrugations;
- u is the perimeter length of a single corrugation;
- I is the moment of inertia of a single corrugation about its neutral axis.

Equation (10.5) is valid for fasteners in every corrugation. In the case of fasteners in alternate corrugations, the left-hand term has to be halved.

e. Failure of the edge members under tension or compressions
The calculation of the axial force in the edge members also has to take into account internal actions due to vertical loads. The edge members have to be sized with a 25% reserve of safety, expecially against the buckling of the compression members.

f. Gross distortion or collapse of the profile at the end of the sheeting
In order to avoid this failure mode the following limitations on shear in a panel should be observed:

$$\frac{0.9\,t^{1.5}\,a\,f_y}{d^{0.5}} > V^* \tag{10.8}$$

in the case of fasteners in every corrugation, and

$$\frac{0.3\,t^{1.5}\,a\,f_y}{d^{0.5}} > V^* \tag{10.9}$$

in the case of fasteners in alternate corrugations.

The sheet tearing at the seam fasteners or shear connector fasteners (failure modes **a** and **b**) is a gradual ductile failure. On the contrary, the other failure modes are not able to provide sufficient ductility and, therefore, are undesirable. For this reason a reserve of safety has been incorporated into the design expressions for these failure modes in order to ensure that they have a considerably greater strength than the lesser of the ones corresponding to the failure modes **a** and **b**.

By considering the strength of all diaphragm components, it is easy to establish the weakest link in the chain and, as a consequence, to recognize the connecting system details to be modified for improving the diaphragm strength.

The total shear flexibility is calculated by summing the following component flexibilities listed below.

a. Flexibility due to sheet distortion

$$c_{11} = \frac{a\, d^{2.5}\, K}{E\, t^{2.5}\, b^2} \tag{10.10}$$

The sheeting constant K takes the value K_1 for sheeting fastened in every corrugation (Table 10.2) or K_2 for sheeting fastened in alternate corrugations (Table 10.3). Equation (10.10) is valid provided that the elementary sheets constituting the panel are able to cover the full depth of the panel.

b. Flexibility due to shear strain in the sheet

$$c_{12} = \frac{2\,a\,(1+v)\,(1+2h/d)}{E\,t\,b} \tag{10.11}$$

c. Flexibility due to slip in the fasteners connecting the sheet to the top and bottom members perpendicular to the corrugations

$$c_{21} = \frac{2\,a\,s_p\,p}{b^2} \tag{10.12}$$

where s_p is the slip per sheet-to-purlin fastener per unit load.

d. Flexibility due to slip in the seam fasteners

$$c_{22} = \frac{2\,s_s\,s_p\,(n_{sh}-1)}{2\,n_s\,s_p + \beta_1\,n_p\,s_s} \tag{10.13}$$

where s_s is the slip per seam fastener per unit load and n_{sh} is the number of sheet widths per panel.

e. Flexibility due to slip in the side fasteners (edge member fasteners)

$$c_{23} = \frac{2\,s_{sc}}{n_{sc}} \tag{10.14}$$

where s_{sc} is the slip per sheet-to-shear connector fastener per unit load. Equation (10.14) is valid provided that all four sides of the panel are fastened.

As the previous component flexibilities are derived for a shear force acting parallel to the corrugations, taking into account that for infill panels for MR frames the shear force acts in the direction perpendicular to that of the corrugations, the total shear flexibility of the panel is given by

Table 10.2   Values of K_1 for fasteners in every trough

θ	h/d	\(d'/d\) 0.1	0.2	0.3	0.4	0.5	0.6	0.7	0.8	0.9
15°	0.1	0.017	0.031	0.040	0.041	0.041	0.047	0.066	0.115	0.241
	0.2	0.062	0.102	0.118	0.115	0.113	0.134	0.209	0.403	
	0.3	0.139	0.202	0.218	0.204	0.200	0.254	0.440	0.945	
	0.4	0.244	0.321	0.325	0.293	0.294	0.414	0.796		
	0.5	0.370	0.448	0.426	0.371	0.396	0.636	1.329		
	0.6	0.508	0.568	0.508	0.434	0.513	0.941			
	0.7	0.646	0.668	0.561	0.483	0.664	1.349			
	0.8	0.768	0.735	0.578	0.527	0.861				
20°	0.1	0.018	0.032	0.039	0.039	0.039	0.046	0.066	0.111	0.276
	0.2	0.068	0.101	0.111	0.106	0.104	0.131	0.221	0.452	
	0.3	0.148	0.193	0.194	0.174	0.176	0.255	0.492		
	0.4	0.249	0.289	0.267	0.230	0.259	0.444	0.931		
	0.5	0.356	0.372	0.315	0.270	0.364	0.725			
	0.6	0.448	0.420	0.326	0.303	0.512				
	0.7	0.509	0.423	0.301	0.356					
	0.8	0.521	0.372	0.259	0.413					
25°	0.1	0.019	0.032	0.038	0.038	0.038	0.045	0.068	0.126	0.313
	0.2	0.072	0.099	0.103	0.095	0.095	0.129	0.236	0.513	
	0.3	0.151	0.178	0.166	0.144	0.160	0.268	0.557		
	0.4	0.238	0.244	0.204	0.176	0.247	0.494			
	0.5	0.306	0.272	0.203	0.204	0.376				
	0.6	0.333	0.248	0.172	0.241					
	0.7	0.300	0.174	0.142						
	0.8	0.204	0.081							
30°	0.1	0.020	0.032	0.037	0.036	0.036	0.044	0.070	0.133	
	0.2	0.075	0.095	0.094	0.084	0.087	0.133	0.256		
	0.3	0.148	0.157	0.135	0.116	0.152	0.291			
	0.4	0.208	0.186	0.139	0.139	0.253				
	0.5	0.226	0.161	0.112	0.176					
	0.6	0.180	0.089	0.093						
	0.7	0.077								

Table 10.2 (continued)

θ	h/d	\multicolumn{9}{c}{d'/d}
		0.1	0.2	0.3	0.4	0.5	0.6	0.7	0.8	0.9
35°	0.1	0.021	0.032	0.036	0.034	0.034	0.043	0.072	0.142	
	0.2	0.076	0.089	0.083	0.072	0.082	0.137	0.281		
	0.3	0.137	0.130	0.102	0.093	0.151				
	0.4	0.162	0.119	0.082	0.120					
	0.5	0.123	0.059							
	0.6	0.032								
40°	0.1	0.023	0.032	0.034	0.032	0.032	0.043	0.075	0.155	
	0.2	0.075	0.081	0.070	0.060	0.077	0.146			
	0.3	0.116	0.096	0.068	0.078					
	0.4	0.100	0.053	0.048						
	0.5	0.024								
45°	0.1	0.024	0.031	0.032	0.029	0.030	0.043	0.079		
	0.2	0.071	0.069	0.056	0.050	0.073				
	0.3	0.086	0.057	0.041						
	0.4	0.032								

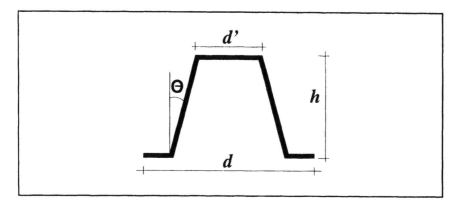

$$c = \frac{b^2}{a^2}(c_{11} + c_{12} + c_{21} + c_{22} + c_{23}) \tag{10.15}$$

Typical values of the design resistance and of the slip per unit load are given in Table 10.4 for the most common types of connectors.

Table 10.3 Values of K_2 for fasteners in alternate troughs

θ	h/d	\(d'/d\) 0.1	0.2	0.3	0.4	0.5	0.6	0.7	0.8	0.9
15°	0.1	0.093	0.142	0.188	0.231	0.271	0.313	0.364	0.448	0.682
	0.2	0.325	0.458	0.586	0.707	0.824	0.953	1.140	1.523	
	0.3	0.703	0.942	1.174	1.393	1.610	1.874	2.316	3.411	
	0.4	1.237	1.602	1.953	2.285	2.624	3.089	3.981		
	0.5	1.937	2.443	2.926	3.379	3.869	4.640	6.256		
	0.6	2.778	3.428	4.058	4.664	5.366	6.581			
	0.7	3.692	4.488	5.273	6.081	7.138	8.902			
	0.8	4.648	5.570	6.516	7.628	9.190				
20°	0.1	0.096	0.144	0.190	0.232	0.273	0.315	0.368	0.459	0.680
	0.2	0.339	0.472	0.597	0.716	0.832	0.966	1.177	1.659	
	0.3	0.743	0.978	1.204	1.416	1.633	1.927	2.481		
	0.4	1.317	1.673	2.009	2.325	2.679	3.246	3.840		
	0.5	2.075	2.559	3.011	3.436	3.993	4.969			
	0.6	3.006	3.625	4.194	4.752	5.588				
	0.7	4.042	4.789	5.494	6.272					
	0.8	5.122	6.013	6.883	7.861					
25°	0.1	0.098	0.147	0.192	0.234	0.274	0.317	0.373	0.475	0.665
	0.2	0.355	0.485	0.609	0.725	0.840	0.983	1.226	1.566	
	0.3	0.784	1.015	1.233	1.437	1.660	2.000	2.589		
	0.4	1.398	1.740	2.057	2.359	2.753	3.427			
	0.5	2.205	2.659	3.064	3.490	4.114				
	0.6	3.199	3.752	4.218	4.796					
	0.7	4.318	4.941	5.480						
	0.8	5.487	6.132							
30°	0.1	0.101	0.150	0.194	0.236	0.276	0.319	0.378	0.495	
	0.2	0.372	0.500	0.621	0.734	0.850	1.005	1.298		
	0.3	0.827	1.051	1.260	1.456	1.697	2.098			
	0.4	1.477	1.801	2.092	2.393	2.830				
	0.5	2.319	2.727	3.075	3.499					
	0.6	3.320	3.738	4.041						
	0.7	4.378								

Table 10.3 (continued)

θ	h/d	d'/d								
		0.1	0.2	0.3	0.4	0.5	0.6	0.7	0.8	0.9
35°	0.1	0.105	0.153	0.197	0.238	0.278	0.322	0.385	0.525	
	0.2	0.390	0.516	0.634	0.744	0.862	1.035	1.329		
	0.3	0.872	1.088	1.284	1.476	1.741				
	0.4	1.553	1.849	2.105	2.412					
	0.5	2.400	2.713							
	0.6	3.278								
40°	0.1	0.109	0.156	0.200	0.241	0.280	0.325	0.394	0.569	
	0.2	0.411	0.538	0.647	0.753	0.878	1.077			
	0.3	0.919	1.122	1.301	1.496					
	0.4	0.614	1.859	2.085						
	0.5	2.376								
45°	0.1	0.114	0.160	0.203	0.243	0.282	0.329	0.409		
	0.2	0.434	0.553	0.661	0.764	0.899				
	0.3	0.965	1.148	1.306						
	0.4	1.634								

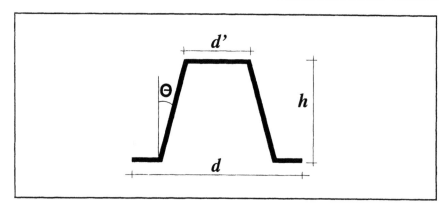

10.4 NUMERICAL SIMULATION

10.4.1 General

The most advanced method to predict the effective shear response of diaphragms, as an alternative to full scale experimental tests, is

Table 10.4 Strength and slip values of fasteners

(a) Sheet-to-edge member fasteners

Fastening	Washer type	Diameter (mm)	Design strength (kN per mm thickness)	Slip (mm/kN)
Screws	Steel washer head	6.3	6.0	0.15
		5.5	5.0	0.15
	Neoprene	6.3	5.0	0.35
		5.5	4.0	0.35
Fired pins	–	3.7–4.5	5.0	0.10

(b) Seam fasteners (no washer)

Fastening	Diameter (mm)	Design strength (kN per mm thickness)	Slip (mm/kN)
Screws	4.1–4.8	2.5	0.25
Steel or Monel blind rivets	4.8	2.8	0.30

computer-assisted finite element analysis. Test procedures have been standardized by the American Iron and Steel Institute (AISI), but their application is very expensive and time consuming. In addition, the results are applicable only to specific combinations of panels, supporting members and connectors.

The essential feature of the finite element method is that a continuous body is divided into a large number of small elements arranged in such a way as to approximate the geometry of the structure. This method is particularly effective for shear diaphragms. The system consists of a large number of individual members, connected at discrete points; for this reason the finite element representation is more realistic than that based on an equivalent continuum. Therefore, such a system can be idealized as a plane stress type problem in which each individual sub-panel becomes a finite element with nodes at points of interconnection with other sub-panels and the supporting frame. Nevertheless, the definition of this model is quite complex and particular care is required for modelling the connecting system and panel geometry [18].

Three different elementary structural systems have to be modelled: the edge members, the panel sheets and the connections. Edge members at the perimeter of the diaphragm are represented as one-dimensional elements, able to resist axial loads as well as bending. If the panel is a continuous flat sheet, it can be modelled as a uniform isotropic elastic plate loaded in its own plane and with elastic constants equal to that of the base material. For corrugated trapezoidal sheets, the behaviour of the panel is interpreted better as an orthotropic plate with different extensional moduli in the two principal directions. In this case, the corrugated sheet is modelled by an equivalent orthotropic thin plate, of uniform thickness equal to that of the actual sheet. Five elastic constants are necessary to characterize its behaviour: two extensional moduli in the direction parallel (1) and perpendicular (2) to the corrugations, the effective shear modulus and two Poisson's coefficients in the same directions as (1) and (2). All these elastic constants have to be derived considering not only the properties of the base material, but also the geometry of the cross-section. In particular, the shear modulus is considerably influenced by the warping effect at the end of the sheeting and depends on the type of edge fasteners, if fastened at discrete locations. With regard to the evaluation of the elastic constants, theoretical formulations are suggested in the technical literature as an alternative to experimental tests. Comparison between the results provided by these formulations and those obtained by means of finite element computations has shown that the theoretical approach strongly underestimates the stiffness of the panel in the transverse direction.

With regard to the connecting system, in both isotropic and orthotropic sheets, a regular discretization for the equivalent plate is adopted, in which nodes are placed according to the position of the connectors. The displacements due to connector slip play an important role in the response of the shear diaphragm. As a consequence, particular care is required for modelling these elements. In the finite element model, the connections are idealized by means of two-directional springs, which interconnect the dual nodal points placed at any connector location. For spring elements, a non-linear force–displacement relationship has to be assumed in order to account for the plastic redistribution of forces among the various connectors. In the whole system a non-linear behaviour is introduced for the connection system only according to experimental investigations, which have shown that this is the only important source of mechanical non-linearity [19].

A typical idealized panel for the application of the method is shown in Fig. 10.4.

10.4.2 Influence of the connecting system

In a shear diaphragm, there are essentially two systems of connectors: seam fasteners between sub-panels and panel-to-beam edge fasteners. A finite element analysis of an ad hoc panel has been performed, using the advanced non-linear code ABAQUS, in order to study the influence of the

connecting system on the shear flexibility [20, 21]. The sheet considered in
the analysis is an isotropic elastic plate of uniform 5 mm thickness; the
connectors consist of two-dimensional spring linkage and the edge
members are rigid elements with pin-ended connections. The response of
the spring elements has been characterized by assuming the experimental
load–slip relationship [19] shown in Fig. 10.5. Several numerical
applications have been carried out in order to calculate the shear
flexibility of the diaphragm by summing the single-component
contributions (Fig. 10.6). The first application is devoted to a single
sub-panel by considering the linear behaviour of spring elements. First,
the shear strain of the plate S is evaluated by assuming an infinite value
for the elastic constants of every spring used to model the panel-to-beam
connections. Then, the flexibility due to the slip in the fasteners in the two
directions of the plane is analysed by three different steps. In the first,
$S + O$, an infinite stiffness value for spring elements in the longitudinal
direction is assumed. For springs in the transverse direction, the initial
stiffness of the load–slip relationship shown in Fig. 10.5 is taken to be $K =$
12440 N/mm.

In the second step, $S + V$, relative displacements between panel and
the beams have been allowed in the longitudinal direction only, by assigning a stiffness of 12440 N/mm to the springs along this direction.
Finally, the spring elements with the same finite stiffness in both directions have been considered, $S + O + V$. The results are summarized in
Fig. 10.6. In the histograms the panel flexibility has been normalized with
respect to the value of the shear strain flexibility of the plate without
connectors S. In the same figure, results for the diaphragm formed by

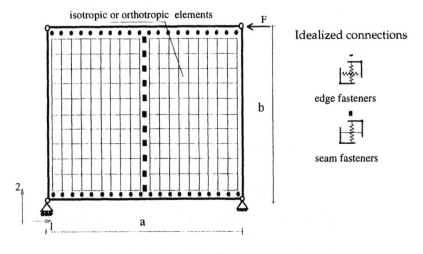

Fig. 10.4   Typical mesh for simulating light-gauge steel panels

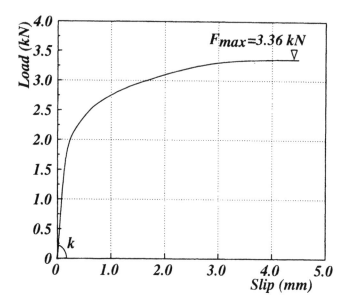

Fig. 10.5   Experimental load–slip relationship of fasteners

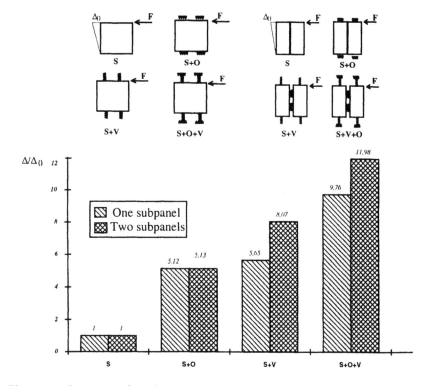

Fig. 10.6   Summary of results

two sub-panels are also shown. In this case, the same fasteners in the longitudinal direction have stiffness $K = 12440 \, \text{N/mm}$, while in the transverse direction an infinite value for K has been assumed to simulate the absence of relative displacements due to the presence of the constraint produced by adjacent sheets. The comparison between the two histograms emphasizes the flexibility contribution due to the seam fasteners.

The shear flexibility of the whole panel under the hypothesis of nonlinear behaviour of the connections has also been investigated, by assuming the load–slip relationship shown in Fig. 10.5. The comparison between linear and non-linear curves (Fig. 10.7) demonstrates that the inelastic global response becomes significant at about 40% of the failure load. In each case, the analyses show that the contribution of the connecting system to the shear flexibility of the panel is always prevalent.

The diaphragm failure load has been defined as the value of the external shear force corresponding to the collapse of the spring connector. In the linear analysis, this occurs for an external shear load of 32.05 kN, which allows the ultimate strength in the corner panel-to-beam connector to be reached. However, when a non-linear approach is followed, a significant increment (up to 52.5 kN) in the diaphragm shear strength is observed, due to the plastic redistribution of stresses along the connectors.

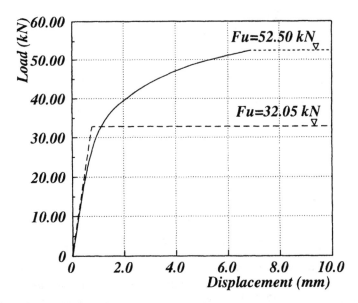

Fig. 10.7   Load–slip relationship

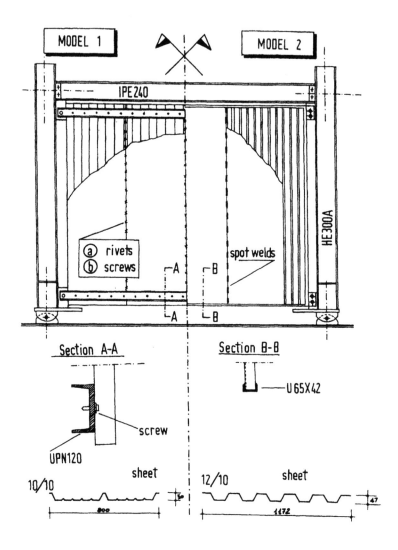

Fig. 10.8   Tested panels

10.4.3 Comparisons with experimental results

In order to calibrate the finite element model, it has been applied to the simulation of monotonic shear behaviour of a panel tested at the University of Pisa [22]. The examined prototype, shown in Fig. 10.8, is composed of three symmetrical trapezoidal sheets, having a cross-section

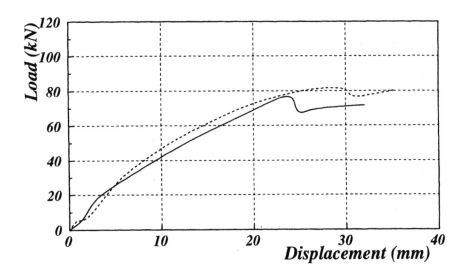

Fig. 10.9   Experimental shear force versus shear displacement response

of depth 47 mm and a thickness of 1.2 mm, assembled by means of aluminium rivets 6.3 mm in diameter at a pitch of 254 mm. The complete panel is connected to the upper and lower UPN 120 chords with 6.3 mm diameter self-tapping screws every 210 mm. The external frame is made of HEA300 columns connected to an IPE240 beam by double angle joints. The experimental shear force versus shear displacement curve under monotonic load shows a non-linear behaviour of the diaphragm (Fig. 10.9). The collapse of the system is reached by means of rivet failure up to the full disconnection of each sheet with consequent loss of load-bearing capacity.

The diaphragm idealization for finite element analysis is shown in Fig. 10.10. The trapezoidal sheets are represented by orthotropic plates of uniform thickness of 1.2 mm. Each elastic constant has been evaluated through a finite element analysis of the actual geometry of the sheet; the values obtained are summarized in the same figure.

The force–displacement relationship of the two systems of connection has been obtained by additional tensile tests carried out at the Engineering Faculty of the University of Naples. The tests used steel sheet elements 1.2 mm thick fastened together by aluminium rivets, and fastened to the beam elements by self-tapping screws. Both rivets and screws have a diameter of 6.3 mm. The average experimental curve has been assumed in order to define the behaviour of the vertical spring

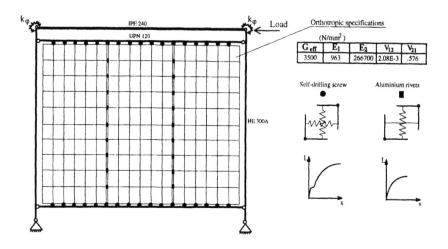

Fig. 10.10   Finite element mesh with semirigid beam-to-column connections

elements of the seam fasteners (Fig. 10.11), while an infinite value of stiffness has been assigned to the horizontal spring elements.

In the case of screwed connections, the tests have shown that failure occurs for bearing and tearing of the sheet. The experimental load–slip relationships obtained for three specimens are shown in Fig. 10.12. The figure also shows the average curve introduced in the numerical simulation to describe both horizontal and vertical spring element behaviour along the chords. The particular kind of beam-to-column connection used in the external supporting frame required modification of the model by introducing two semirigid joints with appropriate values of stiffness K_φ (Fig. 10.10).

In order to overcome the uncertainties over the correct evaluation of semirigid joint stiffness K_φ, a bound criterion has been assumed. According to this assumption, values of K_φ ranging from 0 to $2EI/L$ have been considered, where EI/L is the flexural stiffness of the upper beam (IPE240).

A comparison between the range of simulated curves and the experimental data is shown in Fig. 10.13. The average curve of the scatter band gives a good prediction of panel strength.

For the flexibility, the main influencing parameter is the semirigidity of the beam-to-column joint. In particular, comparison of the slopes at the beginning of the curves indicates that the best correlation with the

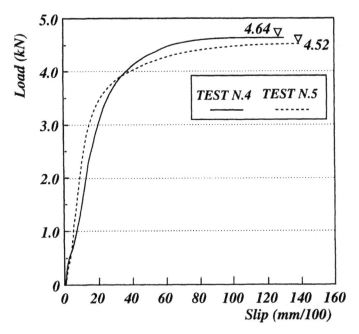

Fig. 10.11   Average load–slip experimental curve for rivets ($d = 6.3$ mm)

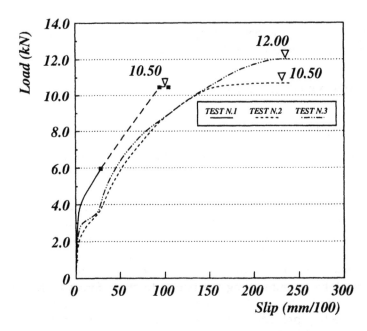

Fig. 10.12   Average load–slip experimental curve for self-tapping screws
($d = 6.3$ mm)

Influence of claddings

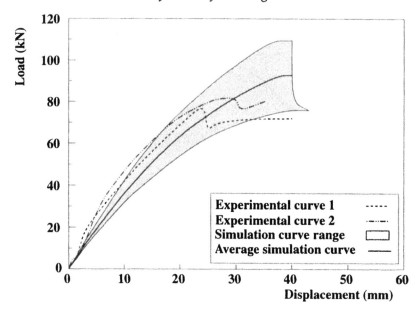

Fig. 10.13   Comparison between the range of simulated curves and the
experimental data

experimental results is obtained by setting $K_\varphi = 2EI/L$, which is considered as an upper bound for the real value.

10.5 EXPERIMENTAL DATA

10.5.1 General

The definition of the analytical model to be considered by the calculations requires particular attention to the type of panel and its connecting system.

The shear forces can be transferred from the frame to the panel by a continuous connection along the perimeter or by a limited number of connections generally placed at the corners of the supporting structure. These connections can be made using bolts, rivets, screws or welds.

Therefore, there is a wide range of situations that need to be studied experimentally by submitting the panels to shear forces both in monotonic and cyclic ranges. In so doing it is most important to establish standard testing procedures for the panels in order to obtain consistent interpretation of the results.

The tests performed to date have been mainly under monotonic loading conditions, although a correct prediction of the seismic response of the whole system, composed of the main structure and the panel system, requires a knowledge of the behaviour of the panels under cyclic loads.

The influence of the connecting system on the cyclic behaviour of cladding panels made of corrugated light gauge sheets has been investigated at the University of Pisa [22, 23]. These tests have demonstrated the importance of connection type on the strength and ductility of panels. It was observed that the extremes of behaviour are given by riveted and welded connections, while the use of other systems (bolts, screws) provides intermediate results. The influence of the connecting system on the seismic response of a pin-jointed structure with bracing panels is evaluated by examination of these extreme cases.

10.5.2 Riveted and screwed panels

The cladding panel is usually fixed by means of mechanical fasteners (bolts, rivets or screws).

The available test results for these types of panels are mainly for monotonic loading conditions. A knowledge of the cyclic behaviour is also required in order to take into account the stiffening effect of panels which is always present and influences the dynamic behaviour of structures.

The data available for the cyclic behaviour of riveted and screwed panels [23] are used here to calibrate the analytical models. A single layer panel composed of four elementary sheets, assembled by rivets, is shown in Fig. 10.8 (MODEL 1a). The complete panel is screwed to the upper and lower chords. For riveted sub-panels, the shear force versus shear displacement curve under monotonic loads (Fig. 10.14a) shows a nonlinear behaviour. The early achievement of the maximum load is followed by discontinuities due to the collapse of the rivets, until the single sheets become disconnected and lose their load-bearing capacity.

The hysteresis loop (Fig. 10.14b) is asymmetrical due to large slippage in the connections (Fig. 10.8, MODEL 1b). The cycles which follow the first are characterized by a reduction of the energy dissipation capacity.

Substitution of the riveted connections between sub-panels with screwed connections provides an improvement in the ductility, as shown by the experimental monotonic behaviour (Fig. 10.15a); a small increase of the energy dissipation capacity under cyclic loads is also obtained (Fig. 10.15b).

Comparison between these two cases (rivets and screws) confirms that rivets lead to the worst behaviour and, therefore, can be taken as a lower bound.

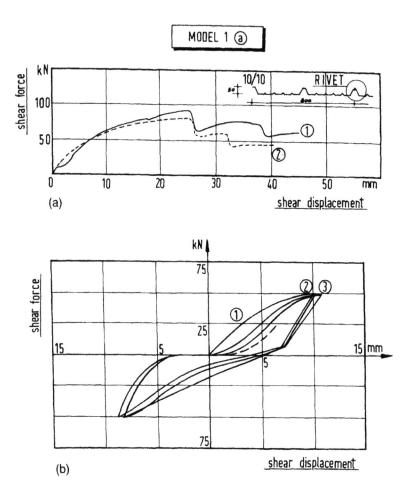

Fig. 10.14　Monotonic and cyclic behaviour of riveted panels: (a) monotonic behaviour; (b) cyclic behaviour

10.5.3 Welded panels

An important increase of strength and ductility with a strong reduction of slips can be obtained by using spot welds for the connections between sub-panels and by inserting the complete panel into a perimeter frame (Fig. 10.8, MODEL 2). A further improvement of strength and ductility could be obtained by adopting continuous welding.

The monotonic and cyclic behaviour of panels composed of sheets connected by spot welds and inserted into a perimeter frame has been investigated [23].

Under monotonic loading conditions, their behaviour is non-linear and the softening branch of the shear force versus shear displacement relationship (Fig. 10.16a) arises from local buckling of the panel, which produces collapse. The hysteresis loop under loads comparable with the collapse static load has shown a very appreciable improvement of the energy dissipation capacity with respect to the previous cases of mechanical fasteners (Fig. 10.16b). It is evident that welded panels provide the best behaviour and, therefore, can be taken as an upper bound.

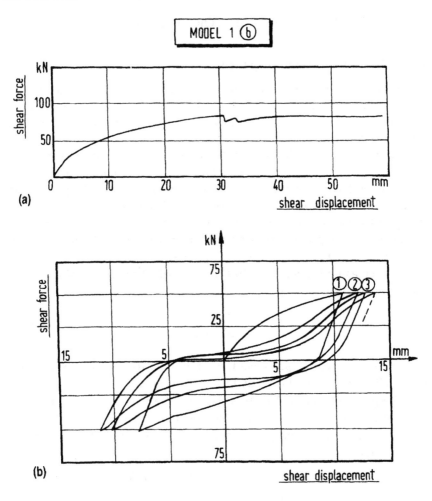

Fig. 10.15 Monotonic and cyclic behaviour of screwed panels: (a) monotonic behaviour; (b) cyclic behaviour

These last results are encouraging, because they confirm that trapezoidal sheet panels are able to contribute to the dissipation of earthquake input energy.

Moreover it is clear that, due to the considerable impact of connection type, prediction of the seismic behaviour of structures with collaborating trapezoidal sheet panels requires the setting up of models calibrated on experimental data. Analytical models corresponding to riveted and welded panels, which can be considered as lower and upper bounds respectively, are now described.

10.6 MATHEMATICAL MODELS FOR CYCLIC BEHAVIOUR

10.6.1 Riveted and screwed panels

The above experimental results have shown that the most important feature of the cyclic behaviour of riveted and screwed panels is represented, in the shear forces versus shear displacement loops, by the presence of large slips due to the collapse of rivets or screws.

In order to predict the seismic behaviour of framed structures taking into account the collaboration of riveted or screwed panels, a simplified model for interpreting their cyclic behaviour has been set up. The proposed simplified model (Fig. 10.17) [2] has been developed in order to interpret the worst behavioural case, in which the energy dissipated during the reloading phase can be partially neglected in a safe interpretation of the testing behaviour of MODEL 1a (Fig. 10.14b).

The increasing branch (Fig. 10.17) is described by means of a curve of the Ramberg–Osgood type:

$$v = \frac{F}{K_o} + \left(\frac{F}{B} \right)^n \tag{10.16}$$

where v is the shear displacement, F is the shear force and K_o is the initial shear stiffness provided directly by the experimental curve.

The parameters n and B characterizing the non-linearity of the loading phase can be obtained by imposing the passage of the theoretical curve through two given points (v_1, F_1) and (v_2, F_2) of the experimental one, to give

$$n = \left[\ln \left(\frac{F_1}{F_2} \right) \right]^{-1} \ln \left(\frac{v_1 - \dfrac{F_1}{K_o}}{v_2 - \dfrac{F_2}{K_o}} \right) \tag{10.17}$$

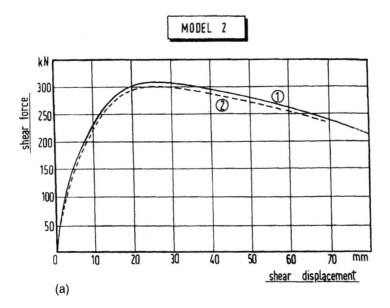

(a)

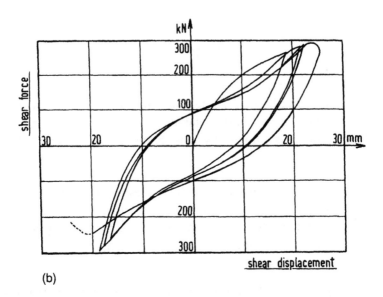

(b)

Fig. 10.16 (a) Monotonic and (b) cyclic behaviour of welded panels

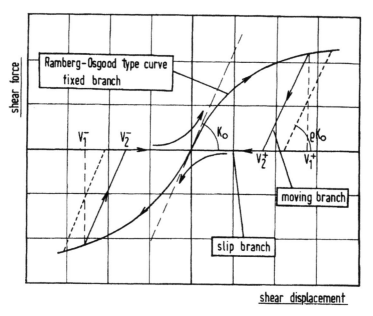

Fig. 10.17   Mathematical model for riveted and screwed panels

$$B = \frac{F_1}{\left(v_1 - \dfrac{F_1}{K_o}\right)^{1/n}} \tag{10.18}$$

The unloading branch is linear and its slope is given by ρK_o, where also the coefficient ρ can be obtained from the experimental curve. The unloading phase is followed by a slip branch corresponding to the release of the total accumulated deformation until reloading in the opposite sense is reached.

It is clear that the proposed model overestimates, in the loading phase, energy dissipated by cycles following the first. This approximation is compensated for after the unloading phase by moving from the slip branch up to the reloading phase, where the energy dissipation is neglected.

In Fig. 10.17 v_1^+ and v_2^+ are values defining the current position of the unloading branch in the positive range, while v_1^- and v_2^- are the analogous values for the unloading branch in the negative range.

Both unloading branches can assume different positions as the number of cycles increases and can therefore be defined as moving branches, whereas the Ramberg–Osgood curve represents a fixed branch.

Considering the case in which v is greater than zero, the following situations can occur.

- If $v < v_1^+$, the point $P(v, F)$, which represents the stress and strain state of the panel, is moving on the fixed (loading) branch; therefore, equation (10.16) provides the current value F of the shear force and the position of the moving branch has to be updated by the values

$$v_1^+ = v \tag{10.19}$$

$$v_2^+ = v - \frac{F}{\rho K_o} \tag{10.20}$$

- If $v_2^+ \le v \le v_1^+$ the point $P(v, F)$ belongs to the moving (unloading) branch, therefore

$$F = \rho K_o (v - v_2^+) \tag{10.21}$$

and the values of v_1^+ and v_2^+ must not be updated.

- Finally, if $v < v_2^+$ the point $P(v, F)$ lies on the slip branch and therefore $F = 0$, while the values of v_1^+ and v_2^+ must not be updated.

The case for $v < 0$ can be studied in an analogous way.

An important feature of the proposed analytical model consists in its easy codification. This benefits its implementation in computer programs for the dynamic inelastic analysis of structures braced with trapezoidal sheet panels connected to the main structure.

10.6.2 Welded panels

From the cyclic point of view, the shear force versus shear displacement relationship can be considered stable, provided that it exhibits the same behaviour as in monotonic tests even when the number of cycles increases. Conversely, the behaviour is considered unstable when its stiffness decreases as the number of cycles increases.

For panels in which sub-panels are assembled by spot welds, the simplified analytical model (Fig. 10.18) [3], neglects the slight slip phase of the loading branches of the cycles following the first (assuming a stable behaviour), while the unloading branch can be either linear or non-linear. In the specific case the use of a linear unloading branch is preferable, because the corresponding approximation provides a reduction of energy dissipation compensating the overestimate corresponding to the loading branch.

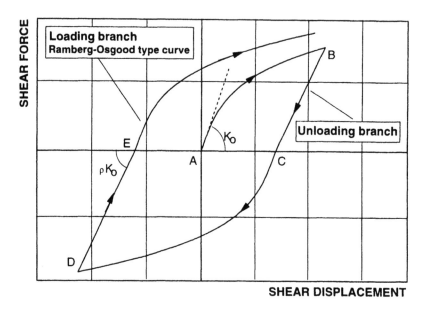

Fig. 10.18 Mathematical model for the cyclic behaviour of welded panels

The loading curve can be represented by means of the following equations, which are still of Ramberg–Osgood type:

$$v = v_0^- + \frac{F}{K_o} + \left(\frac{F}{B}\right)^n \qquad \text{for} \quad F > 0 \qquad (10.22)$$

and

$$v = v_0^+ + \frac{F}{K_o} - \left|\frac{F}{B}\right|^n \qquad \text{for} \quad F < 0 \qquad (10.23)$$

The numerical coefficients of equations (10.22) and (10.23) and that describing the unloading branch can be obtained by minimizing the difference between the energy dissipation corresponding to the analytical model and the one corresponding to the hysteresis loops, as given in Fig. 10.16.

The point $B(F^+, v^+)$ the highest shear displacement experienced in the loading range; the unloading curve in the positive range can be provided by the relationship

$$v = \frac{F - F^+}{\rho K_o} + v^+ \qquad (10.24)$$

while that for the negative range is given by

$$v = v^- + \frac{F - F^-}{\rho K_o} \qquad (10.25)$$

It is important to realize that the proposed hysteretic model represents a simplification and an adaptation to this problem of a more general model which was originally proposed for the simulation of beam-to-column joint behaviour under cyclic loads [24, 25].

The model in its original form (Chapter 3) takes into account either the slippage phenomena in the loading branch or the non-linearity of the unloading branch. Furthermore, the original model also allows for stiffness degradation as the number of cycles increases.

In this case the assumed simplifications are justified for the application of the model in a computer program for the dynamic inelastic analysis of structures.

10.7 APPLICATIONS

10.7.1 Structures with bracing panels

The first application of the models described above was for a pin-jointed steel structure braced by trapezoidal sheet panels, shown in plan and elevation in Fig. 10.19. Its transversal behaviour under seismic loads has been examined for riveted panels and for welded panels. The storey weight is 1120 kN for each storey.

The structure has been examined by dynamic inelastic analysis, performed using a ground motion generated starting from the elastic design response spectrum given in the new proposal of the Italian seismic code (CNR-GNDT) for soil type S_1, with peak ground accelerations of 0.15 g, 0.25 g and 0.35g for low-, medium- and high-seismicity zones, respectively [26].

The structural typology under examination can have economic advantages, particularly in low-seismicity zones, because it provides the maximum economy in cost. In order to investigate the possibility of using cladding panels, in the form of trapezoidal sheets, as a bracing system in zones of higher seismicity, it is of interest to evaluate the different seismic performances provided by riveted and screwed panels. This section examines the main differences in seismic response for pin-jointed structures braced by riveted panels and screwed panels.

Dynamic inelastic analyses have been performed for the two cases of riveted (or screwed) and welded panels, using the analytical models described in sections 10.6.1 and 10.6.2 respectively. The computations were carried out by using a computer program [2] in which bracing panels are introduced by means of an equivalent couple of diagonals. It is

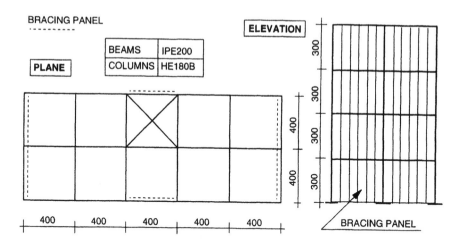

Fig. 10.19   Analysed structure braced with light-gauge steel infill panels

assumed that these diagonals have a cyclic behaviour corresponding to that provided by the proposed analytical models.

The program allows evaluation of all time histories, such as those for nodal displacements, stress and strain of members and panels.

Three zones considered were:

- zone 1, low seismicity
- zone 2, medium seismicity
- zone 3, high seismicity.

Their maximum displacements (Fig. 10.20) and maximum interstorey drifts (Fig. 10.21) are compared as significant behavioural parameters. The results of Fig. 10.21 are very important in view of the control of ductility demands for bracing panels.

In fact, the required interstorey drift has to be compared with the available ductility of panels, which is given by the experimental data (Figs. 10.14–10.16). From this comparison, it can be concluded that riveted and screwed panels can be used in low-seismicity zones only, while welded panels are strictly necessary in zones of higher seismicity. The above result confirms how the connecting system plays an important role if pin-jointed structures are to be used in seismic zones, where cladding panels made of corrugated sheets are expected to provide the ductility required for their bracing function.

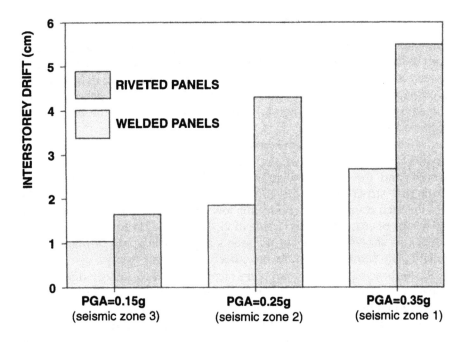

Fig. 10.20   Maximum interstorey drift for different seismic zones

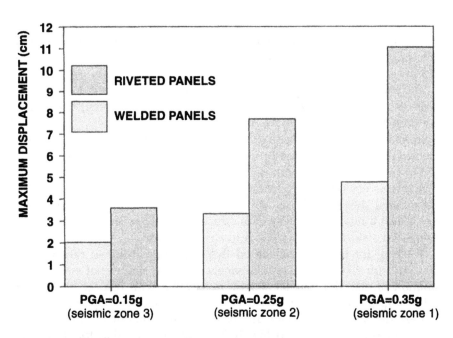

Fig. 10.21   Maximum top sway displacement for different seismic zones

10.7.2 Comparison with the X-braced scheme

Pin-jointed structures braced by collaborating trapezoidal sheet panels can be a convenient structural typology, expecially in low-seismicity zones, due to the reduction in overall weight. The convenience of this typology in seismic zones can be demonstrated from a comparison with the equivalent most common and simple structural type, the pin-jointed X-braced scheme.

The alternative structural solution, based on the concentrically X-braced scheme, is shown in plane and elevation in Fig. 10.22. The storey weight and the number of bracings is the same as for the previously examined structure and the corresponding design base shear force (54 kN for each transversal bracing) for low-seismicity zones can be resisted by means of cross-diagonals made of 2L 25x4 angles in Fe430 steel. Figure 10.22 also shows the sizes of diagonals required in zones of higher seismicity. The same dimensions have been kept in all storeys.

These bracing diagonals are very slender and their energy dissipating capacity under compression can therefore be neglected. Consequently, for diagonal bracings the hysteresis loop can be assumed as in Fig. 10.23, corresponding to members which buckle in compression.

Dynamic inelastic analyses have been carried out starting from the ground motions already used in order to examine the seismic response of the equivalent structure braced with trapezoidal sheet panels.

The comparison between the seismic behaviour of pin-jointed steel structures braced with trapezoidal sheet panels and X-braced structures gives significant data to grasp the possibilities of use of the former type in seismic zones.

The main differences between the behaviours of the examined structures can be pointed out with reference to the first storey, because it undergoes the most severe damage.

In all cases the interstorey drifts of the X-braced structure are always greater than those required by the diaphragmed pin-jointed structure.

The most important result is that the ductility demand in the X-braced structure is considerably greater than that required for structures with trapezoidal sheet panels, as can be observed from Fig. 10.24 where the maximum interstorey drifts of the structure braced with infill panels are compared with the ones of the X-braced structure.

Looking for the reasons behind this important result, it can be seen that (and for riveted or screwed panels also), the manner of energy dissipation of cladding panels is very favourable with respect to that for X-bracing. In fact, comparison between the hysteresis loops shows that trapezoidal sheet panels can dissipate a given amount of energy for a lower ductility value. From Fig. 10.25 it can be seen that for X-bracing the energy corresponding to the area A–B–C–D is dissipated only in the

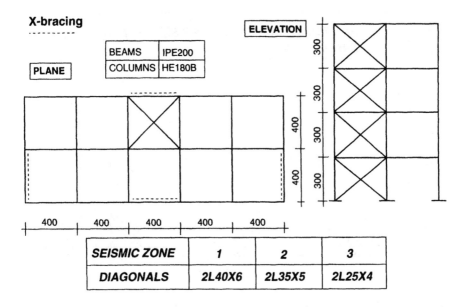

Fig. 10.22 Analysed X-braced scheme

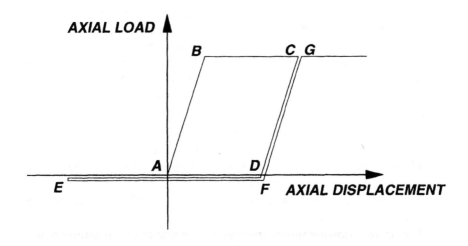

Fig. 10.23 Simplified cyclic behaviour of slender bracings

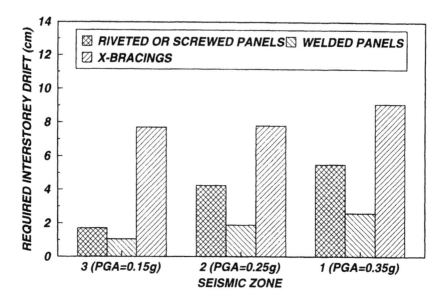

Fig. 10.24   Main results of dynamic inelastic analyses

first cycle, while in the case of a bracing panel (Fig. 10.26) the energy corresponding to the area G–B–C is dissipated not only in the first cycle but also in all the following cycles. As a consequence, for a fixed value of the ductility demand, the bracing panel is able to dissipate in the cyclic range a greater amount of energy than the X-bracing.

In the case of X-braced structures the ductility demand is considerably higher, for all seismic zones, so that the choice of the connecting system for the bracing members is determinant. Using Δ = 10 cm as the design value of the interstorey drift, where for the examined structure $\cos\alpha$ = 0.80 (α is the angle between the diagonal and the horizontal axis), the lengthening of the tensile diagonal is δ = $\Delta\cos\alpha$ = 8 cm; therefore the tensile strain is given by ε = 8/500 = 0.016. In addition ε_y = 0.0013 is the yield strain for Fe430 steel, therefore the connecting system has to provide at least a ductility ratio ϕ = $\varepsilon/\varepsilon_y$ = 12, which can only be obtained using welded connections according to the experimental results discussed elsewhere [27, 28].

The above considerations are based on equalizing the ultimate accumulated ductility ratio in one direction under cyclic loads and the ultimate ductility ratio under monotonic loads.

It is also of interest to analyse the different behaviour of the X-braced solution and the diaphragmed solution in terms of interstorey drifts.

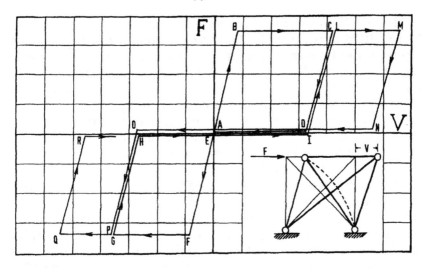

Fig. 10.25 Cyclic behaviour of an X-braced storey with slender diagonal bracings

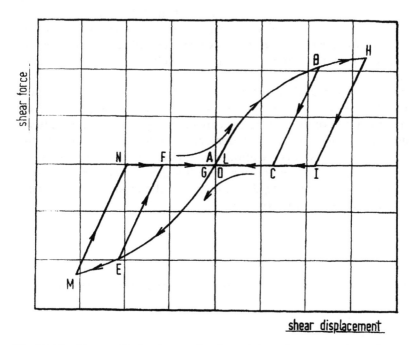

Fig. 10.26 Energy dissipation mechanism of riveted or screwed panels

From the time histories of the first-storey interstorey drift, given in Figs. 10.27 and 10.28 respectively for the X-braced and for the diaphragmed solutions, it can be seen that, in the first case, the seismic response is asymmetrical. This behaviour is unsatisfactory, because it produces, for a fixed amount of energy dissipation, an increase of the required ductility. Conversely, for the structure braced with corrugated sheet panels, the seismic response is quite symmetrical, with values of the interstorey drifts of the same magnitude in both positive and negative directions. This last characteristic provides a high energy dissipation without increasing the ductility demand.

Finally, in the case of structures braced by means of cladding panels, in low-seismicity zones (in seismic zone 3 the required shear displacement is less than 2 cm) both panels composed by riveted and screwed connections can be used (Figs. 10.14 and 10.15); in higher seismicity zones, however, only the panel with welded connections and inserted into a perimeter frame is able to provide the required ductility (Fig. 10.16).

In conclusion, the foregoing comparison has shown that pin-jointed structures with bracing panels can be used in zones of low and moderate seismic activity because they have a cyclic behaviour better than that of traditional X-braced structures with slender diagonals. It can be seen that this result depends on the manner of energy dissipation which requires lower values of ductility for a fixed value of dissipated energy.

In particular, the examined structural typology is economically advantageous in low-seismicity zones because, as all members can be pin-jointed, it provides the best savings in structural weight and erection processes.

Moreover, it is clear that further investigation is required to fully understand the seismic behaviour of this unusual typology. From the experimental point of view, different connecting systems, both traditional and innovative, must be tested under cyclic loading conditions. On this basis, refined analyses must be applied to modelling the cyclic behaviour of trapezoidal panels by correlation with the type of connection system, with the purpose of verifying the influence of the models on the results of dynamic analysis of the structure.

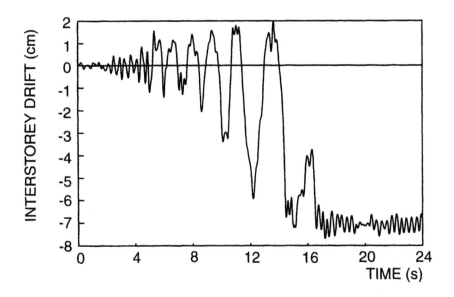

Fig. 10.27 X-braced structure: interstorey drift time history of the first storey

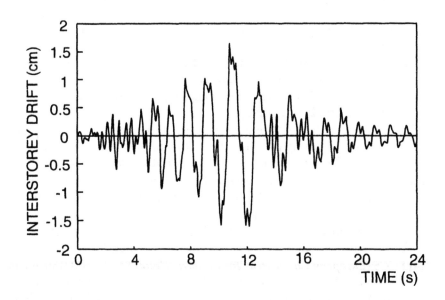

Fig. 10.28 Stucture braced with panels: interstorey drift time history of the first storey

10.8 REFERENCES

[1] *F.M. Mazzolani, F. Sylos Labini:* 'Skin-Frame Interaction in Seismic Resistant Steel Structures', Costruzioni Metalliche, N.4, 1984.

[2] *F.M. Mazzolani, V. Piluso:* '"Skin-effect" in pin-jointed Steel Structures', Ingegneria Sismica, N.3, 1990.

[3] *F.M. Mazzolani, V. Piluso:* 'Influence of Panel Connecting System on the Dynamic Response of Structures composed by Frames and Collaborating Claddings', Second International Workshop: 'Connections in Steel Structures: Behaviour, Strength and Design', Pittsburgh, Pennsylvania, April, 10–12, 1991.

[4] *R. Landolfo, F.M. Mazzolani:* 'Il Comportamento Strutturale delle Lamiere Grecate di Terza Generazione', Acciaio, N.2, 1988.

[5] *W.M. El-Dakhakhni, J.H. Daniels:* 'Frame floor-wall system interaction in buildings', Fritz Engineering Lab Report No. 376.2, Lehigh University, April, 1973.

[6] *C.J. Miller:* 'Analysis of Multistorey Frames with Light Gauge Steel Panel Infills', Department of Structural Engineering, Report No. 349, Cornell University, August, 1972.

[7] *I.J. Oppenheim:* 'Control of Lateral Deflexion in Planar Frames using Structural Partitions', Proceedings of ICE, Vol. 55, Pt 2, pp.435–445, June, 1973.

[8] *E.R. Bryan, J.M. Davies:* 'Stressed Skin Diaphragm Design', published in 'Constructional Steel Design: An International Guide', edited by P.J. Dowling, J.E. Harding, R. Bjorhovde, Elsevier Applied Science Publishers, London, 1992.

[9] *J.M. Davies, E.R. Bryan:* 'Manual of Stressed Skin Diaphragm Design', Granada Publishing Ltd, London, 1982.

[10] *E.R. Bryan, J.M. Davies:* 'Steel Diaphragms Roof Decks', Granada Publishing Ltd, London, 1981.

[11] *Steel Deck Institute:* 'Tentative Recommendations for the Design of Steel Deck Diaphragms', Steel Deck Institute, 9836 West Roosevely Road, P.O. Box 207, Westchester, Illinois, 60153, October 12, 1972.

[12] *U.S. Army:* 'Seismic Design for Buildings', U.S. Army Technical Manual 5-809-10, Department of the Army, Washington, D.C., April 17, 1973.

[13] *D. Huang, L.D. Luttrell:* 'Theoretical and Physical Approach to Light Gage Steel Shear Diaphragms', Civil Engineering Report, West Virginia University, Morgantown, West Virginia, December, 1979.

[14] *ECCS (European Convention for Constructional Steelwork)* 'European Recommendations for the Stressed Skin Design of Steel Structures', Committee 17: 'Cold-Formed Thin-Walled, Sheet Steel in Building', ECCS-XVII–77–1E, published by CONSTRADO, 1977.

[15] *P.E. Liedtke, D.R. Sherman:* 'Comparison of Methods predicting Dia-

phragm Behaviour', Sixth Specialty Conference on Cold Formed Steel Structures, University of Missouri, Rolla, 1982.

[16] *J.M. Davies:* 'A General Solution for the Shear Flexibility of Profiled Sheets I: Development and Verification of the Method. II: Applications of the method.', Thin Walled Structures, Vol. 4, pp. 41–68, pp. 151–161, 1986.

[17] *J.M. Davies, J. Fisher:* 'End Failures in Stressed Skin Diaphragms' Proceedings of Institute of Civil Engineers, Part 2, March, 1987.

[18] *A.H. Nilson, A.R. Ammar:* 'Finite Element Analysis of Metal Deck Shear Diaphragms', Journal of the Structural Division, Vol. 100, No. ST4, pp. 711–726, 1974.

[19] *E. Atrek, A.H. Nilson:* 'Nonlinear Analysis of Cold-Formed Steel Shear Diaphragms', Journal of the Structural Division, Vol. 106, No. ST3, pp. 693–710, 1980.

[20] *G. De Matteis:* 'La modellazione del Comportamento a Taglio dei Pannelli in Lamiera Grecata', Graduation Thesis, Istituto di Tecnica delle Costruzioni, Faculty of Engineering, University of Naples, 1994.

[21] *R. Landolfo, F.M. Mazzolani:* 'Shear Behaviour of Steel Corrugated Panels', First International Workshop and Seminar on Behaviour of Steel Structures in Seismic Areas, Timisoara, Romania, 26 June –1 July, 1994.

[22] *L. Sanpaolesi, L. Biolzi, R. Tacchi:* 'Indagine Sperimentale sul Contributo Irrigidente di pannelli in Lamiera Grecata', IX Congresso C.T.A., Giornate Italiane della Costruzione in Acciaio, Perugia, pp. 251–264, 1983.

[23] *L. Sanpaolesi:* 'Indagine Sperimentale sulla Resistenza e Duttilità di Pannelli-Parete in Lamiera Grecata', Italsider, Quaderno Tecnico N.7, 1984.

[24] *A. De Martino, C. Faella, F.M. Mazzolani:* 'Simulation of Beam-to-Column Joints Behaviour under Cyclic Loads', Costruzioni Metalliche, N.6, 1984.

[25] *F.M. Mazzolani:* Mathematical Model for Semi-Rigid Joints under Cyclic Loads, in 'Connections in Steel Structures: Behaviour, Strength and Design', Elsevier Applied Science, 1987.

[26] *CNR-GNDT:* 'Norme Tecniche per le Costruzioni in Zone Sismiche', Dicembre, 1984.

[27] *P. Zanon:* 'Resistenza e Duttilità di Angolari Tesi Bullonati', Costruzioni Metalliche, N.4, 1979.

[28] *G. Ballio, F.M. Mazzolani:* 'I Collegamenti nelle Strutture Sismo-Resistenti di Acciaio', Italsider, Quaderno Tecnico N.11, 1980.

[29] *R. Landolfo, F.M. Mazzolani, V. Piluso:* 'L'Influenza dei Pannelli di Chiusura sul Comportamento Sismico dei Telai in Acciaio', VI Congresso Nazionale, L'Ingegneria Sismica in Italia, Perugia, pp. 785–794, 1993.

Index

Milton Keynes UK
Ingram Content Group UK Ltd.
UKHW021917071024
449327UK00022B/1674